Praise for

AS GODS

"A gripping, bawdy tale of science fiction morphing into business history. Exhaustively researched and beautifully written, *As Gods* provides the histories of recombinant DNA, biotech, GMOs, gene therapy, and cloning in a single lively, accessible account."

—Nathaniel Comfort, John Hopkins University

"A lucid and vigorously insightful account of the pitfalls and triumphs of the twenty-first century's most ethically challenging and potentially world-changing technology."

—Paul McAuley, author of *Fairyland*

"A superb account of genetic engineering in life and culture, in all its myriad anxieties and exhilarations. Should we be scared? Read this book and you'll have a sense of the answer."

—Adam Roberts, author of *It's the End of the World*

"Powerful gene technologies, long foreseen, are finally with us. Taking the measure of this daunting prospect calls for historical acumen, technical appreciation, and a clear-eyed view of human foibles. As this book attests, Matthew Cobb has all three."

—Jon Turney, author of *The Rough Guide to the Future*

"A superb guide to the global history of the dreams, fears, and science of genetic engineering, and why it matters for tomorrow."

—Jon Agar, author of *Turing and the Universal Machine*

"Rich, vivid, and well-documented, this is a wonderful introduction to genetic technology with the take-home message—just because we can doesn't mean we should."

—Michel Morange, author of *A History of Molecular Biology*

"Cobb is a great storyteller of science, a tapestry of intriguing and enlightening ideas thoughtfully and entertainingly told."

—Robin Ince, author of *The Importance of Being Interested*

"Faced with a new round of genetic dreams and nightmares, Cobb skillfully sifts the truth from the hype in this thrilling and alarming account of our most dangerous and exciting technology."

—Paul Mason, author of *How to Stop Fascism*

"Cobb's riveting analysis warns that in a world beset by poverty, inequality, and climate catastrophe, chasing apparently dazzling technofixes is rarely cost effective or morally justified."

—Georgina Ferry, author of *A Computer Called LEO*

AS GODS

ALSO BY MATTHEW COBB

The Idea of the Brain: A History

Life's Greatest Secret: The Race to Crack the Genetic Code

The Egg and Sperm Race: The 17th-Century Scientists Who Unravelled the Secrets of Sex, Life and Growth

Smell: A Very Short Introduction

The Resistance: The French Fight Against the Nazis

Eleven Days in August: The Liberation of Paris in 1944

AS GODS

A Moral History of the Genetic Age

MATTHEW COBB

BASIC BOOKS
New York

Cover design by Ann Kirchner

Cover image © Zita / Shutterstock.com

Basic Books
Hachette Book Group
1290 Avenue of the Americas, New York, NY 10104
www.basicbooks.com

Printed in the United States of America

Originally published in 2022 by Profile Books Ltd. in Great Britain

First US Edition: November 2022

Published by Basic Books, an imprint of Perseus Books, LLC, a subsidiary of Hachette Book Group, Inc. The Basic Books name and logo is a trademark of the Hachette Book Group.

Typeset in Palatino by MacGuru Ltd

Library of Congress Control Number: 2022945141

ISBNs: 9781541602854 (hardcover), 9781541602847 (e-book)

LSC-C

Printing 1, 2022

CONTENTS

In memory of my PhD co-supervisor, Barrie Burnet (1935–2020), Reader in Genetics at the University of Sheffield, and of my University of Manchester colleague, Roger Wood (1934–2021), who, like me, studied both insect genetics and the history of science.

INTRODUCTION

Ever since the early 1970s the world has been living through a scientific and technological revolution. New genetic techniques have transformed science, providing profound insights into the whole tree of life and allowing us to investigate a huge range of organisms with astonishing precision. These discoveries have been turned into novel technology with far-reaching implications, putting food in the fields and healing bodies. Vast fortunes have been made as whole sectors of the pharmaceutical industry have been transformed. To give just one example: if you take insulin, it was produced in a genetically engineered microbe.

Built upon dreams, genetic engineering has equally provoked nightmares throughout its history. The impact of the new genetics on popular culture and our global mindset over the last fifty years has mirrored the promises and doubts about nuclear power during the post-war decades. Nuclear physics lost its innocence in the searing white heat of Hiroshima and Nagasaki; genetic engineering has yet to suffer a similar fate but that is not for want of trying – from the very beginning, hidden from history, the new science has been used to create terrifying new weapons. The potential for the nightmares to become reality grows with our increasing mastery and our endless appetite for ever more audacious applications.

With every new development there have been promises of new sources of food or medicine, or new cures for diseases, which have been rapidly counterbalanced by fears of plagues, of genetically manipulated humans, or of the inadvertent or deliberate release of dangerous organisms. Genetically modified plants promised to transform agriculture, but opponents feared ecological catastrophe. Neither happened. The genetic dreams and nightmares have tended to recur: the promises are rarely fully realised, the worst fears never come to pass and after a while the whole issue subsides, only to re-emerge a few years later following new discoveries and applications.

This cycle has occurred repeatedly over the last half-century. Now, after a long period in which the technology began to appear commonplace, three recent developments have brought the dreams and nightmares back into our waking lives, providing amazing opportunities and raising the real possibility of catastrophe:

- In 2018 we stepped into the brave new world of heritable human genome editing when Chinese researcher He Jianqui used CRISPR gene editing in a botched experiment that mutated three healthy embryos, with unknown consequences for the resulting children. Despite a global outcry, there is no agreed way of preventing this from happening again.
- We can now transform whole ecosystems through a process known as a gene drive, which is essentially a genetic chain reaction. This could eradicate malaria mosquitoes, but as the scientists working on these systems have warned, it could also wreak havoc on the ecosystem.
- Well-meaning scientists trying to gain insight into the potential shape of future pandemics have deliberately produced new variants of lethal pathogens that are even more dangerous than before. These experiments were not behind the COVID-19 pandemic, but such research could inadvertently lead to an outbreak of a terrifying new disease.

These are not fantasies nor absurd predictions; this is where science has brought us.

My motivation in writing this book has been to explore my own fears about these three areas. Each of them worries me in different ways, but I recognise that many of my concerns are similar to those expressed by people faced with previous applications of genetic engineering, most of which turned out to be either exaggerated, or at least to be controllable by careful regulation and strict safety procedures. I needed to decide if these most recent developments of genetic engineering are truly novel and full of real threat, or if they, too, will turn out to be overblown in terms of both promise and peril. For the moment, I remain deeply concerned – we are indeed faced with new and serious threats and great care will be needed to negotiate the coming years. Above all, the key lesson I have learned is that to understand what these developments mean and how we should respond to them, we need to understand how we got here.

*

The history of nuclear power shows that dangerous technology can be used safely, despite its inherent potential for accidental or deliberate disaster. The rise and fall of the atom as a driver of culture and of popular anxiety also shows that as dangerous technology is safely employed, it loses some of its emotional power. Over recent years, that seems to have happened with genetic engineering too.

Many technologies have followed a similar arc, but in one extremely significant respect genetic engineering is unique. From the very outset, potential dangers led scientists to impose a temporary halt to experimentation while safety concerns were addressed.

No group of researchers, apart from geneticists, has ever voluntarily paused their work because they feared the consequences of what they might discover. Extraordinarily, this has happened not just once but four times – in 1971, in 1974, then again in 2012 and most recently in 2019.

There were some precedents for this. In the 1930s, Leo Szilard, who conceived the nuclear chain reaction, argued for the details to be kept secret; following the subsequent discovery of nuclear fission he tried to persuade his colleagues to omit certain key bits of information from their publications to avoid helping Nazi Germany. Later, while working on the Manhattan Project, Szilard and other researchers opposed the continuing development of the atom bomb after the defeat of Germany in May 1945. But Szilard's protests were of limited effect and the result was the horror of Hiroshima and Nagasaki.

Geneticists have been more decisive in response to potential threats. In 1971, when the first experiment mixing two very different kinds of DNA – from a virus and a bacterium – was proposed, researchers raised fears that it might go horribly wrong, causing an outbreak of cancer. There were private discussions about the dangers and eventually some experiments were quietly abandoned by the handful of scientists involved. But new discoveries soon made the technique much simpler and within two years dozens of laboratories around the world could master it. The question of what might happen was posed once again, but this time on a global scale. That led to the first publicly declared research moratorium, which was announced in July 1974 and lasted about eight months while scientists argued about the issue.

The culmination of this process was a conference held in February 1975, at Asilomar in California, at which scientists came up with safe ways of performing their experiments but notably refused to consider the social or political consequences of what they were doing. Over the subsequent decades the self-regulation that was proposed at Asilomar has repeatedly been held up as an example of how science can act responsibly. And indeed, the great virtue of Asilomar was that the meeting took the potential dangers of genetic engineering very seriously indeed, thrashing out protocols that would protect researchers, the public and the environment and insisting that even with these safety measures, some experiments involving pathogens remained too dangerous to be carried out under any circumstances. But the debates were focused entirely on these biosecurity issues, with the sole objective of establishing safety criteria that would allow the moratorium to be lifted.

The organisers ruled out any discussion of moral issues, or of the potential military uses of what was known as recombinant DNA. And yet the political and social issues that have dogged the last fifty years of the development and application of genetic engineering were precisely those that were ruled off the Asilomar agenda. Two key issues that shaped subsequent decades – the commercial exploitation of genetic engineering and the terrifying threat of new bioweapons made with recombinant DNA – were both being actively developed at the time of Asilomar but were not discussed. They were known to only a handful of privileged delegates, and no one was aware of both developments. Had there been an open debate about these matters, subsequently involving the whole world, later events might have turned out rather differently.

The reasons why molecular geneticists took this unprecedented step of halting their research until it could be made safe, while scientists in other controversial areas such as nuclear weapons research did not, probably lies as much in the people and the times as in the degree of existential threat. Genetic engineering was the right subject, in the right place, at the right time, with the right scientists involved, for such a stance to be taken.

Asilomar occurred in a period of doubt about science and its role in society, centred around the upheavals of 1968 and the long wave of global unrest and social uncertainty that surrounded those events, all of which was framed by the Cold War. Most of the organisers of Asilomar had been involved in campus protests against military research and against the US war in Vietnam and Cambodia. Furthermore, this was a field that involved a small number of scientists, working in groups of only a handful of researchers with no military or government involvement, which gave them a degree of autonomy.

The third pause in genetic engineering took place more recently, in 2012, when a few dozen scientists became alarmed at the direction being taken by their research on the extraordinarily dangerous H5N1 bird flu virus, which they were manipulating in order to prepare for future pandemics. That self-imposed moratorium lasted about eight months and, as at Asilomar, was similarly resolved through the adoption of new safety procedures, which arguably saved us from an accidental lab-leak pandemic that would have dwarfed COVID-19. However, those procedures were not globally binding – different countries have different biosecurity standards, some of which may lead to disaster in the future.

Both the recombinant DNA and the H5N1 research pauses were widely accepted and observed. The most recent call for a research moratorium, focused on heritable human gene editing, has not been met with such unanimity. In 2015 leading researchers and scientific authorities around the world declared it would be irresponsible to try and edit a human embryo using CRISPR, but there was no call for a pause on research. Indeed, those same researchers and authorities sought to chart what they called a 'prudent path' to heritable human genome editing. Then, in November 2018, He Jiankui announced to a stunned world that he had carried out such an experiment on three normal baby girls, with potentially disastrous results. A few weeks later, at the beginning of 2019, a group of leading geneticists called for a five-year moratorium on editing humans, but others, including Nobel Prize-winning leaders of the field, have rejected this approach. This time around, there is no consensus. Despite the widespread revulsion at the CRISPR babies experiment, there is no guarantee that it will not be repeated tomorrow.

These examples show the singularity of genetic engineering in the history of science and technology – its strong connection with questions of social responsibility and a sense of doing the right thing. That is why, although this book surveys the last half-century and more, the central issues it addresses revolve around the current potential of the new science and how we can control it and prevent catastrophe. In a sense, the past, present and future of genetic engineering are on trial in these pages, and you are part of the jury.

*

To some readers, the term 'genetic engineering' will have a rather old-fashioned feel. This is partly because the name by which this technology is known has morphed and shifted down the decades. From genetic engineering, through recombinant DNA, gene stitching (that one did not catch on), gene cloning and gene splicing, to genetic modification and most recently gene editing, different terms have embodied various subtleties to both scientists and the public. Sometimes these names have been adopted because they present a new application of the science as somehow less threatening than previous, contested versions (this is clearly the case with 'gene editing' which seems simple and rather domestic). Some terms refer to a much broader field – for example, 'biotechnology' or 'synthetic biology', both of which incorporate genetic engineering as a base technology but tend to have different ambitions and outlooks. What all these approaches have in common is the ability to deliberately and precisely introduce new changes into the genes of an organism – genetic engineering.

One of the key consequences of the new ability to manipulate genes that appeared in the 1970s was that, right from the outset, this technique was turned into a technology and was given practical application. This in turn led to fears and protests as new organisms made the transition from the laboratory to the factory or the field. These challenges form an integral part of the history of genetic engineering – they not only shaped public attitudes to the science, they also led to a series of regulations designed to assuage fears and to allow the safe application of the technology.

Nevertheless, faced with the ability to mix genes from wildly different species, many people continue to feel uneasy. Writing in 1997, as global suspicion of genetically modified (GM) crops was about to reach a paroxysm, pioneer molecular biologist François Jacob reached deep into our collective psyche to identify what he thought was the fundamental problem:

> The notion of recombinant DNA is tied to the mysterious and the supernatural. It rekindles the terror associated with the hidden meanings of monsters, the revulsion engendered by the notion of two beings merged in defiance of nature.[1]

The results of genetic engineering can be unsettling, and not simply because it can seem a bit weird when put starkly (for example, in my research I use flies that have genes from jellyfish in them, controlled by other genes from yeast). But a far greater problem than any unease we might feel is that, as with any technology, the outcome of genetic engineering is not guaranteed in advance. Its essence is the ability to produce inherited changes, and change may not always go in the direction we intend.

*

This book is not simply an account of how a revolution in science and technology has taken place – how brilliant men and women have developed theories, dreamed up experiments and imagined applications – and how the dangers have been perceived and countered. It also shows how those discoveries have formed part of a broader cultural, political and economic history, shaping our present day.

One significant feature of this story is the growing internationalisation of genetic engineering. In its earliest years it was largely US-based, now every continent is touched by this science and its application, and every inhabited continent is covered in the pages that follow. There is one specific change in the geographical focus of this story which is of fundamental importance and will increase in significance in the years to come – the growing influence of Chinese research. This is a product of both the growing economic and scientific power of China and the Chinese government's long-standing commitment to the application of genetic technology. China was the first country in the world to approve a GM crop and it now has the largest number of approved gene therapy protocols. And yet China is as vulnerable to genetic nightmares as anywhere else – the prospect of GM rice divides the Chinese Communist Party leadership and has led to a strikingly open debate in the country. Genetic engineering has the power to excite and disturb right across the planet.

Throughout its history, this technology has been intertwined with broader cultural and political changes – from the US counterculture of the 1960s and early 1970s, through the get-rich-quick deregulation years of the 1980s and 1990s, the collapse of the Soviet Union and the end of the Cold War, the growing opposition to globalisation around the turn of the century, the global responses to the 9/11 attacks, right up to the fears of a new pandemic which were tragically realised at the beginning of 2020. Genetic engineering and its applications have played a significant role in local, national and global politics, shaping and threatening our future as surely as the atom shaped the post-war world.

Although the pages that follow include plenty of experimental detail, this is not an academic account covering every conceivable aspect of the subject. The impact of genetic engineering on virtually every part of biology over the last half-century has been so overwhelming that a book detailing those effects would turn into a history of modern biology itself. As a consequence, many scientific, technical and social developments that flowed from, or coincided with, the appearance of genetic engineering are deliberately not covered in any detail here – IVF, stem cell biology, embryo research, biotechnology, mammalian cloning, synthetic biology, genomics, DNA sequencing, transhumanism and many others. Some experts will undoubtedly find that their favourite technique, their favourite experiment or their favourite researcher has been missed out for reasons of space. My apologies for any disappointment, but rather than covering all the twists and turns of the history or every conceivable implication of the technology, I have preferred to focus on the key applications of genetic engineering and how it has acted as a cultural force, prompting creators to produce works that have inspired, stimulated and amused us.

In some cases, these creations have proved so powerful that they have almost replaced the old references to *Frankenstein*. Like Mary Shelley's influential book, many of these works – from *Jurassic Park* to untold numbers of B-movies, from high-concept novels by Nobel Prize winners to clunky thrillers, from pop songs and comic books to sculptures and pieces of conceptual art – focus on the nightmares, shaping attitudes to the scientific and technological moments we have been living through for the last few decades. Like news media, TV and radio programmes, cultural and artistic artefacts have their place in the story of genetic engineering as they have refracted the science and politics of our times and left lasting traces. This broad focus makes this book much more than an inward-looking history of science.

*

Nevertheless, there is quite a bit of science in this history. I have tried to keep the technicalities to a minimum and to explain each step only as much as is strictly necessary, but if you are unsure about the basics of genetics, the following three paragraphs will give you the key information you need to begin (there is also a glossary at the back).

Our genes are made of DNA, a molecule that has two strands – the famous double helix. Each strand carries a sequence of four chemical structures called bases, which are known by their initials, A, C, G and T. The shape of these bases means that the two strands of DNA are complementary – if there is an A on one strand, the base on the opposite strand has to be a T, and vice versa. The same is true of the C and G bases. Genes consist of unique sequences of bases and, in general, a gene codes for a protein (a string of amino acids), which is produced by the cell following the instructions in the gene. Each group of three bases, called a codon, corresponds to an amino acid. Amino acids are strung together by the cell to form proteins, which can do an infinite variety of things in the organism – in particular they can be structural (for example, hair) or they can alter physiology (for example, enzymes or hormones).

When a gene is activated, the double helix is unravelled by enzymes in the cell and the gene is transcribed: the DNA on the strand that carries the gene is used to produce a molecule called messenger RNA (mRNA), which is a complementary copy of the gene, just like the DNA on the other strand, except that in RNA the T base is replaced by another base called U. This mRNA molecule is then used by cellular structures called ribosomes to turn the genetic message into a protein. Each three-base codon is read by the ribosome, and the corresponding amino acid is found within the cell and brought to the ribosome by transfer RNA (tRNA) molecules, where it is attached to the amino acid preceding it to form a protein. Ribosomes are made of RNA and protein; like everything else, they are encoded by genes. In some cases, RNA or protein produced from a gene's DNA is used by the cell to control the activity of other genes.

Viruses have played a key role in genetic engineering. They are replicating parasitic molecules, made of either DNA or RNA, concealed in a protein coat (the instructions to make that coat are encoded in the virus's DNA or RNA). The sole function of a virus is to penetrate a cell and hijack its internal mechanisms, using the cell's biochemistry to reproduce copies of the virus's DNA or RNA, and then to turn that into more viruses. There are untold numbers of different kinds of virus on the planet; most are harmless, some are lethal. The beginning of the genetic engineering revolution occurred when scientists set out to harness the ability of viruses to introduce DNA into a cell. Their ultimate objective was to alter a virus so that it carried well-understood bacterial DNA and to transfer that DNA into a mammalian cell, thereby shedding light on the mysteries of gene function in multicellular organisms.

That is all you really need to know to start reading.* You will know a lot more by the time you have finished.

*

This book takes you through the changing science of genetic engineering, explaining the links between science and politics, ethics, business and culture, rooting our developing knowledge in the changing world of the last half-century and showing how fears and protests have been there from the very outset. This history carries lessons for all branches of science where discoveries might produce dangers, perpetuate inequalities or otherwise damage society. It highlights the importance of the general population being informed about and involved in decisions about science and its application from the very beginning.

Such social questions are not an optional add-on, something that can be left at the door of the laboratory. They are there, in embryo, in every genetic engineering experiment. They are part of the politics implicit in this revolutionary technique which has transformed science, medicine and agriculture. These issues are present in genetic engineering in a way that is not the case in other branches of genetics because, as my good friend the historian of science Michel Morange has put it, the whole point of the field is to transform molecular biology 'from a science of observation into a science of intervention and action'. When you intervene and act upon the world, rather than simply observing, when you create things that have never existed before, you run the risk of things happening that you did not intend,

* If you want to know how we discovered all this, I would immodestly recommend my 2015 book, *Life's Greatest Secret: The Race to Crack the Genetic Code* (London, Profile).

or which you may desire but others do not. Resolving these issues is a political question – science and society are intertwined.

This interventionist, creative aspect of genetic engineering was first imagined centuries ago, long before the discovery of genetics. When he died in 1626, the English thinker Sir Francis Bacon left an unfinished fragment of fiction known as *The New Atlantis*, which purported to be an account of life on the imaginary island of Bansalem, somewhere in the Pacific Ocean. A key feature of the island was Salomon's House, a kind of research institute that investigated all aspects of the natural world with a view to 'enlarging the bounds of human empire', developing not just knowledge, but also application and control. At a time when heredity was so profound a mystery that the word had no biological meaning, Bacon's fictional institute was able to manipulate plants and animals in a precise and desired way, mixing different species and creating new organisms for the benefit of all:

> We finde Meanes to make Commixtures and Copulations of diverse Kindes; which have produced New Kindes, and them not Barren, as the generall Opinion is. We make a Number of Kindes, of Serpents, Wormes, Flies, Fishes, of Putrefaction; Wheroff some are advanced (in effect) to be Perfect Creatures, like Beastes or Birds; And have Sexes, and doe Propagate.[2]

Despite the weird spelling and understandable lack of detail, Bacon was describing our world. Four centuries later, we have exactly the kind of power over nature that Bacon dreamed of. The idea of mixing the characteristics of different species goes deep into mythology – half-human, half-animal creatures are a common feature of many cultures – but Bacon's proposal that we might deliberately create such hybrids to exploit their characteristics was both novel and profoundly significant, heralding the shift in attitudes to the natural world that took place with the development of science, technology and industry over subsequent centuries. In a way, this book is about the realisation of Bacon's dream.

I hope you will be enthralled, amused, moved and alarmed by what you read here – I did not know all of this history and there were things that I discovered that made my blood run cold. Above all, you should be informed – decisions on using the latest versions of this technology need to be in the hands of every citizen on the planet. By understanding the past and the present we can be more confident in our ability to control the future, or at least to limit the damage that might occur there. As you will see, this is too important to be left to the scientists.

Four hundred years ago, Francis Bacon hoped we would enlarge the bounds of human empire, subduing the forces of nature. To an extent, genetic engineering has helped us do that. But technology is not neutral – it changes the way we behave. In 1872 the German philosopher-revolutionary, businessman and honorary Mancunian Friedrich Engels explored the consequences of our mastery of the forces of nature through technology. As he put it, those forces avenge themselves by imposing a veritable despotism upon us. In other words, technology shapes our social organisation, it requires us to behave in a particular way. In the case of genetic engineering, this has shaped our view of life itself – some organisms have become pieces of machinery that can apparently be controlled and their behaviour predicted. But sometimes, our predictions are poor, our control is inadequate and discovery creates danger.

There is no way of unlearning what we have found out, and equally we cannot escape the implications of what we have created – we have to meet the potential threats that we now face. To do that, we must first understand them. As Francis Bacon also said, 'knowledge is power'. That is the point of this book.

Manchester, March 2022

– ONE –

PRELUDE

Humans have been changing genomes for millennia. From our earliest days in Africa, hundreds of thousands of years ago, we have inadvertently altered the genes of the animals and plants we eat, acting as a force of natural selection just like other predators. Some animals and plants were able to adapt to our attentions; others could not and went extinct, in particular the megafauna – mammoths, woolly rhinoceroses, giant sloths and so on. Then, with the slow development of agriculture around 10,000 years ago, we began to systematically domesticate animals and plants, deliberately breeding those types that suited our needs.

The results could be dramatic. Genomic analysis shows that all modern horses are descended from a small group of animals that were domesticated in the Western Eurasian steppes around 4,000 years ago.[1] Strong and docile, they rapidly replaced the other breeds of horses that we had tamed. We can see this process in the horse genome – our ancestors selected for behavioural and physiological characteristics that allowed the animals to be ridden for long distances and made them more placid, but underlying this process were unseen molecular genetic changes that we can now understand. It is even possible that, over hundreds of thousands of years, we domesticated ourselves, intuitively selecting against aggressive behaviour and in favour of all sorts of cooperative characteristics. With the development of agriculture, we also began to use the simplest form of biotechnology, unwittingly harnessing the activity of microbes to make bread, cheese, beer and wine, inadvertently selecting the varieties that best suited our purpose.[2]

That activity has also led to insight – understanding mating and pollination became of great importance once agriculture had begun. There is a rough synchronicity between when people around the world realised how reproduction works and when they domesticated plants and animals: the growth in our knowledge of the natural world and our increasingly detailed attempts to assert control over it have gone hand in hand.[3]

Despite this deep history, 1972 marked a real qualitative change in our ability to change genes – blind tinkering became precise and deliberate manipulation. This happened through the publication of the work of a group of researchers at Stanford University in Palo Alto, California, in what is now known as Silicon Valley (not much of a valley, to be honest). The researchers, led by 45-year-old Paul Berg, took a mammalian virus called SV40 and added to it DNA from a bacterium, *Escherichia coli*. Berg's idea was to use the virus to introduce the well-understood genetic material from *E. coli* into a mammalian cell and thereby gain the first insight into how genes work in multicellular organisms. The following year the approach was simplified, allowing the fusion of DNA from virtually any organisms. Once the procedures were shown to be safe and controllable, this ability to produce what was called recombinant DNA led to the explosion of the biotechnology industry, the development of GM crops and gene therapy, massive advances in our scientific understanding of the whole of biology and, ultimately, the current excitement over CRISPR gene editing.

The techniques used during this revolution have changed – Berg's pioneering but primitive genetic engineering and today's gene editing are radically different in their detail – but in outlook and in underlying approach all the methods used over the last half-century trace their origin back to Stanford in 1972. As with any revolution, to understand the key moment and the events it unleashed, we need to explore what came before and what came afterwards. We need to begin, not at the beginning, but before the beginning. And before there was fact, fiction could outline what might be possible, enabling thinkers to explore both the promise and the perils of future developments centuries before science could make them a reality.

*

The most powerful fictional portrayal of the dangers of science is surely *Frankenstein*, written by the teenager Mary Shelley in 1816 and published two years later.[4] Influenced by Greek mythology and the story of Prometheus, the Greek god who gave humanity fire and suffered terribly as a consequence ('The New Prometheus' is the subtitle of the book), by the Jewish myth of the Golem – a man-made creature that would do its creator's bidding – and by the German story of Faust and his pact with the Devil, Shelley's book has been taken as a warning tale about the potential dangers of the new way of knowing – science – and in particular the risks of creating something profoundly unnatural.

Towards the end of the nineteenth century, the slow discovery of the mechanisms of heredity led to this new phenomenon becoming the focus of fiction. H. G. Wells conjured the horror that might occur if modern medicine were able to create grotesque hybrid animals through vivisection in his 1896 novel *The Island of Doctor Moreau*. Eight years later, Wells imagined what would happen if food additives changed our heredity. In his now-forgotten novel *The Food of the Gods*, farm animals and children grow to giant size after consuming a substance known as Boomfood; most significantly, they pass their gigantism to their offspring, with terrible social and political consequences. *The Food of the Gods* was published three years after the rediscovery of Mendel's laws of genetics, which were first established in 1865.

The twentieth century was the century of genetics. For its first few decades, it was also the century of eugenics – the desire to deliberately manipulate human genes by selective breeding. This was widely applied, notably in the United States and Sweden, mainly by sterilising those who were deemed unfit – in particular the poor and the disabled. This reached its foul culmination in Nazi Germany, where selective breeding became systematic murder.[5] The most influential fictional account of the consequences of eugenics is Aldous Huxley's 1932 novel *Brave New World*, in which humans are bred in artificial wombs using Podsnap's technique and Bokanovsky's process and have a genetically determined role in society. Although Huxley made up the sciencey-sounding genetic technology employed in his dystopia, he did not have to imagine what eugenics might look like – it was around him, infesting society, irrespective of politics (the socialist Wells was one of many left-leaning eugenicists).

The horror of the Holocaust dampened the appetite for eugenics in the post-war world but growing scientific interest in the nature of genetic material and what might be done with it percolated into fiction. In 1951, the pulp science fiction writer Jack Williamson published an unremarkable novel, *Dragon's Island*, which revolved around the fictional science of what he called genetic engineering – 'a process for creating new varieties and species at will' by 'directing mutation'.[6] The result was the appearance of 'Not-men' – 'superhuman monsters ... Hiding among mankind, and waiting to overwhelm us.'[7] By the end of the story the protagonists prophetically decide to set up a company to spread the benefits of the new technology and make pots of money.*

Williamson might have written a clunker (in best pulp tradition, the dramatic title and the lurid cover had little to do with the story), but he had detected the glimmer of scientific and pecuniary potential contained in a discovery that had been made in 1944 by Oswald

* Although Williamson believed he had coined the term 'genetic engineering', the phrase had already appeared in *Science* in 1949 referring to genetic counselling and eugenics. Furthermore, an earlier usage, closer to today's meaning, was made in 1934 by Nikolaj Timoféeff-Ressovky, who described the creation of mutations by radiation as a type of genetic engineering. Two years before that, a talk entitled 'Genetical Engineering', referring to the selective breeding of crops and farm animals, was given at the Sixth International Congress of Genetics. Nevertheless, Williamson was the first to use the term to describe deliberate, directed mutations. Stern, K. (1949), *Science* 110:201–8; Timoféeff-Ressovky, N. (1934), *Biological Reviews* 9:411–57; Crowe, J. (1992), *Genetics* 131:761–8.

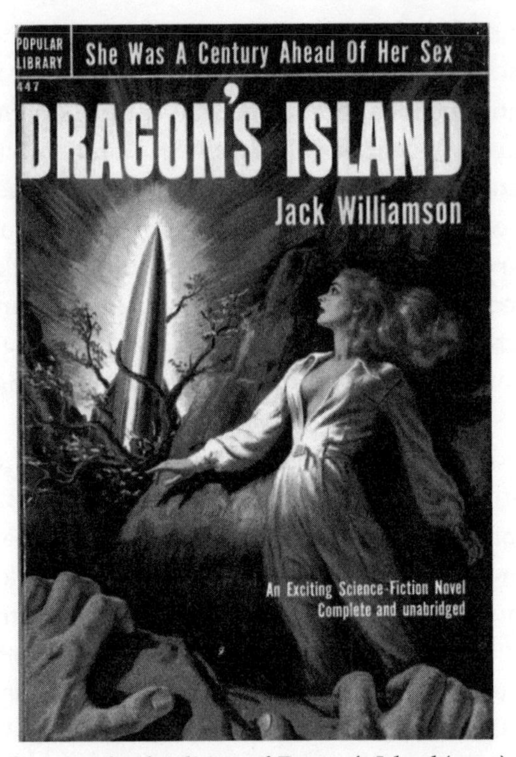

The paperback edition of *Dragon's Island* (1952).

Avery at the Rockefeller Institute in New York. Avery showed that pneumonia bacteria could be transformed from infectious to benign, and vice versa, by adding an extract of the other kind of microbe. The earliest interpretation of Avery's work suggested that transformation induced a mutation in the receiving bacteria – it looked as though the process produced a specific genetic change. [8] This would have been revolutionary – existing methods of creating mutations with X-rays or chemicals were essentially random, while selective breeding, which had been the basis of agriculture for millennia, was a slow, hit-and-miss process.

But this interpretation was wrong. Avery and his colleagues had in fact discovered something much more fundamental: the nature of the genetic material itself. To their surprise, the material they transferred between bacteria – they called it the transforming principle – turned out to be made of deoxyribonucleic acid, or DNA.

This substance had been thought to have a purely structural role in chromosomes, which carry genes – it was seen as a kind of scaffold, while genes were generally assumed to be made of proteins. Through a series of careful experiments, Avery and his colleagues, Colin MacLeod and Maclyn McCarty showed that, in bacteria at least, genes were made of DNA. This opened the road to decades of discoveries in molecular biology, the great new science of the second half of the twentieth century.

During the 1950s, speculation about transferring DNA from one organism to another, or using enzymes to change DNA sequences, moved from science fiction into science itself. In December 1958, Ed Tatum gave a lecture in Stockholm, to mark his share in half of the 1958 Nobel Prize in Physiology or Medicine for showing with George Beadle that genes can produce enzymes (the other half of the prize had gone to Joshua Lederberg, who had turned to bacterial genetics after reading Avery's 1944 paper on DNA). Tatum was speaking a little over five years after Jim Watson and Francis Crick, using data from Rosalind Franklin and from Maurice Wilkins, had proposed that DNA has a double-helix structure. At this time the role of DNA as the hereditary material in all life was still no more than a working hypothesis. It had taken over a decade for Avery's discovery to be widely accepted, and there was still no decisive proof that all genes were made of DNA. Scientists could be reasonably sure of the genetic function of DNA only in bacteria and viruses; the situation in more complex organisms remained unclear until the late 1960s.[9]

The closing part of Tatum's speech – a mere 400 words entitled 'Predictions' – looked into the future and foresaw our present:

> Perhaps within the lifetime of some of us, the code of life processes tied up in the molecular structure of proteins and nucleic acids will be broken. This may permit the improvement of all living organisms by processes which we might call biological engineering.[10]

The kind of engineering Tatum had in mind involved synthesising DNA molecules with desired characteristics and introducing these molecules into an organism by injection or by using engineered viruses. The aim of this work, he argued, would be to cure genetic defects and to create more productive and disease-resistant animals and plants. He predicted that a full understanding of the interplay of nature and nurture would lead to a new Renaissance 'in which the major sociological problems will be solved and mankind will take a big stride towards the state of world brotherhood and mutual trust and well-being envisaged by Alfred Nobel'.

Although Tatum's description of the direction of science was remarkably accurate, you may have noticed that it did not lead to his utopia. Tatum, like many who predict the future, underestimated the difficulties in applying scientific discoveries. Understanding needs to be turned into reliable and scalable technology and there must be cultural acceptance of both the techniques involved and their application. That acceptance can be affected by all sorts of contingent factors: in the years that followed Tatum's speech, political and social developments profoundly altered global views on science, building on a growing suspicion of the power of the atom that dominated the Cold War world of the 1950s. By the end of the 1960s a new, pessimistic attitude crystallised, colouring and shaping responses to the coming revolution in genetics.

*

In January 1960, *Time* magazine named US scientists as its 'Men of the Year'.[11] The fifteen men who represented the profession were mainly physicists, but there were two molecular geneticists on the list – Beadle and Lederberg – and the *Time* article hailed the 'glittering opportunities' of molecular biology, in particular the hope that in the near future 'it should become possible to treat and correct genetic diseases, now mostly incurable'. For *Time*, science was 'at the apogee of its power for good or evil'.

A mere twelve months later the balance seemed to have shifted towards the negative. In January 1961, as President Eisenhower handed over power to his successor, John F. Kennedy, the old soldier broadcast to the nation, alerting the American people to the growing influence of the military-industrial complex, criticising the links between science and government, and predicting that 'public policy could itself become the captive of a scientific-technological elite'. Eisenhower's gnomic warning – soon amplified by fears of nuclear annihilation during the Cuban Missile Crisis of October 1962 – connected to a slow growth in pessimism about science that was taking place across Western societies.[12]

These doubts and suspicions were principally driven by the menace of nuclear weapons, and focused not only on the immediate threat of thermonuclear annihilation, but also on the danger of radioactive fallout from nuclear tests that was increasingly recognised as a potential trigger for a possible epidemic of cancers and mutations. In 1954, US nuclear tests on Bikini Atoll caused a scandal when they produced unexpectedly high levels of fallout, which affected hundreds of people on nearby islands and on a Japanese fishing boat. In Nevil Shute's best-selling 1957 novel *On the Beach*, which was made into a film two years later, nuclear weapons cause the death of humanity through a massive cloud of fallout that rolls around the planet. By the mid-1950s there were alarming numbers of atmospheric tests (over 250 had taken place by 1958) and calls grew for a ban on atmospheric and undersea tests. Eisenhower was initially supportive of a 'moratorium' on tests – this term became widespread with regard to nuclear weapons and would eventually be used repeatedly down the decades in debates about the dangers of genetic engineering.

Eventually, after a great deal of politicking on both sides of the Iron Curtain and a lot of campaigning by protestors, including chemist Linus Pauling, who won the Nobel Peace Prize in 1962 for his role, the Partial Test Ban Treaty came into effect at the end of 1963. That did not mean that the threat of fallout went away. In the United States there were repeated attempts to develop the peaceful use of nuclear weapons through Project Plowshare, in which nuclear bombs would gouge out new harbours in Alaska and Australia or even blast a new sea-level Panama Canal that would directly link the Atlantic and Pacific Oceans (this would have required anywhere between 185 and 925 explosions, depending on the route).[13] Devised by Manhattan Project veteran and 'father of the hydrogen bomb' Edward Teller, Project Plowshare carried out twenty-seven test detonations between 1957 and 1973 (it was finally defunded in 1975), leaving

100-metre-deep craters and ejecting debris nearly five kilometres into the atmosphere. From the outset, the project was dogged by hubris, as grandiose plans turned out to be unfeasible, unwanted or hideously expensive, while there was repeated opposition from the public and from local authorities who were concerned that underground explosions might produce dangerous levels of fallout. Which is what happened. Radioactive isotopes were repeatedly found in the environment – including in neighbouring towns and on farmland, causing widespread protest and weakening governmental support for a project that looked increasingly out of step with reality.

Hollywood picked up on these fallout fears in the crudest way – 1954 saw the release of *Them!*, a monster movie in which radiation from atomic tests led to the appearance of two-metre-long giant ants that threatened Los Angeles. When the ants were finally defeated by conventional weaponry, one of the scientists concluded: 'When Man entered the Atomic Age, he opened the door to a new world. What we may eventually find in that new world, nobody can predict.' The power of the atom and the power of the gene became intertwined in the popular imagination. Media aimed at the younger generation was particularly interested in the question – key Marvel Comics characters created in the early 1960s, such as the Hulk, Spider-Man and the X-Men, derived their superpowers from the effects of radioactivity on their DNA.* Similarly, when the Daleks, the iconic baddies in the BBC TV science fiction series *Doctor Who*, first appeared in 1963

*Despite its apparent triviality, the changing origin story of the X-Men nicely illustrates one of the themes of this book – the shift in public fears from radiation to genetic engineering and how these were reflected in culture at all levels. At their creation, the X-Men gained their powers through mutations that were implicitly caused by radiation, with this link being explicit in the case of The Beast (his father was a nuclear engineer, as explained in *X-Men* #15, from 1965). But in 1980, as fears of radiation receded and worries about the power of genetics grew, a new layer of Marvel mythology was added, involving hokey von Däniken-esque aliens known as the Celestials. According to the new origin backstory, a million years ago the Celestials visited the Earth and carried out a series of genetic engineering experiments on pre-humans (*What If...* #23). One experiment, carried out by a Celestial called Oneg the Prober, produced the human lineage, replete with a 'latent gene' that when mutated would later produce superpowered individuals.

they were revealed to be mutants created by a centuries-long atomic war on the planet Skaro.

Both the bomb test moratorium and the test ban treaty required the public to trust their political and military leaders at a time when confidence in authority – including in science – was becoming increasingly fragile. Intellectuals began to question science's claims to authoritative knowledge, in particular following the publication of Thomas Kuhn's *The Structure of Scientific Revolutions* in 1962. Kuhn explored how scientific disputes were also struggles for power within science; to his dismay, some readers concluded that truth was a purely relative thing and that ultimately science was just another story. More concretely, there were growing worries about the inadvertent effects of chemicals, as highlighted by the thalidomide scandal, when a drug prescribed to relieve morning sickness caused tens of thousands of children to be born with deformities. The side effects of chemicals were also at the heart of Rachel Carson's prophetic 1962 book *Silent Spring*, which described the unintended environmental effects of insecticides. Meanwhile, the fallibility of technology and the potential for a technical incident to produce annihilation were portrayed with savage black humour in Stanley Kubrick's 1964 film, *Doctor Strangelove*.

The creeping American involvement in Vietnam was a key contributory factor to a growing mistrust of government in the United States.[14] Gradual awareness of the use of chemical weapons such as napalm, Agent Orange and dioxin in Vietnam reinforced widespread suspicions about the involvement of scientists in government-sponsored projects.[15] * Throughout the 1950s and 1960s, fiction writers explored the inadvertent consequences of chemical and microbial research, for example in John Christopher's novel *The Death of Grass* (1956), in which a pesticide causes a virus to mutate and kill all grasses and related plants, leading to a global famine. In the United Kingdom, concerns about bioweapons were amplified in 1962 when a scientist at the Microbiological Research Establishment at Porton Down died after being accidentally infected with bubonic plague.

---

* Ironically, leading scientists, such as the pioneer molecular geneticist Matthew Meselson, played a key role in exposing the use of these weapons.

Fears of a bioweapon escaping from the laboratory were a popular trope that featured, for example, in Alistair MacLean's 1962 thriller *The Satan Bug*, which was turned into a blockbuster movie in 1965. In one of the most famous post-apocalyptic novels, John Wyndham's 1951 *The Day of the Triffids*, the main character speculates that the triffids – motile, venomous, oil-producing plants with a malevolent streak – were deliberately created in a Soviet biolab before being accidentally released.[16] In 1963, Kurt Vonnegut's satirical novel *Cat's Cradle* described the creation by the military of a special form of ice – *ice-nine* – which froze at room temperature. It inevitably escaped and spread its characteristics, in a terrifying runaway fashion, to every drop of water on the planet.

Fears of hidden forces using science to manipulate the world were expressed in fiction by Thomas Pynchon in novels such as *V* (1963) and *The Crying of Lot 49* (1966) and by films like *The Manchurian Candidate* (1962), which soon nourished conspiracy theories over the assassination of President Kennedy a year later. A year after publishing *Dune*, science fiction writer Frank Herbert combined the themes of *Brave New World* with the new sensibility of suspicion in *The Eyes of Heisenberg* (1966) in which an immortal elite used genetic manipulation to control the behaviour of the human population. Non-fiction also pursued similar alarming themes. *Science and Survival*, published in 1966 by cell biologist Barry Commoner, who had played a key role in the opposition to Project Plowshare, helped to lay the basis for the US environmentalist movement, with its critique of industrial and technological development, frequently spiced with urban fantasies about the attractiveness of country living. Books such as Barbara Ward's *Spaceship Earth* (1966) sought the solution to the ills they diagnosed not in more growth or smarter technology but in fundamental changes to the way in which society worked. The times they were a-changin'.

*

In the late 1960s, the global public's alarmed fascination with the sometimes macabre implications of modern biology reached its height with the publication of a series of popular books with dramatic titles, each of which offered a deeply disturbing vision of the near future. These books sold millions, influencing the public, the media and scientists themselves.[17]

In 1968, US biologist Paul Ehrlich published *The Population Bomb*, which opened with the terrifying assertion that overpopulation and lack of resources would lead to hundreds of millions of people starving to death in the coming decade. A few months later, *The Biological Time Bomb* by British science journalist G. Rattray Taylor appeared. This focused on imminent biological and medical developments such as IVF, genetic engineering and the apparent possibility of creating life. The book contained disturbing predictions, including the claim that scientists would shortly be able to create 'combinations of characteristics which have never previously existed in nature' and that they would even 'with the aid of artificial gestation, go on to manufacture completely novel organisms'.[18] Rattray Taylor's conclusion was bleak:

> Man now possesses power which is so extreme as to be, at most, godlike. ... It is precisely because we cannot see, in detail, the consequences of using the new technology that they constitute dangers. The fact that they might be used for benign purposes or so as to benefit man is not the point, for history shows us that man is far more likely to use power wrongly rather than rightly.[19]

The media responded to *The Biological Time Bomb* with alarm. As a UK newspaper put it:

> Mr Rattray Taylor has drawn a diagram of the bomb. This must now be used to defuse it, use its components carefully and build in the safeguards for our survival. The alternative is inaction that will set the world on fire. It is not too late to confiscate the matches and put them in the hands of responsible, politically independent far-seeing people.[20]

Some scientists took these concerns seriously. In October 1968, Francis Crick gave a lecture in which he surveyed the future of biology, taking as his starting point Rattray Taylor's book.[21] Although Crick considered that the two key steps required for genetic engineering to become a reality – altering genes in a desired fashion and transferring genes between organisms – looked rather difficult, he felt that the implications needed to be addressed immediately. 'We must start to think about these problems now', he told his audience. 'Many of us need a totally new outlook to deal with them.'

A year earlier, Marshall Nirenberg, who cracked the genetic code in 1961 and was shortly to receive a Nobel Prize for his work, had accurately predicted the 'reprogramming' of DNA in mammalian cells, embracing the view of life as information that had become increasingly widespread in the 1960s: 'my guess is that cells will be programmed with synthetic messages within twenty-five years. If efforts along those lines were intensified, bacteria might be programmed within five years.'[22] Worried that we would be able to manipulate our own genes long before we had resolved the concomitant ethical and moral problems, Nirenberg proposed that once this power became real, humanity should refrain from using it. In other words, he proposed there should be a moratorium on such genetic experimentation. Other scientists disagreed. Joshua Lederberg responded by saying that experimental bans, whether they were imposed by individuals or by the state, would inevitably hamper scientific and medical progress.[23] The future battle lines surrounding genetic engineering were being drawn, even before the science was a reality.

Some people were enthusiastic at the prospect of manipulating genes. In 1967, Brian Richards and Norman Carey, two UK research scientists at the pharmaceutical company Searle, wrote a confidential document in which they outlined how the hypothetical technology might be used: curing genetic diseases and cancer, enabling organ transplantation and improving agricultural animals and plants. The pair even suggested that genetic manipulation might be a route to enabling agriculture in the ocean and in extraterrestrial environments. They do not appear to have considered any potential downsides.[24]

Despite the alarming tone of much of the popular and scientific coverage in this period, there were few suggestions about how to respond to the looming possibility of manipulating genes. Most proposals revolved around action by the United Nations, by nation states or simply by scientists themselves – there was no sense that the broader public should be involved. And yet in the world outside the laboratories, profound political and cultural changes were taking place that had serious implications for science and how it was viewed by the population, in particular the young.

– TWO –

TOOLS

In the closing years of the 1960s, tumultuous events shook the whole world. In the United States, the United Kingdom, Mexico, Germany, France, Japan, Czechoslovakia and elsewhere, student unrest, a partial consequence of the recent massive global growth in higher education, fused with protests over civil rights, the Vietnam War and the perception of technocratic rule. There were occupations, protests and, in France in May 1968, the largest general strike the world has ever seen as over ten million workers downed tools. These movements challenged existing power structures at every level of society and culture, radicalising millions of young people, including the younger generation of scientists.[1]

This inchoate movement was deeply suspicious of technology, in particular where it had links with industry and the military.[2] The most striking expression of this was seen a few years later in the critical response of many activists to what still seems to have been the height of US technological power – the Apollo moon landings.[3] In the words of Gil Scott-Heron's 1970 poem 'Whitey on the Moon':

> I can't pay no doctor bill.
> (but Whitey's on the moon)
> Ten years from now I'll be payin' still.

(while Whitey's on the moon)
The man jus' upped my rent las' night.
('cause Whitey's on the moon)
No hot water, no toilets, no lights.
(but Whitey's on the moon)

In the United States, university departments that undertook military research became a focus of protest as radical young scientists set up a loose group called Science for the People. One of their main targets was the staid American Association for the Advancement of Science (AAAS) – from 1969 to 1971, Science for the People members regularly disrupted AAAS meetings, holding alternative sessions, street theatre events and generally aping some of the features of Chairman Mao's Cultural Revolution. As one Science for the People leaflet from 1970 put it, referring to AAAS attendees: 'They are not here to educate us. We're here to educate them.'[4] For some of those protestors, power came from the end of a test tube.

One key event took place on 4 March 1969 when scientists and students at the Massachusetts Institute of Technology (MIT) organised a day-long research strike in protest against the growing power of the military-industrial complex. One MIT group released a statement highlighting the areas in which science posed an existential threat to humanity: nuclear weapons, pollution-induced climate, ecological changes and genetic manipulation. One MIT debate, which involved students and academics from all the neighbouring institutions, discussed the lessons to be learned from the failure of the nuclear scientists in the 1940s to prevent the use of the atomic bomb.[5] The overwhelming view of the protestors was that such a mistake should not be repeated – scientists should act. In response, the chairman of President Nixon's Science Advisory Committee, Lee DuBridge, insisted on the right of researchers to study whatever they wanted, without any social control. DuBridge accepted that government might restrict research that was 'harmful or dangerous', but that was not the fundamental point. 'Scientists', he argued, 'must be free to pursue the truth wherever they can hope to find it.'[6]

On the other side of the country, the very nature of science itself was challenged by Stanford historian Theodore Roszak in his highly influential book, *The Making of a Counter Culture*. Roszak argued not only that the younger generation was opposed to science and technology, but also that scientific objectivity itself was an illusion.[7] In what now looks like a profound contradiction, Roszak's anti-science book nestled on young scientists' shelves next to Jim Watson's recent best-selling novelised account of the race to discover the structure of DNA, *The Double Helix*.* Watson's book described events that occurred fifteen years previously but the breakthrough was only now bearing fruit. Furthermore, his lively first-hand descriptions of scientists and their sometimes base motivations seemed to capture the imagination of the young generation: although Watson's scornful treatment of women, in particular his depiction of Rosalind Franklin, was straight out of the 1950s, *The Double Helix* made molecular biology seem modern, attractive, rule-breaking and sexy.

These complex and contradictory cultural, political and psychological changes that were shaking the world had consequences for how genetic discoveries were viewed. This can be seen in the strikingly contrasting responses to two important findings that took place on either side of the pivotal year of 1968.

In December 1967 Arthur Kornberg and his colleagues at Stanford and Caltech reported they had been able to synthesise viral DNA in a test tube. Furthermore, the DNA was functional – if the molecule was introduced into a bacterial cell, the microbe reproduced the virus, following the molecular instructions in the synthesised DNA.[8] A key step in this work had been described a month earlier, when Kornberg reported that an enzyme called a ligase could cause the ends of two molecules of DNA to fuse together (to ligate, in the jargon). In these essential steps towards making genetic engineering a reality, Kornberg's experiments showed that bits of DNA could be synthesised, assembled and would function in a cell.[9]

A carefully organised press conference announced the discovery to enormous global excitement. James Shannon, director of the US National Institutes of Health (NIH), provocatively described the work as involving 'the creation of one form of life in a test tube'.[10] President

* Well, on my bookshelf at least. What both books evoked in me was a numinous sense of the power and immanence of discovery.

Johnson hyped up the result in his Texan drawl, calling it 'an awesome accomplishment ... [one of] the most important stories you ever read or your daddy read or your grandaddy ever read'.[11] Scientists piled in to highlight the implications – one bacteriologist suggested 'it may be only a few years before scientists will be able to take selected genes from humans, make large amounts of genetic material, then get it into cells where it can alter the heredity of the cells'.[12] On the front page of the *Los Angeles Times*, Kornberg's co-author Robert Sinsheimer spoke of a forthcoming 'second Genesis' when humanity would make a completely new cell from scratch, while in *The Encyclopedia Britannica Yearbook*, Joshua Lederberg praised the experiment as 'a cardinal event in the history of biology', claiming that 'the historic importance of the event was unchallengeable'.[13]

Two years later, after 1968 had convulsed the world, another step towards control over DNA got a completely different reception. In November 1969 the scientific journal *Nature* published a study by a group of young researchers from Harvard Medical School. They were led by 31-year-old Jon Beckwith, a member of the Boston chapter of Science for the People who had been involved in the 4 March events, and included James Shapiro, at 26 years old a recently minted PhD. Beckwith's youthful group described how they had isolated a gene known as *lac* from *E. coli* – this gene encodes an enzyme that enables the bacterium to break down the sugar lactose.[14] *lac* was of particular interest because it is only activated under certain circumstances – the environment in which the bacterium is reared affects whether the enzyme is produced.

When viruses infect bacteria – these viruses are called bacteriophages, or 'phages' – they often end up with fragments of their host's genome in their own DNA. Beckwith and his group allowed phages to repeatedly infect *E. coli*, thereby gathering up bits of bacterial DNA and eventually producing a virus that contained the whole bacterial *lac* gene. The degree of control involved was exquisite and unprecedented; Beckwith emotionally described his feelings about the experiment: 'we derive a great deal of pleasure from the type of work we do. The manipulations of genes, practically at will, has been a lot of fun. It is a constant temptation for me to spend all my waking hours thinking and working in this area.'[15]

But when the research was announced at a press conference there was none of the hype and triumphalism that had accompanied Kornberg's work two years earlier. Instead, the young researchers created an international scandal by emphasising the potential dangers that flowed from their discovery. Beckwith told journalists:

> I feel the bad far outweighs the good in this particular work. I feel that it is far more frightening than hopeful. It is obvious that it raises the possibility of genetic engineering.[16]

Shapiro declared that their method could lead to 'bad consequences over which we have no control'.[17] A few weeks later the youngest member of the team, Lawrence Eron, who had been involved in the project as a summer student, outlined the lurid possibilities the trio were concerned about – possibilities that in fact were far removed from their work:

> For example, it might be possible at some future date to inject viruses containing a sterility gene into the water supplies. ... You could thereby eliminate future generations of blacks or Indians. ... It might also be possible for dictators to use this technique in the distant future to eliminate dissent by injecting genes into humans which make their behaviour more placid, more assenting.[18]

Although Beckwith soon admitted that their statements were misleading and awkward, the world's press immediately jumped on the story, ramping up the fear factor.[19]

'Genetic "Bomb" Fears Grow' proclaimed London's *Evening Standard*, while the *Los Angeles Times* inadvertently conjured up an odd image when it warned its readers 'Test-tube Man Feared'.[20] In London, an editorial in *The Times* drew parallels between the huge leaps made by molecular biology and the link between modern physics and nuclear power:

> So may there not be a biological equivalent of the atomic bomb? ... In molecular biology, as in nuclear physics, the step

> between the test-tube and the production plant is not merely uncertain but long. Vigilance is for the time being sufficient. Foreboding is not yet called for.[21]

Within weeks, Shapiro and Beckwith created further waves when they were interviewed on NBC's *Today* programme. Shapiro doubly scandalised viewers, first by not wearing a tie and then by announcing that he was going to give up science because of his fears about the misuse of research. As he put it priggishly in a newspaper interview:

> I am dropping out of science because it is simply being exploited by the people who run this country to serve their own ends. To work in a laboratory is futile at the present time. The only useful mode of life I can imagine now is to challenge the political system.[22]

Surprisingly, Shapiro's declaration that he was going to drop out and become a political activist (he would survive on his inheritance, he said) led to a very positive article in *Science*. Salvador Luria told the journal he supported the move – 'I think it is important that there are scientists like Shapiro who point out the misapplications of science.' On the other hand, when Luria was asked if he thought Shapiro's decision represented a loss to science, he laughed and said 'There are enough scientists as it is …'[23]

A couple of months later, Beckwith received an award for his research from the US pharmaceutical company Eli Lilly; in keeping with the spirit of the times, he warned the prize ceremony that 'science in the hands of the people who rule this country and who run our industries is being used to exploit and oppress people all over the world' and went on to denounce the US drug industry for its vast profits, its manipulation of patents and its pernicious links with the medical profession. In a final snub Beckwith announced that he was going to give his $1,000 prize money to the Black Panther Party, a radical African American organisation.[24]

Nature responded to the Beckwith group's statements with a disdainful editorial that expressed bewilderment about the growing tendency of the public 'to seek sombre consequences for scientific discoveries'.[25] A month later the editor of *Nature*, John Maddox, tried to rally scientists against what he called the 'doomsdaymen', calling on his readers to stamp out 'heresies' – negative views of science and technology.[26] As to all the recent fuss, it could be dismissed – 'it is not merely irrelevant but mischievous to raise false spectres like genetic engineering', he said.*

This was becoming a theme for *Nature*, which earlier in the year had despaired that apparently radical young people were technological conservatives, almost Luddites, but was unable to understand why, nor to explain why these feelings had become stronger in the 1960s.[27] Raging against pessimism – this became something of a hobby horse for Maddox, who soon dashed off a best-seller on the topic – would not do the trick.[28] The growing gulf between the scientific establishment of journal editors and Nobel Prize winners on one side and the younger generation of scientists and concerned members of the public on the other was not simply caused by a difference in mood.[29] Deeper changes in global society were responsible.

*

Some senior scientists were more in tune with the new mood. In April 1969, Maurice Wilkins, who in 1962 had won the Nobel Prize with Watson and Crick for his work on the structure of DNA, helped set up the British Society for Social Responsibility in Science (BSSRS – pronounced 'bissriss'); I joined when I was a student in the middle of the 1970s.[30] Wilkins had been a member of the Communist Party until 1939 and had retained his left-wing sympathies – the British security services routinely opened his mail until 1953, and in October 1962 the MI5 Scientific Officer, Peter 'Spycatcher' Wright, questioned Crick about Wilkins's political reliability shortly after he won the Nobel Prize along with Watson and Crick.[31] BSSRS was hardly a

* To bolster his argument Maddox compared concerns about genetic engineering to the recent suggestion that using fossil fuels would lead to increased CO_2 levels and rising temperatures. There was no need to worry about CO_2, he insisted: 'whatever effects there may be are likely to be at once slower and smaller than the more gloomy prophets say.'

product of the counterculture: it was launched at the Royal Society's new headquarters at Carlton House Terrace in the heart of the British Establishment and had the backing of the UK's left-leaning intellectual aristocracy – among those who signed a statement of support were Eric Hobsbawn, Aldous Huxley, J. D. Bernal and Bertrand Russell.

One of BSSRS's first initiatives was a London meeting on 'The Social Impact of Modern Biology', a kind of extended version of the 4 March debates at MIT, which focused on the issues raised by recent discoveries and the worries highlighted by Beckwith, who was one of the key speakers.[32] Held at Friends House opposite Euston Station on 26–28 November 1970, with an audience of up to 800 crammed in each day, the meeting saw a procession of scientific superstars, from Jacques Monod to Jim Watson, as well as mathematician Jacob Bronowski (yet to make his influential *Ascent of Man* documentary series), historian Bob Young and sociologist Hilary Rose (the only woman invited).

In his opening remarks, Wilkins focused on 'the crisis in science', as evidenced by widespread student unrest and a perceptible change in how scientists viewed their work: he felt that some of his colleagues were beginning to lose confidence in the value of science.[33] In general, the meeting did not take up the challenge of exploring this question – the lectures tended to be heavy on the biology and light on the social impacts. Above all there was a mismatch between the views of the older generation, as represented by Jacques Monod's defence of science's quest for objective knowledge, and those of younger speakers, such as Young and Rose, who emphasised not only that 'all facts are theory-laden' but that, even more provocatively, in Young's words, no biological fact 'is immune from the ideologising influences of its social content'.[34] The old guard were firm believers in science as a neutral, objective force; the young radicals understood that science takes place in a social context and has political implications, in particular when it is applied.

The contrast between these views was particularly stark in the final two talks, from Beckwith and Bronowski. Beckwith called for scientists to 'build the ties with the rest of the people' in order to 'bring science to the people'. He denounced the structure of the

BRITISH SOCIETY FOR SOCIAL RESPONSIBILITY IN SCIENCE

PROGRAMME AND ABSTRACTS

The Social Impact of Modern Biology

NOVEMBER 26, 27 & 28 1970
Friends House, Euston Road, London WC1.

Supported by the Council for Biology in Human Affairs, SALK Institute, San Diego, California.

TWO SHILLINGS.

Programme of the November 1970 BSSRS meeting on 'The Social Impact of Modern Biology'.

meeting, with its Fellows of the Royal Society (FRSs) and Nobel Prize winners – even his own presence, he explained, was simply a consequence of the press-sponsored notoriety he had acquired. In contrast, Bronowski focused on the need to make science independent from the influence of the military and the state by creating an international umbrella organisation to oversee funding. This staid view of social responsibility in science was not well received by the more rebellious parts of the audience. In the discussion, psychologist Tim Shallice – then a representative of the radical wing of BSSRS, now an FRS – dismissed Bronowski's proposal as 'a prize example of liberal claptrap'. There were even occasional interjections of 'balls' and 'bullshit' from the audience at various points.[35] Writing in the US magazine *Science for the People*, Beckwith later repeated Jim Watson's suggestion at the meeting that scientists should begin a dialogue to educate the world's citizens, but followed Watson's words with a classic piece of seventies radical-speak: 'Right on! Science for the People!'[36]

Looking back, it is striking that there was very little discussion at the meeting about how to respond to the possibility of genetically engineered organisms. Although there was a brief debate about their potential use in biological warfare, the most focused exploration of the question came from Bronowski, who – like Wilkins – referred to the horrified realisation of the physicists in the 1930s that their science could destroy the world. (Wilkins had worked on the Manhattan Project and abandoned physics for molecular biology after the bombs were dropped on Hiroshima and Nagasaki.) Like Nirenberg three years earlier, Bronowski argued that scientists should voluntarily agree not to pursue potentially menacing discoveries. As he put it:

> If science is to express a conscience, it must come spontaneously out of the community of scientists.[37]

This lack of engagement with what was soon to become a key issue is all the more surprising because one of the talks at the meeting described recent progress in creating genetically modified plants. In his lecture on 'Molecular Biology and Agricultural Botany', Arthur Galston of Yale University presented his unpublished research showing that the introduction of foreign RNA into tobacco plant tissue could stop the production of a particular enzyme. He also described Dieter Hess's claim from the previous year that DNA could be transferred between plants, transforming a white petunia into a red-flowering version, much like Avery had done with bacteria in the 1940s. There had been a number of such reports – in 1968 the Belgian researcher Lucien Ledoux published an article in *Nature* showing that bacterial DNA could be integrated into the DNA of germinating barley seeds.[38] Galston was a leading campaigner against the United States' use of Agent Orange in Vietnam, well aware of the potential dangers of science, and yet neither he, nor anyone else at the meeting, raised any ethical, environmental or safety concerns about the suggestion that nucleic acids could be transferred between very different organisms.[39] The potential implications were all seen as positive. There were no triffids or death of grass to give the audience sleepless nights; instead Galston hoped that genetic manipulation of plants might 'yet permit man to gain time in his struggle against starvation, pollution and waste'.[40] No one batted an eyelid.

These reports of the genetic modification of plants do not form part of the usual story of genetic engineering, or even of molecular biology.[41] Such histories are always focused on viruses and bacteria – they are coloured the grey of the Petri dishes in which scientists grow their microbes, not the green of plants. The reason for this apparent bias is simple: there were growing doubts about the validity of all these experiments that claimed to show transfer of genetic material in plants. Researchers found it difficult or impossible to replicate the results and less dramatic explanations could be found for the effects.[42] Confidence in the claims gradually waned until they were forgotten, even by historians.[43]

*

Despite these supposed breakthroughs with plants, by the end of the 1960s molecular biology appeared to be in the doldrums. The excitement associated with the fundamental discoveries of the 1950s and of the race to understand the genetic code and its role in protein synthesis had all evaporated. Leading researchers such as Francis Crick, Sydney Brenner and Seymour Benzer had shifted their attention, arguing that all the interesting molecular stuff had been discovered and that the key intellectual challenge was now to understand the nervous system, while Jim Watson was concentrating on cancer and running Cold Spring Harbor Laboratory. In 1968 phage geneticist Gunter Stent captured the mood in an elegiac account of the field that appeared in *Science*: 'That Was the Molecular Biology that Was.'[44] French researcher François Gros, who had worked with Watson at the beginning of the 1960s, later recalled that 'by the beginning of the 1970s the whole discipline went into crisis … research was treading water and the heart had gone out of it'.[45]

Not everyone agreed. In September 1970 British molecular biologist and *Nature* journalist Benjamin Lewin – he would soon jump ship and create *Cell*, an enormously influential and highly profitable journal – wrote a long review proclaiming 'The Second Golden Age of Molecular Biology'.[46] But despite Lewin's confidence-boosting title, the results he highlighted merely suggested that the models of gene function and protein synthesis developed in bacteria were indeed applicable to multicellular organisms, as everyone had expected. However, Lewin's article also described the recent discovery of the key tools that would soon make it possible to carry out genetic engineering, although he dismissed the possibility as far-fetched: 'The idea that the day will soon come when extra pieces of DNA can be synthesised and added at will to the genome of higher organisms to repair their defects is not only naive but also wrong-headed.'[47]

As well as the ability to synthesise DNA in the test tube and to stitch bits of nucleic acid together to make a gene, Lewin highlighted the surprising discovery that some RNA viruses were able to make an infected cell copy their RNA into DNA, thereby enabling it to integrate into the cell's genome, where it would produce more viruses. This process was known as reverse transcription and involved an enzyme called reverse transcriptase, which had been simultaneously described in 1970 by Howard Temin and Satoshi Mizutani at the University of Wisconsin and by David Baltimore at MIT.[48] This enzyme was of decisive importance for the future of genetic engineering because it provided a way to turn messenger RNA – produced from the cell's DNA as a gene functions – back into a DNA sequence (this became known as complementary DNA, or cDNA). Eventually, scientists were able to transfer cDNA molecules between organisms.

In 2021, Baltimore told me it took him only a couple of days to do the experiments that led to the demonstration of the mechanics of reverse transcription and then to an unusually rapid Nobel Prize a mere five years later. He also revealed how, for his 32-year-old self, science and politics were intertwined, and one of the most significant discoveries of the decade was made amid campus protest. When Baltimore began doing the experiment in spring 1970, much of MIT – staff and students – was on strike against the recent US invasion of Cambodia. Baltimore – who in 1969 had helped organise the 4 March events at MIT – told me that after he had completed his experiment, 'I closed up the lab. I didn't tell anybody what I'd done. I was out on the streets, with everybody else, trying to keep my students from ending up in jail. And we did that for the next three or four days. Then I went back to the lab.'

Another key set of genetic engineering tools that was discovered at this time, but which were not highlighted by Lewin, were what are known as restriction enzymes. They take their name from the fact that some strains of bacteriophage cannot survive in some species of bacteria – bacterial enzymes restrict the ability of the virus to reproduce by cutting viral DNA into pieces. This ability of bacteria to defend themselves against viral infection had been discovered in 1953 and had been intensively studied by Werner Arber and Daisy Roulland-Dussoix in Geneva. In 1968, Matthew Meselson and Robert Yuan at Harvard isolated the first of these enzymes and two years later Hamilton Smith at Johns Hopkins University discovered a restriction enzyme that chopped up viral DNA into stretches that always began and ended in the same three bases, suggesting that the enzyme somehow recognised a specific sequence.[49] This finding was of such significance that in 1978 Smith was awarded the Nobel Prize for his work, along with both Arber and Daniel Nathans, who was rewarded for his research in the 1970s on the topic.

Scientists had now assembled the key tools for manipulating DNA. They had found enzymatic scissors for cutting up DNA at precise points, identified enzymes that could stick two bits of DNA together, developed methods for replicating nucleic acids in the test tube and learned how to move genetic information from one kind of nucleic acid (mRNA) into another (DNA) using reverse transcriptase, and from one bacterium to another using a virus as an intermediary. The possibility of turning knowledge into action, of transforming biology into engineering, was slowly materialising. As to the potential dangers, any concerns were repeatedly slapped down by the leaders of the scientific community.[50] In the summer of 1971 the editor of *Science*, Philip Abelson, condescendingly dismissed such worries:

> Talk of the dire social implications of laboratory-related genetic engineering is premature and unrealistic. It disturbs the public

> unnecessarily and could lead to harmful restrictions on all scientific research.

This same point had been highlighted a few weeks earlier by Lee DuBridge, who used the broader context of the culture wars that were wracking US society to repeat his argument of the previous year, claiming science as an essential part of the American dream:

> Make no mistake, a limitation on experimentation in whatever cause is the beginning of a wider suppression. When we fail to experiment, we fail. In failing, we bring the best part of American society as we know it today to a halt. … If these trends progress, our society will become dull, stodgy, and altogether stagnant.

The coming conflict over genetic engineering was going to extend far beyond the benches of the molecular biology laboratory.

– THREE –

BIOHAZARDS

On 28 June 1971, Bob Pollack made a telephone call from his office at Cold Spring Harbor Laboratory on Long Island. Pollack was a keen young researcher whose Amish-style 'chin curtain' beard with its bare upper lip seemed to go with his earnest, somewhat reserved character.* He had just got wind of a planned experiment to introduce DNA from a virus that caused tumours in mammalian cells into *E. coli*, a bacterium found in the human gut. This had made him unusually agitated. He was concerned this proposed experiment might cause an epidemic of cancer and wanted to convince the researchers involved to abandon their project. His telephone call would set in motion a train of events that eventually changed the world.

Cold Spring Harbor is on the north coast of Long Island, about an hour's train journey from New York. The laboratory buildings sit on the west side of an inlet, nestling in a massive variety of trees, some native to the region, many others brought in to form a fabulous collection. Converted from a disused whaling station into a laboratory at the end of the nineteenth century, the research institute

* In 2022 Pollack told me that he adopted the beard in honour of the Soviet dissident author Alexander Solzhenitsyn.

was a centre of eugenics before its shift to the genetics of bacteria and viruses in the 1940s. In the 1950s and 1960s, the Cold Spring Harbor Laboratory annual symposium and training courses became an essential part of global science as molecular biologists raced to crack the genetic code and then tried to understand gene function. Over the decades, tens of thousands of researchers from around the world have played volleyball on the sand bar, held late-night beach parties, attended the inevitable lobster banquet and studied hard in these idyllic surroundings. In 1968, Jim Watson became director and reoriented the laboratory towards the study of cancer. Pollack, who was only thirty years old, had recently been recruited by Watson to work on mammalian cell genetics, with new facilities being built especially for this research.

In that warm June, Pollack was running an intensive three-week summer course to initiate researchers in the arcane techniques of cell culture, with a particular focus on cancerous cells. There were twenty students, from Europe and the United States; one of the four women was Janet Mertz, a 21-year-old PhD student from New York.[1] After completing a dual degree in engineering and life sciences at MIT (only 5 per cent of MIT students were women at the time), in January 1971 Mertz crossed the continent to work in Paul Berg's laboratory in the Department of Biochemistry at Stanford Medical School.[2] Berg was impressed with her intellect – 'smart as hell', he later recalled – but he also found that 'in the beginning she was a pain in the butt'.[3]

A couple of years earlier, Berg had begun working on SV40, a monkey virus that causes tumours when injected into newborn hamsters. SV40 was used to explore the cellular processes of cancer, in particular the then-fashionable idea that all cancers were viral diseases.[4] Many bacterial viruses end up adding bits of their host DNA to their own genome – this was the feature that had been exploited by Beckwith and his colleagues when they synthesised the *lac* gene. Berg began to wonder if mammalian viruses might do the same thing; if so, it might be possible to use this effect to transfer DNA from one mammalian cell to another, opening the road to the precise genetic manipulation of mammals.

Berg's initial idea was to use SV40 to transfer a bacterial gene of known function, such as *lac*, into a mammalian cell line. This would enable researchers to study gene function in mammals, about which virtually nothing was known. During laboratory discussions about this project Berg's group also considered the possibility of performing the opposite experiment – transferring SV40 DNA into an *E. coli* bacterial cell, after first integrating it into phage-derived plasmid DNA that naturally infects *E. coli*. This side experiment would not directly advance any of Berg's main interests, but it would provide insight into viral action and might ultimately contribute to the aim of using SV40 as a tool to probe how human cells work. This relatively minor project was given to Janet Mertz, months before she headed east to Cold Spring Harbor for Pollack's training course.

At the beginning of the course, each student summarised their current research. When Pollack then gave a seminar on 'Safe Work in a Safe Laboratory', there was discussion about Mertz's proposed experiment, which continued throughout the course. Pollack was incredulous: 'Do you really mean to put a human tumour virus into *E. coli* – a gut bacterium?' he asked her.[5] Inserting viral DNA that could potentially cause cancer into a bacterium that is naturally present in the human gut appeared to Pollack to be reckless. SV40 was already intensively studied without using bacteria that could infect a human; putting SV40 DNA into *E. coli* seemed to be simply asking for trouble. Eventually, after a lot of thinking, Pollack picked up the telephone on that Monday afternoon and got through to Berg in Stanford.

The conversation started off badly – Pollack blurted out 'Why are you doing this crazy experiment?' while Berg was irritated by Pollack's audacity in calling him at all. In 2021, Pollack told me that Berg replied: 'Who the *bleep* are you?' (Pollack, who is still a quiet and reserved man, refused to say what the missing word was.) Berg in turn said he could not recall his exact words but that he probably told Pollack to go to hell.[6]

Berg had already heard from Mertz that the proposed experiment had caused a row on the course and he had his responses ready: the chances of the manipulated bacterium escaping from the lab and causing problems were negligible, and anyway he would use a strain of *E. coli* that would die if it escaped from the lab. Pollack was unmoved and pressed his case, even inviting to fly Berg to the east coast so he could give a closing talk at the end of the course in four days' time (Berg declined, saying he was too busy). Eventually, Berg agreed to think about the problem and rang off. As he later admitted, he was 'really quite annoyed', or as he put it on another occasion, perhaps more frankly: 'My first reaction was, this is stupid; this is bunk. I thought it was outrageous.'[7]

Part of the reason behind Berg's irritation was that as departmental chair he had paid a great deal of attention to the potential dangers associated with SV40 – he had responded to concerns amongst researchers and technicians by building a laboratory with filtered air and special fume cupboards; the facility was maintained at negative pressure so that air would only flow into the room, trapping any escaped microbe.[8] But now it appeared that there was an additional problem he had not foreseen.

There was an element of culture clash in the exchange between the younger Pollack, a radically minded postdoctoral researcher on a temporary contract at an east coast research institution, and the established Berg, a tenured professor in a prestigious university. In his mid-forties, recently elected Chair of the Department of Biochemistry, Berg was definitely part of the square generation. Nevertheless, he was politically on the left – he had been involved in various teach-ins against the war in Vietnam and had travelled to Washington to protest against Nixon's bombing of Cambodia.[9] Further proof that Berg was no fuddy-duddy can be seen in a scientific countercultural event that he organised a few weeks before the telephone call from Pollack. Hundreds of writhing, stoned students and some talented dancers and musicians turned the molecular processes of protein synthesis into a performance.

This happening, which involved a fourteen-piece funky acid-rock band (organ, saxophone, bongos and the rest), took place on a Stanford playing field and was filmed.[10] Released a few months later under the title 'Protein Synthesis: An Epic on the Cellular Level', with a biochemical pastiche of *Jabberwocky* read over the soundtrack, the thirteen-minute film opens with a slim, staid Berg – short hair, tie, white short-sleeved shirt with two pens in the pocket – explaining protein synthesis in front of an office chalkboard. The scene then abruptly switches to long-haired, brightly clothed (or semi-naked) students dancing in the sun, acting out the steps involved in protein synthesis to the sound of pulsating music (Janet Mertz – completely sober – was part of a ribosome). The film has had over a million views on YouTube.[11] *

After Pollack's course was over, Mertz flew back to California, taking with her a bundle of documents, including a six-page document on safety and ethics, drawn up by Pollack and his friend and colleague, Joe Sambrook. This included a section entitled 'Are there any good experiments using human cells and viruses that should not be done?' which focused on the possibility of changing eggs or spermatozoa, or introducing DNA from humans or from viruses that infect humans into bacteria – exactly the experiment Berg envisaged. The document concluded:

> You ought to ask yourself if the experimental results are worth the calculable dangers. You ought to ask yourself if the experimental techniques are over the boundary and amount to experimentation with people. Finally, you ought to ask yourself if the experiment *needs* to be done, rather than if it *ought* to be done, or if it can be done. If it is dangerous, or wrong, or both, and if it doesn't need to be done, don't do it.[12]

In a particularly prescient passage, Pollack and Sambrook warned their students how scientists could be seduced by the attractions of mass communication:

> No one should be permitted the freedom to do the first, most messy experiments in secret and present us all with a reprehensible and/or dangerous *fait accompli* at a press conference.[13]

* One of the students involved has said: 'Nobody who participated in that dance remembers anything about it, due to pharmacological events out of our control.' Bohannan, J. (2010), *Science* 330:752. There have been a number of attempts around the world to repeat the performance, but none have been anywhere near as successful, lacking the numbers, talent, resources or perhaps simply the drugs that were available at Stanford.

On the same day that Pollack made his phone call to Berg, he and Sambrook produced a condensed version of their document, in the form of a letter to the editors of *Science* and *Nature*, which they hoped would 'initiate some discussion, in print and in private'.[14] In the end, they decided not to send the letter, wary of provoking the ire of laboratory director Jim Watson who might have felt attacked. Had the document been sent and published, the course of history might have been, if not different, at least substantially accelerated.

Mertz returned to California uncertain about how to proceed, her confidence shaken by the discussions that had taken place. As she recalled a few years later:

> Coming from a radical background, I figured, 'Well, even if there's only a 1 in 10^{30} chance that there's actually something dangerous that could result, I just don't want to be responsible for that type of danger.' I started thinking in terms of the atomic bomb and similar things. I didn't want to be the person who went ahead and created a monster that killed a million people. Therefore, pretty much by the end of the week I had decided that I wasn't going to have anything further to do with this project.[15]

Berg was also mulling over the criticism from Pollack and over the next few months he discussed the issue with leading researchers, including Joshua Lederberg and David Baltimore, both of whom thought the risks were extremely small, but could not find a good reason to proceed. At a July 1971 conference in Sicily, Berg described his experiment over beer one night to a group of several dozen young researchers. To his discomfort, many of them focused on the ethical and moral questions raised by his proposed experiment and expressed their disapproval. Francis Crick, who was also present, uncharacteristically had nothing to say.[16]

One of the many informal discussions Berg had in that summer of 1971 was with his close friend Maxine Singer, at a dinner party she organised with her husband, the lawyer Dan Singer. One of the guests was Leon Kass, who worked on bioethics at the NIH. Berg's proposed experiment worried everyone, with Kass and Dan Singer raising ethical and legal points that Berg had not previously considered. Although Berg maintained that the experiment was safe, the criticisms from friends and colleagues eventually wore him down – all the more so because the experiment that was causing the stink was not fundamental to his research. In the autumn of 1971, he telephoned Pollack and announced his decision:

> We are not going to do that part of the SV40 hybrid experiment, that worried you. You were right about it. We had not fully considered the possibly hazardous consequences.[17]

The first pause on research in genetic engineering had been decided and virtually no one knew.

*

Berg not only told Pollack he would not be doing the experiment, he also explained that he now fully appreciated the significance of biosecurity issues in this kind of research, and urged Pollack to join him in organising a conference on the question. As Pollack put it a few years later: 'typical of the man, he didn't just give me a brush-off; he used judo on the whole argument – just flipped the whole thing over and instead of being the guy who is the *problem*, he became the guy who is ostensibly the *solution*'.[18] Berg's judo flip was even more astute than appeared – having convinced Pollack that a conference would be a good idea, Berg deftly handed the tedious task of organising it to Pollack and two colleagues.

The conference, 'Biohazards in Biological Research', involved around 100 delegates (all but two of them from the United States) and took place in January 1973 at the Asilomar Conference Centre, a location habitually used by Berg's department for its academic retreats. Situated on the California coast just north of Santa Monica, the Asilomar centre is surrounded by pine trees, with the waves of the Pacific crashing on the beach a few hundred metres away. The key word in the title of the meeting – biohazards – was relatively new. The NIH and the National Cancer Institute had begun to use the term in the mid-1960s to describe various viruses and

Biohazard symbol, as designed in 1966 by
Charles Baldwin of Dow Chemical.

bacteria, and in 1966, Charles Baldwin of Dow Chemical designed a warning symbol to denote containment facilities that contained such hazardous biological material. Its fluorescent orange-red circles – redolent of the three rays of the radiation symbol – soon became widespread.[19]

The 1973 Asilomar conference focused on procedures for dealing with existing threats, in particular SV40. The issue that had so preoccupied Pollack – the introduction of tumour virus DNA into a widespread bacterium – was not mentioned. Indeed, there was no discussion of genetic engineering at all. At various points the discussion revealed the dangerous practices that existed in many laboratories – one researcher sheepishly admitted to having accidentally swallowed SV40 while mouth pipetting (a largely abandoned procedure that involves sucking up material into a pipette, hoping you do not suck too hard and that the aerosol components are not inhaled, or even worse, the liquid is not ingested, which inevitably happens sometimes).[20]

Some attendees felt that the whole biohazards business was a fuss about nothing. This led to a spiky clash between Jim Watson and Francis Black of Yale Medical School. Black adopted a coldly utilitarian approach, emphasising the significance of the breakthroughs that might come by studying cancer viruses:

> Even if, as has been suggested, five or ten people were to lose their lives, this might be a small price for the number of lives that would be saved.

Watson fired back:

> I'm afraid I can't accept the five to ten deaths as easily as my colleague across the aisle. They could easily involve people in no sense connected with the experimental work, and most certainly not with the recognition and fame which would go to the person or group that shows a given virus to be the cause of a human cancer.

Watson feared that unless laboratory security improved, the wider population might be exposed to real dangers, which he likened to a catastrophic accident at a nuclear power plant – the parallels between the potential dangers of genetic research and the perils of the atom were repeatedly brought up at this time.

As to what should be done, the proposals adopted by the meeting were limited to calls for a long-term cancer screening programme for research workers (this never happened), the publication of a biohazards newsletter and an insistence on the need for increased funding for better containment facilities. There was no mention of Pollack and Sambrook's argument that there might be experiments that ought not to be done. The potential consequences of mixing DNA from different species were not raised at all. This was surprising because the question was no longer hypothetical. Berg's original starting point – the project of fusing viral and bacterial DNA – had already happened.

*

In autumn 1971 Berg abandoned the idea of introducing DNA from a mammalian virus into a bacterium, but he continued with the central part of his project, which involved introducing bacterial DNA into the SV40 virus. Other people had already begun thinking along the same lines, completely independently. This is not unusual in science

– in general, scientific discovery is so overdetermined that even if researcher X fell under a bus on her way to the laboratory where she was about to make a particular discovery, the history of science would generally not be much perturbed, because researcher Y would have made a similar discovery at around the same time. No matter what scientists might imagine, in the long term the impact of any given individual is generally of little significance.

On 6 November 1969, shortly before Jon Beckwith denounced his own research at that press conference about his article in *Nature*, Peter Lobban submitted a proposal for his PhD research to a panel at Stanford. Lobban was a student in Berg's Department of Biochemistry working in Dale Kaiser's laboratory, which focused on how viruses were able to insert themselves into the DNA of their bacterial victim. In his nine-page proposal Lobban described how he aimed to use viruses to collect DNA from animals and plants and then introduce those sequences into bacteria that could be much more easily manipulated. According to the jargon, these bacteria would become 'transductants', capable of expressing a gene from another organism. But in contrast to Berg, who wanted to create a similar tool for the relatively narrow project of understanding human gene function, Lobban's bold idea was to study a range of animals and plants and 'to produce a collection of transductants synthesising the products of genes of higher organisms'.[21] With the eventual approval of the members of his thesis committee – some of whom, along with other people he consulted, were convinced the experiment would not succeed, but decided to let him learn the hard way – Lobban set to work at the beginning of 1970.

There was also a group of private-sector researchers who were working on a similar project. In 1971, scientists at the International Minerals and Chemical Corporation in Illinois started trying to add cow DNA to the T7 phage. However, there was something wrong with the protocol used by the Illinois group, and the enzyme that was supposed to bring the bits of DNA together – the ligase – did not do its job. No new molecule was created and the group dispersed without pursuing the topic any further.[22]

Throughout 1971 and the first half of 1972, the Stanford researchers tried to solve the problem of getting two pieces of DNA to fuse together, using what were called 'sticky ends'. Phage DNA is composed of the usual double helix, but at either end one of the strands sticks out an extra twelve bases.* These twelve-base sequences at each end are complementary, meaning that the two ends can fuse together – the shape of the DNA molecule means that A can only bind with T and C can only bind with G. The result is a chain of viral double helixes, or even a circular molecule. These sticky ends would be used as pieces of molecular Velcro that would allow larger molecules to be assembled.

David Jackson, a post-doctoral researcher in Berg's lab, was working on Berg's original problem of trying to get bacterial DNA into a mammalian cell. Some of the enzymes that were needed for his experiments and for Lobban's were made available by the department's founder and resident Nobel Prize winner, Arthur Kornberg, and were shared by the different groups in what Berg later described as a very 'giving and open' atmosphere – 'It wasn't secretive. It wasn't competitive', he said. As a result, he put it nonchalantly, 'the actual accomplishment was quite straightforward'.

Whether it was straightforward or not, as a result of cooperation between the different researchers, both Jackson's and Lobban's projects came to fruition virtually simultaneously. Indeed, another Stanford researcher, Vittorio Sgaramella, a postdoctoral researcher in Nobel Prize winner Joshua Lederberg's laboratory in the Department of Genetics, also reached the same objective of ligating DNA by a different route at about the same time. Nearly all of these findings were published in a flurry in autumn 1972 in the *Proceedings of the National Academy of Sciences*. The exception was the work of Peter Lobban. Given how closely Lobban and Jackson had worked together, Berg hoped their papers would appear at the same time. But either because Lobban wanted to publish a longer article in a different journal, or because his supervisor wanted him to do more work (accounts differ), or perhaps both, the article was delayed. By

---

*The cohesive sequences at either end of the phage lambda double helix were GGGCGGCGACCT and AGGTCGCCGCCC. The two strands of a DNA molecule point in different directions; you can see how the last T of the first sequence matches the first A of the second sequence, and so on.

the time the luckless Lobban eventually published his research in mid-1973 – based on his own brilliant, far-reaching idea from 1969 – the main finding was already old hat and used a method that was out of date before it appeared.[23]

The first paper in the set, which contained the big breakthrough, was from Jackson, Berg and Robert Symons, a visiting Australian researcher.[24] Jackson had been able to fuse genetic material from more than one source – in fact he fused DNA from three species. Once a phage had taken up four genes in the *gal* operon from *E. coli*, Jackson was then able to insert that DNA molecule into SV40, using the techniques for creating and ligating sticky ends that he had jointly developed with Lobban. The result was what they called 'a trivalent biological reagent' composed of DNA from two viruses (SV40 and phage) and a bacterium. This could, in principle, be used to investigate the molecular biology of SV40 and of the mammalian cells that it could infect, by seeing how the well-understood *gal* operon behaved in a mammalian cell. Scientists now had a powerful tool for directly investigating mammalian molecular genetics. As the article stated in its bold opening sentence: 'Our goal is to develop a method by which new, functionally defined segments of genetic information can be introduced into mammalian cells.'[25]

For Michel Morange, the historian of molecular biology, this article was of the utmost significance:

> The article by Berg and his coworkers thus has the same founding value as Jim Watson and Francis Crick's first 1953 article. In it the authors united a series of elements that might have been dispersed over a large number of experiments and articles. They grouped together a project, a series of techniques, and a result that was important in and of itself. The article was a scientific work of art.[26]

Berg himself considered that if one paper was the basis of his 1980 Nobel Prize, this was it.[27]

The other articles from autumn 1972 were all related to the technical question of how to get DNA molecules to fuse.[28] The most significant of them, by Janet Mertz and Ron Davis, both from

Stanford, used a restriction enzyme, EcoR1 (Eco for *E. coli*, R1 because it was the first restriction enzyme identified in this organism). This had recently been isolated by Herb Boyer's group at the University of California San Francisco (UCSF). When this enzyme cut the SV40 double helix, it left short sticky ends which Mertz and Davis used to fuse two different lengths of DNA.[29] Instead of employing six enzymes with a low yield as Jackson and Berg had done, Mertz's method needed only two enzymes and was extremely effective. As she told me in 2021, this new approach took the technique from being 'a protocol that no one outside the Stanford biochemistry department could do, to a protocol that possibly a bright high school student could do'.[30] This changed everything.

The final paper was also from the Bay Area, but not from Stanford. Boyer and his UCSF colleagues immediately took the insights of Mertz and Davis, rushing out an article that showed that the sticky ends left by EcoR1 were four bases long (AATT/TTAA).[31] * Both papers drew the same conclusion: in principle, this method made it possible to bypass Jackson and Berg's complicated method, and ultimately, in the words of the Mertz and Davis paper, 'to generate specifically oriented recombinant DNA molecules'.[32]

This was one of the earliest uses of a phrase that for decades came to be synonymous with genetic engineering: recombinant DNA. Although strictly speaking both your DNA and mine is 'recombinant' – it is the result of the mixing of genetic material from two parent individuals (this was the way that the word had originally been used) – the novel meaning attached to it here was that recombinant DNA came from more than one species.

Nature soon highlighted both the excitement and a potential danger associated with the Jackson, Symons and Berg paper. The journal emphasised that the method 'commands universal admiration', but pointed to exactly the problem that had preoccupied Pollack over a year earlier – if the construct could infect mammalian

* EcoR1 recognises the sequence GAATTC. Because of the way the two strands of DNA are oriented, and the existence of complementary base-pairing, the enzyme cleaves the DNA between G and A on both strands, producing a molecule with two complementary sticky ends: AATT on one strand, TTAA on the other.

cells through its SV40 component, perhaps it could also infect *E. coli* through its phage element, and thence be transferred to a human gut:

> This possibility, remote though it may seem to be, can hardly be ignored, and it will be most interesting to learn what criteria the group adopts when it decides whether or not the scientific information that might be obtained by continuing the experiment justifies the risk. Perhaps those involved will decide that the game is not worth the candle.[33]

Apart from *Nature*, no one else, neither scientist nor journalist, picked up on the breakthrough. In a world preoccupied with the end of the Vietnam War and the growing Watergate scandal, there were no articles in the press on either side of the Atlantic, no scary popular science books, no ominous television programmes, no press conferences with or without prophesies of doom from the scientists involved, and there was not even a mention of this work at the Asilomar meeting on biohazards that took place a few weeks later. And yet this was the revolution that so many had been worrying about for so long – DNA from different species could now be mixed, at will, and it could function. The age of genetic engineering had begun, and no one seemed to care.

Within six months, this state of apparent indifference was transformed into global alarm.

*

In November 1972, microbiologists from America and Japan converged on Honolulu for a conference on bacterial plasmids. Plasmids are small circular pieces of DNA that are found in bacteria and are separate from the main bacterial chromosome. They generally contain genetic material that is associated with adaptations, such as resistance to antibiotics, and were of growing interest to microbiologists. Late one evening, a group of conference attendees met in a kosher deli near Waikiki Beach to excitedly discuss the papers on recombinant DNA that had just appeared and to brainstorm what experiments could now be done. Among them were Herb Boyer, who had supplied the Berg laboratory with the EcoR1 restriction enzyme, and Stanley Cohen, from the Department of Medicine at Stanford. Cohen had been studying how certain plasmids made bacteria resistant to an antibiotic called tetracycline and had shown that these plasmids could be transferred from one bacterium to another, rendering the recipient resistant, too.

As Cohen tells it, over warm corned beef sandwiches and cold beer, Boyer, Cohen and their colleague Stanley Falkow furiously scribbled on paper napkins, devising an experiment that would use EcoR1 to manipulate a plasmid. Cohen's interest lay primarily in using the enzyme to chop the plasmid into pieces, to facilitate investigation of the genetic basis of resistance. Boyer could see much further. He realised that it would be possible to use the plasmid to make recombinant DNA with EcoR1 and then to copy that DNA by inserting the altered plasmid into a bacterium, which would then reproduce both its own DNA and the plasmid, increasing the number of copies of the new DNA molecule with every round of bacterial reproduction. The recombinant plasmid could also be introduced into other bacteria – because these would be exact copies or clones of the original recombinant molecule, this technique became known as gene cloning, or just 'cloning'. This moment, which was the origin of commercial biotechnology, became well known in Honolulu, and was marked in a local newspaper cartoon when the deli was demolished about fifteen years later (Boyer is under the N, Cohen under the A; their arms equally form each of the letters).[34]

Within five months, Boyer and Cohen had used EcoR1 to introduce a DNA fragment that produced resistance to one antibiotic into a plasmid known as pSC101 (p for plasmid, SC for Stanley Cohen, 101 because it was the first plasmid that he named) which already conferred resistance to another antibiotic, tetracycline. The recombinant plasmid was then introduced into *E. coli* where it enabled the bacterium to resist both antibiotics. Like the Berg laboratory a year earlier, they had created recombinant DNA using Mertz's relatively accessible method. Importantly, they had also shown that the new hybrid DNA molecule was functional.

They had also created a strain of *E. coli*, a common cause of illness

Cartoon showing Herb Boyer (under 'N'), Stanley Cohen (under 'A') and others in the kosher deli coming up with the idea of cloning in November 1972. By Dick Adair, published in the *Honolulu Advertiser* in 1987 to mark the demolition of the deli.

in humans, that was now resistant to two key antibiotics. Alarm bells should have rung.

When the research appeared in the *Proceedings of the National Academy of Sciences* in November 1973, a year after the initial recombinant DNA papers, the article contained one of the first applications of a new technique that has since become an essential tool of molecular biology. This involves staining DNA with ethidium bromide, which causes it to fluoresce under ultraviolet light, thereby enabling DNA fragments of different lengths to be distinguished when loaded onto an electrophoresis gel.[35] The glowing bars that featured in the article have since become emblematic of genetic research.[36] In 2021, Boyer told me that when he first saw the simplicity and the beauty of the fluorescent DNA in a gel from this experiment, he was overcome and wept.

Cohen and Boyer's paper concluded by highlighting the relative simplicity of their approach, and the massive significance of what they had done:

> The general procedure described here is potentially useful for insertion of specific sequences from prokaryotic or eukaryotic chromosomes or extrachromosomal DNA into independently replicating bacterial plasmids.[37]

As Berg put it a quarter of a century later: 'the Cohen–Boyer recombinant DNA breakthrough made it possible for *any*body to do *any*thing'.[38]

In June 1973, as the finishing touches were being put to the article, Boyer attended a Gordon Research Conference held in New Hampshire. Gordon Conferences are small, select, off-the-record meetings, designed to facilitate informal discussions; Cohen had urged the ebullient Boyer not to say anything about their work, but he could not resist describing their breakthrough. After his talk, a number of the younger attendees discussed the implications of Boyer's work in private. They were concerned that this technique would allow the creation of potential biohazards, in particular biological weapons. Two of these young researchers, Edward Ziff and Paul Sedat of Cambridge University, expressed the collective alarm to the organisers of the meeting, Maxine Singer and Dieter Söll.

Singer – who had discussed these issues two years earlier with Berg – organised a brief debate at the conference, which revealed that the unease of the younger researchers was widely shared. She and Söll subsequently drafted a letter to the National Academy of Sciences, warning of the potential dangers that flowed from the breakthroughs of the previous months and asking for guidance. This letter was then sent to all the attendees, asking for their views; a non-response, Singer said, would be taken as agreement with the contents of the letter. About half the 140 delegates responded, all in support of the letter, but split on whether it should be made public, with twenty delegates saying the issue should remain confidential.[39] On 17 July, the letter was sent, warning of the creation of 'new kinds of hybrid plasmids or viruses, with biological activity of unpredictable nature' that 'may prove hazardous to laboratory workers and to the public'. Singer and Söll called on the National Academy of Sciences and the US Institute of Medicine to set up a study group to investigate the problem and come up with guidelines for action.

The letter appeared in *Science* in September 1973 and began to attract attention.[40] A month later, Ziff published an article in *New Scientist* in which he described the discussions that took place at the Gordon Conference and the reasons for the Singer–Söll letter. In clear language he outlined the technology and, as the title of the article put it, highlighted the 'benefits and hazards of manipulating DNA'.[41] The editor of *New Scientist*, Bernard Dixon, applauded the Singer–Söll letter, but also expressed the widespread unease that was felt: 'New knowledge cannot be "unlearned", but such socially responsible warning of potential hazards offers the best hope that it can be controlled.'[42]

A few weeks later, following the publication of the proceedings of the Asilomar biohazards meeting that had taken place at the beginning of the year, journalist Nicholas Wade wrote a confused and alarming article in *Science*, mixing up the dangers of biological warfare, the potential for a new pandemic like the 1918 flu outbreak, the Singer–Söll letter and the issues originally raised back in the summer of 1971 over Berg's proposed experiment. Wade concluded with a perceptive comment from Andrew Lewis, of the National Institute of Allergy and Infectious Diseases, who considered that the issue was too important to be left to the scientists:

> If the public feels the scientific community is acting irresponsibly, there will be an immediate reaction and the freedom of research will be curtailed. If we don't exercise due caution we are heading for trouble.[43]

At the same time as these concerns were being expressed, another fundamental scientific step was being taken, one which felt even more dramatic. In the summer of 1973, John Morrow, who was finishing his PhD in Berg's laboratory, offered to supply Cohen and Boyer with some frog DNA that they could try to clone in their system – it coded for ribosomal RNA (ribosomes are cellular structures, made largely of RNA, which are involved in protein synthesis). This would provide Cohen and Boyer with the possibility of showing that their technique would also work for vertebrate DNA, even if the nucleic acid they were cloning related to a fundamental cellular component shared by all life, rather than something specific to animals. The experiments – largely conceived and carried out by Morrow – were completed in the autumn of 1973 and published in May 1974. Not only did the frog ribosomal DNA end up in the plasmid as intended, but after the plasmid was introduced into a bacterium, the gene apparently worked: when the team isolated RNA from the bacteria, some of it had been produced by frog DNA. As they concluded: 'The procedure reported here offers a general approach utilizing bacterial plasmids for the cloning of DNA molecules from various sources.'[44]

What they did not state explicitly was that the method would enable researchers to duplicate mammalian DNA by inserting it into a bacterium that could then be grown in large quantities in culture, enabling the DNA to be extracted and studied. As Lobban had first suggested, it might even be possible to use the mammalian DNA to synthesise a mammalian protein inside a bacterium. This in turn could open the road to discovery and profit, by easily and cheaply synthesising proteins used in medicine. The Stanford University press release that announced the paper picked up on this last point, declaring to the world that recombinant DNA 'may completely change the pharmaceutical industry's approach to making biological elements such as insulin and antibiotics'.[45]

When Berg heard about what Morrow had done – going behind his back, working with another group without telling him – he was livid: 'I almost kicked him out of the lab, I was so furious', he recalled. However, his temper soon cooled:

> Because I thought the experiment was so terrific, I called him and invited him to come and give a talk to the department on this experiment. … I regard it as one of the critical experiments in the whole evolution of the DNA cloning technology.[46]

Despite the evident excitement and concern that was expressed at the time, some leading geneticists did not seem to have the slightest idea of what was happening. In April 1974, eighteen months after the first recombinant DNA papers were published, pioneer molecular biologist Sydney Brenner wrote a think-piece in *Nature*, 'New Directions in Molecular Biology', to mark the twenty-first anniversary of the publication of the double-helix structure of DNA.[47] Looking back, the article seems uncharacteristically out of touch. The new directions Brenner described were the basic academic problems of how organisms develop, the organisation of the genetic material in chromosomes, the structure of the cell surface and the futile search for new laws of nature. His article gives no indication that molecular biology was metamorphosing into a science of action and made no reference to any of the work on recombinant DNA.[48] Despite his brilliance, vision and his many connections with the global scientific community, Brenner gave no sign of what was bubbling up in the minds of researchers. Perhaps he was constrained by the requirements of his *Nature* article, but in August 1973 another molecular biologist of his generation, Gunter Stent, had declared to the *New York Times*:

> Genetics is kind of dead. The problems we were interested in have been solved so it is not so interesting anymore. For the romantics – for those who want to measure themselves against the impossible – this is not such a good field anymore.[49]

In fact, the previous eighteen months had opened up astonishing new possibilities of genetic manipulation in which the impossible became possible; these discoveries had immense scientific and political implications which would soon alarm the general public and preoccupy scientists around the world – among them Brenner – for years to come.

- FOUR -

ASILOMAR

Over three years after Bob Pollack's telephone call to Paul Berg, concerns about recombinant DNA reached a decisive stage. Those worries took the shape of a letter signed by Berg, Boyer, Cohen, Watson and seven others, which in July 1974 appeared simultaneously in *Science*, *Nature* and the *Proceedings of the National Academy of Sciences*. What became known as 'the Berg letter' called on scientists around the world to defer all experiments involving genetic engineering 'until the potential hazards of such recombinant DNA molecules have been better evaluated or until adequate methods are developed for preventing their spread'.[1] For the first time in history, scientists had publicly decided to stop doing a particular kind of experiment until they were sure it was safe.

This letter represented the decisive step on the road to one of the most intensively studied moments in twentieth-century biology – the Asilomar conference of February 1975 and the global debate about the regulation of recombinant DNA research that it ignited. Scores of books, PhD theses, articles and memoirs have been published about the period, which saw protests, furious rows, complex legal regulation and eye-popping financial speculation about potential riches from the patented products of genetic engineering. Asilomar now forms part of the mythology of how science should respond to potential dangers and is regularly invoked by scientists, politicians and historians as an example of how science – and scientists – can take the moral high ground and self-regulate.

In fact, the conference was a far narrower and more confusing event than is generally thought. As historian Jonathan Moreno has quipped, 'Asilomar has become for biology what Woodstock has become for youth culture – a mythology that's grown but that obscures how muddy the event itself was at the time.'[2]

That parallel is more than a mere joke, however. Woodstock can now be seen as the end of the US countercultural wave – the hippie dream curdled less than five months later at Altamont in California, when members of the Hell's Angels motorcyle gang stabbed a man to death in front of the Rolling Stones. Similarly, Asilomar came after the peak of the post-1968 contestation of science in the United States. Although that movement laid the basis for the decision to hold the meeting and provided the context for some of the sharp arguments that took place, by the time that the molecular biologists met on the Californian coast in 1975, the radical steam had largely evaporated. What occurred at Asilomar was relatively staid – things might have been very different had the recombinant DNA breakthrough been made five years earlier, at the height of US protests around science and its social implications.

*

In summer 1973, when the US National Academy of Sciences received the Singer–Söll letter about the recombinant DNA discussions at the Gordon Conference, they asked Berg, as an Academy member and one of the key protagonists, to produce a position document on the safe use of recombinant DNA. Berg in turn invited a small number of colleagues, including David Baltimore and Jim Watson, to work on the question. Watson's involvement was inevitable, given both his prestige and his position as head of the Cold Spring Harbor Laboratory, but despite his outspoken contributions about the hazards of recombinant DNA research at the first Asilomar meeting, he was beginning to change his tune. A few years later, Watson claimed that his initial response to the Singer–Söll letter was one of condescension:

> It seemed just another irrational leftish outburst against genetic engineering to which those of us who had been in Boston had become allergic after Jon Beckwith and Jim Shapiro in 1969 called a press conference to denounce themselves for their isolation of the lac operon DNA.[3]

Whatever the case, he did not express such feelings at the time.

For a while Berg mused about inviting the firebrand Beckwith to join the group (that would have been fun), but eventually he decided someone more moderate would fit in better, so he called on the bioethicist Dick Roblin. Maxine Singer was supposed to be involved, but she could not attend because one of her children was sick on the day the group met in David Baltimore's office at MIT in April 1974.[4] Despite his supposed misgivings, Watson recognised the urgency of the matter. As he recalled in 1977: 'that morning we felt in a hurry, since if anything was clear, it was that recombinant DNA procedures needed only a child's mind to master.' One of the reasons for this haste was that the group had got wind of the still unpublished Morrow–Cohen–Boyer experiment that expressed frog DNA in *E. coli.* During the discussion, one participant, Norton Zinder of Rockefeller University, said:

> If we had any guts at all, we'd tell people not to do these experiments until we can see where we are going.[5]

This spontaneous outburst gradually gelled into the idea of calling for scientists to 'defer' their recombinant DNA experiments. To put a clear end date to the call, Watson suggested organising a conference to decide what to do, at the beginning of 1975.

Over the next few weeks the letter went through four versions as the group sought to focus on technical, biosecurity issues. The result was simultaneously a tightening and a weakening of the scope of the debate. For example, the authors removed a section on bacteriological warfare (this was seen as a distraction) and drafted, and then removed, a clear statement that 'some of these artificial recombinant DNA molecules could prove hazardous to man'.[6] Meanwhile, Stanley Cohen and others joined the signatories. At first Cohen had not been included in the select group, notionally on the basis that everyone involved had to be doing cancer research. It seems more likely that the growing personal and scientific tensions between Cohen and Berg played a role in that initial exclusion. But given the transformational impact of the Cohen–Boyer method for cloning recombinant DNA, the cloners – Stanley Cohen and Herb Boyer, along with Ron Davis and David Hogness, who had recently fused DNA from the geneticists' friend, the *Drosophila* fly, with Cohen's plasmid – were also invited to sign the document.

The Berg letter was published with the imprimatur of the National Academy of Sciences, reinforcing its impact – it not only had the backing of the leading researchers in the field, it also had the support of the US scientific establishment. Its most striking part was the declaration that the signatories would defer experiments involving the creation of recombinant DNA, including the introduction of animal viruses into plasmid or phage DNA that could infect bacteria.

The letter also called on scientists around the world to carefully weigh the implications of experiments fusing animal DNA with plasmids or phages, given that the outcome would be 'new recombinant DNA molecules whose biological properties cannot be predicted with certainty'. Next, they called on the NIH – the main federal funder of US biomedical research – to create an advisory committee that would investigate the potential biological and ecological hazards of these new forms of DNA, develop procedures to 'minimise the spread of such molecules within human and other populations' and draw up guidelines for researchers. Finally, the signatories announced their intention to organise a meeting at the beginning of 1975 to review progress in recombinant DNA and 'to further discuss appropriate ways to deal with the potential biohazards of recombinant DNA molecules'.[7] The whole question was posed in terms of human safety; there was nothing about the ethical dimension of these experiments, no reflection on whether they were the right thing to do. Despite the left-leaning sympathies of many of the signatories, the Berg letter was firmly focused on biosecurity. There was no mention of military applications, of the possibility of human genetic manipulation or of the impact of commercial exploitation. The agenda for Asilomar was set.

Officially released at a press conference at the Washington headquarters of the National Academy of Sciences on 18 July 1974, nearly a year after the alarm bells had first rung at the Gordon Conference, the Berg letter immediately created a stir in the US media.[8] The *San Francisco Chronicle* put the story on the front page with a headline that read 'A Danger in "Man-made" Bacteria'.[9] The *New York Times* had a more neutral 'Genetic Tests Renounced over Possible Hazards',[10] while the *Washington Post* conjured up images of Hiroshima but reassured its readers that the scientists knew best:

> Scientific research and experimentation is surely not a matter for the police to control. The best we can hope for is that the collective conscience of scientists themselves asserts itself to weight the risks in each specific instance.[11]

This was all in keeping with previous press coverage – earlier in the year both the *New York Times* and *Newsweek* had heralded the work of Annie Chang and Stanley Cohen when they introduced a resistance gene from the *Staphylococcus* bacterium into *E. coli*, and of Boyer, Cohen and Morrow, but neither publication gave the slightest hint that there might be anything problematic about these experiments.[12] After the initial flurry of articles around the appearance of the Berg letter, for the rest of the year leading US media outlets were silent on the possibilities – positive and negative – offered by the new genetic engineering.

While the mainstream media soon turned away from the issue, within days of the publication of the Berg letter, scientists began to express their views in letters to scientific journals such *Science*, *Nature* and *Genetics*, and in private correspondence. Some were in favour of what was soon known as the moratorium on recombinant DNA research (the 'm word' was not used in the Berg letter, but these things have a way of catching on). For example, in August, the microbiologist Roy Curtiss – a supporter of the genetic engineering of microbes, who with his beard and long, centre-parted wavy hair looked like a kitsch representation of Jesus – sent a sixteen-page open letter to around 1,000 scientists in which he called for a more detailed set of containment criteria and a more widespread pause in research.

Curtiss's concerns, coming from someone in the field, had a significant impact on the scientific community.[13]

Other scientists were deeply hostile to the idea of a moratorium, dismissing the potential dangers and seeing the deferral of experiments as an attack on their academic freedom. In many respects this was a repeat of arguments that had taken place a few years earlier when US scientific activists had protested against nuclear weapons research, in particular around the shadowy Jason Program, which was revealed in the 'Pentagon Papers' leak in the *New York Times* in 1971.[14] Scientists should be free to study whatever they wanted, went the libertarian argument.

Alongside this flurry of exchanges, leading US specialists were recruited to be part of three working groups in preparation for the conference, each of which focused on a different aspect of recombinant DNA biohazard issues (plasmids, eukaryotic DNA and viruses), and debated the issues with varying degrees of success in the second half of 1974. Their reports informed the position of the Asilomar organising committee, which was presented to the opening of the meeting in February 1975.[15]

Some companies in the commercial scientific sector rallied to the call for a suspension of research. For example, Biolabs – one of the leading suppliers of restriction enzymes – offered to cease production of these molecules in order to help make the moratorium bite.[16] Within a few weeks of the appearance of the Berg letter, Maxine Singer was pleased to see that her order for restriction enzymes was accompanied by a letter stating that the molecules were sold on the basis that they would not be used for any experiments that were forbidden by the Berg letter.[17] In private, Berg was sceptical about the usefulness of such initiatives, as he explained in a letter to his co-signatories:

> I know of an instance where a person with weird ambitions was probably stopped by the unavailability of the enzymes and his unwillingness or inability to make them. But good labs will not be limited in what they do by the lack of commercially available enzymes![18]

The essence of the problem of regulation was posed quite clearly. Even if some sort of pause in recombinant DNA research was adopted, it was hard to see how it could be imposed or controlled, given the relative simplicity of the technology, the potential profits that it might produce and the intellectual excitement that was fizzing about the topic at the time.

As if to accompany these debates, Hollywood returned to the potential threat of genetic mutation. After a series of mutation-based B movies at the end of the 1950s (these included *She Demons*, *The Killer Shrews* and *Konga*, in which the targets of the geneticists' skills were humans, shrews and a gigantic ape, respectively), by the 1960s genetics had gone out of fashion in the cinema.[19] As the science once again became alarming, mutation began to make a comeback as a source of cinematic fear, although with little originality. The year 1973 saw the US release of George Romero's low-budget film *The Crazies*, which recycled the usual trope of a man-made virus escaping from a laboratory, on this occasion causing death and insanity in a small Pennsylvania town. In the second half of 1974, *The Mutations* (also known as *The Freakmaker*), featured Donald Pleasance as a geneticist intent on mixing up human and plant DNA for unspecified reasons. The results – and the film – were predictable, as was the dash of female nudity included to enliven the pedestrian plot. The *New York Times* noted that the 'far-out experiments' portrayed on the screen were 'as up-to-date as genetic research', but the storyline was 'as basically familiar – or convincing – as Boris Karloff going berserk in an underground laboratory'.[20]

*

On both sides of the Atlantic, politicians and administrators turned their attention to the implications of recombinant DNA research. The NIH set up a committee to develop biosecurity protocols – what became known as the Recombinant DNA Advisory Committee or RAC – rapidly implementing one of the Berg committee's proposals. In December 1974, the US House of Representatives received a report entitled *Genetic Engineering: Evolution of a Technology*, which summarised the latest breakthroughs and put them in chilling context: 'for the first time since the development of nuclear weapons, the scientific community will be examining the need for imposing a voluntary deferral of selected types of basic biomedical research.'[21] It also raised the prospect of federal intervention should the forthcoming discussions at Asilomar not prove satisfactory.[22] The threat of legal restrictions on the academic freedom so prized by many US researchers was very real.

In the United Kingdom, matters moved unusually swiftly. News of the Berg committee's meeting at MIT in April 1974 soon circulated, along with copies of the various drafts of the letter.[23] Behind the scenes, the scientific establishment and the civil service began preparing their response, perhaps sensitised by a smallpox outbreak that had occurred in London the year before. Two people died after a lab technician at the London School of Hygiene and Tropical Medicine was accidentally infected and spread the disease to a visitor of another patient on her hospital ward. A potential plague in a capital city was narrowly avoided.

Less than a month after the Berg letter was published, the UK's Advisory Board for the Research Councils set up a Working Party to report on the experimental manipulation of the genetic composition of micro-organisms. Chaired by Lord Eric Ashby, a botanist with an interest in environmental matters, the committee was composed of twelve leading scientists and administrators, including Maurice Wilkins and Oxford geneticist Walter Bodmer. Evidence was heard in private, and the committee's proceedings were absurdly covered by the Official Secrets Act, which can hardly have reassured the doubters.[24]

Ashby was no doubt seen as a safe pair of hands. Master of Clare College, Cambridge, the noble Lord had made his views on the control of scientific research abundantly clear. In 1971, amidst the excitement over the launch of BSSRS, Ashby had given a lecture to the Royal Society on 'Science and Anti-science' in which he lambasted the 'zealots of the New Left' who 'cultivated an ideology of anti-science'. This was not, he growled, the aberration of a few hippies or the view of querulous eccentrics but instead a significant movement in society that represented the stuff of which fascism is made! As he put it incredulously: 'we are at a point in time when it is being seriously suggested that science ought to be practised under some sort of public scrutiny and restraint.'[25]

Three years later, scrutiny and restraint were exactly what the Berg letter was asking for. At the beginning of September 1974, similar calls were made in the United Kingdom by Sir John Kendrew – Nobel Prize winner and Secretary-General of the European Molecular Biology Organisation (EMBO), which had expressed its support for the Berg letter.[26] Quoted in *The Times*, Kendrew argued that the potential dangers of recombinant DNA 'may lead us to question scientists' common and generally unspoken assumption that the acquisition of new knowledge is always an absolute good, requiring no justification, no ethical sanction'. [27] An editorial in the same issue backed Kendrew's suggestion that there should be some form of monitoring, highlighting the differences with nuclear weapons:

> Unlike nuclear physics molecular biology does not require elaborate laboratories or expensive raw materials so that external control is virtually impossible. Voluntary monitoring by the scientists themselves seems the only answer and the chance to establish a credible system should be taken at the international conference to be held early next year.

In an indication that middle England was also concerned, the *Daily Mail* waded into the debate:

> If the scientist insists on playing at God, for the artificial creation of life must be regarded in this context, then it is crucial that his work should undergo the meticulous scrutiny of society. And that means all of us.[28]

At about this time, Berg came over to London to debate the issue of recombinant DNA at the Royal Institution. The discussion was chaired by Nobel Prize winner Sir George Porter and was broadcast on BBC2 under the title 'Controversy: Certain Types of Genetic Research Should Be Suspended'.[29] Berg trod a fine line between his genuine concern about the problems that might emerge and his recognition of the potential gains that might be made. The benefits were clear, he said, perceptively and accurately distinguishing things that were possible in the near future from others that remained far distant:

> Why can't these simple organisms become the factory for producing some of society's most needed supplies – antibiotics – hormones – and even a source of food. And for those who enjoy real speculation there are of course the dramatic possibilities for introducing new genes into human cells and thereby trying to cure certain genetic diseases.[30]

The British scientists who appeared with Berg embraced the positive side of what he had to say but seemed complacent about the potential dangers. *New Scientist* reported wearily:

> The viewer was left with the feeling that British microbiologists were not much bothered, a sneaking suspicion that perhaps they ought to be, and a slight sense of mental exhaustion.[31]

In a private letter, Ashby congratulated Berg for his 'masterly handling' of the debate and assured him that 'opinion in our committee is consistent with your views'. The committee had barely met, but Ashby already knew what it would decide. Sydney Brenner also wrote to congratulate Berg, although, typically, he was more direct:

> I thought you carried it off extremely well considering the number of morons that were participating.[32]

Amidst these media alarums and excursions, the Medical Research Council imposed its own UK-wide moratorium on recombinant DNA research, pending the outcome of the Ashby committee. At the end of September, the working party heard expert evidence from Brenner, which was to prove extremely influential in the months that followed.[33] He told the committee that recombinant DNA

> has produced a qualitative change in this field. It cannot be argued that this is simply another, perhaps easier, way to do what we have being doing for a long time with less direct

> methods. For the first time, there is now available a method which allows us to cross very large evolutionary barriers and to move genes between organisms which have never had genetic contact.

In terms of benefits, Brenner highlighted the suggestion that it would be possible to use bacteria to produce, say, insulin on an industrial scale. However, he also accepted that for the moment this prospect was purely hypothetical: it would require the bacterial cell to express the unknown DNA sequences involved in controlling the vertebrate gene – something most molecular biologists thought was unlikely. In other words, one of the most hyped practical reasons for doing these experiments hinged on a complete unknown.

Like the Berg letter, Brenner emphasised the significance of tumour viruses and the possibility of accidentally creating a dangerous hybrid molecule. As he put it with his usual humour:

> The essence is that we now have the tools to speed up biological change and if this is carried out on a large enough scale then we can say that if anything can happen it certainly will. In this field, unlike motor car driving, accidents are self-replicating and could also be contagious.[34]

Brenner opposed an outright ban on recombinant DNA research, suggesting instead that existing physical containment protocols might suffice, in particular if they were combined with the possibility of engineering the microorganisms so they would be unlikely to survive outside the laboratory. He proposed that a scale of hazards and appropriate protection should be drawn up and administered and monitored by the UK Research Councils. As he put it: 'This is clearly our responsibility and we should shoulder it.'[35] While this sounded noble, it could have been translated as: 'We know best, give us the money and we will deal with it.'

The Ashby Committee produced its report in record time, releasing it in January 1975. As Ashby had made clear from the outset, it was never going to come to any radical conclusion; it did not even follow the Berg letter in arguing for a temporary deferral of experimentation. Instead, in keeping with his lecture of four years earlier, Ashby put his confidence in the good sense of the scientific community – 'prompt and open publication of results' would 'help in dissuading people from conducting irresponsible or unnecessarily hazardous experiments', the report concluded, before outlining a set of underwhelming recommendations.

The most radical of these were that recombinant DNA work should be done by trained researchers (!), laboratories should have the basic equipment required for containing ordinary pathogens and that each institution should have a safety officer. Taking Brenner's advice, the Ashby Group also encouraged scientists to develop biological containment – 'genetical devices to "disarm" the organisms they use'.[36] This idea had been discussed by Berg's group, but had then been dropped and did not appear in the July letter – they may have thought this approach was unlikely to succeed, or that it would not reassure either the public or lawmakers.[37] Despite Ashby's endorsement of the suggestion and Brenner's secondment to the Asilomar organising committee in November, there was no slot on the Asilomar conference agenda for any discussion of this approach.

The Ashby report left *New Scientist* unimpressed. The magazine complained that Ashby and his colleagues had dodged the key ethical questions relating to potential military and human applications: 'No embargo. Not even a voluntary pause while such practical safeguards are developed.'[38] In a rather meandering editorial, *Nature* wondered whether it was practical to expect researchers and technicians skilled in molecular genetics to suddenly adopt the strict biohazard protocols of infectious disease laboratories, pointing out that even this would be no guarantee of safety, as shown by the 1973 London smallpox outbreak. Furthermore, calling for safety officers was insufficient and a system of inspection would be required – the creation of disarmed microbes that could not survive outside the laboratory seemed the best bet.[39] The rest of the British press simply passed over the Ashby report, although *The Times* did publish a brief factual account on page 14, beneath an article from the Sale Room Correspondent about the latest auction prices.[40]

Recombinant DNA was an international issue, not confined to the United States and the United Kingdom. In October 1974, 200 researchers met in Davos, Switzerland to discuss the ethics of genetic engineering. Berg was there, too, as part of an exhausting globe-trotting tour on which he explained to scientists and to the media what he was proposing and why. In his presentation, Berg insisted that his whole approach was technical, not ethical. 'It is simply a public health problem,' he said. And when asked if he would stop doing his experiments, he replied: 'I'd stop if there was a sound practical reason, but not if it were an ethical judgement.'[41] While this position met with approval by many at the meeting, it did not impress journalist Roger Lewin, who was reporting on the conference for *New Scientist:*

> Scientists should not perform their ritual act of retreating behind their religion of 'the unhampered search for the truth' in order to escape their responsibilities. Asserting that ethics have nothing to do with facts is simply another way of claiming that scientists have an absolute right to continue their research unhindered by public influence. This is dangerous to the public and the scientists alike.[42]

This issue should have been at the heart of the debates over recombinant DNA, which instead were deliberately focused on technical, biosecurity issues.

In France, researchers planned to fuse the human myosin gene with phage so it could then be transferred into a bacterium. On hearing of the proposal, Nobel Prize winner François Jacob urged the scientific authorities to intervene. As a result, two national committees were set up at the end of 1974, one dealing with ethical issues, the other with safety criteria.[43] Although Jacob was critical of experiments such as the original Berg–Mertz proposal, he generally supported the application of the new technology.[44] In Germany, researchers and civil servants discussed the possibilities raised by cloning and recombinant DNA, but agreed that it would be prudent to wait until after the Asilomar meeting.[45]

This was wise. Whatever the Europeans and the British might think would ultimately prove irrelevant. The new genetic engineering had been devised in the United States, the inevitable consequence of that country's predominance in molecular biology. As Brenner had put it in his *Nature* article of April, referring to the first revolution in molecular biology: 'Watson and Crick may have invented it but Uncle Sam certainly fuelled it.'[46] This would prove even more true of the early decades of this second molecular revolution, that of recombinant DNA.

*

In drawing up the programme for the 1975 Asilomar meeting – officially entitled the International Conference on Recombinant DNA Molecules – Berg and his fellow-organisers concentrated on allowing time to discuss the latest research and whether any of the vectors or reagents involved, in particular *E. coli* and the SV40 virus, represented biohazards. Only one session was devoted to ethical and legal considerations, crammed into an hour or so on the final evening. From the outset, the fundamental objective of the organisers was to agree the conditions under which experimentation could recommence.

The 140 attendees at the invitation-only meeting, which took place on 24–27 February, were mostly from the United States, but around fifty were from another sixteen countries (Berg had made a conscious effort to include researchers from outside the United States, but they were all his choices), including five from the USSR, among them eminent, if apparently out of touch, members of the Soviet scientific establishment.[47] * Notably, there were four researchers from big US companies, including three from leading drug manufacturers, who were expected to show an interest in applying the new technology sooner or later.

Organising an international conference is a pain in the backside at the best of times. Organising an international conference with the world's press clamouring to attend must be hell. In the days before the meeting began it became apparent that this would not be a confidential discussion. The organising committee recognised

* In other respects, the meeting was as undiverse as you might expect, given the composition of the scientific community at the time. There were only four women present, and no Black researchers.

that the meeting might have historic significance – all the sessions were recorded, with the tapes stored at the MIT library and embargoed until 2025 (expect a flurry of articles, books, podcasts and programmes when they are released) – but they were less keen on press involvement. Eight reporters were initially accredited, with the proviso that no articles could be filed before the end of the meeting, but such was the demand from the media that more places had to be added. Finally, twenty-one journalists managed to attend, some of them by just turning up on the day and wheedling their way in.[48] Among those who crept under the wire was Michael Rogers, a science journalist working for the distinctly countercultural magazine *Rolling Stone*, who produced the best journalistic account of the meeting – a nine-page article entitled 'The Pandora's Box Congress'.[49]

One person who was not at Asilomar was the man who started it all, Bob Pollack. He turned down the invitation because he did not work on recombinant DNA and felt it would be inappropriate to give lessons to colleagues who did. Instead, he suggested his place be given to Science for the People activist Jon Beckwith, to provide an alternative point of view. Beckwith in turn declined, refusing to play the role of the token critic. No other member of Science for the People was considered both appropriate and available, so no radical dissenting voice was heard. Science for the People issued an open letter to the meeting signed by Beckwith and others, but it contained little that would have altered the views of those attending, nor did it make any useful suggestions about how to proceed, beyond arguing for more involvement of laboratory staff and non-scientists, and more discussions.[50]

As scientists and journalists assembled at the Pacific Grove conference centre – once a YWCA facility, with a converted chapel, all golden wooden beams, serving as the conference hall – the weather took a turn for the better. Michael Rogers described the scene lyrically:

> The early spring air at Asilomar was crisp and the sky still a cloudless Kodachrome blue. Inside the chapel, however, the curtains were drawn, the air was still and heavy, and only an occasional shaft of sunlight managed to penetrate, striking a

> bald head here, a greying one there. The texture and colour in the rustic chapel were early Rembrandt.[51]

The tone of the conference was a chaotic mixture of a particularly cantankerous academic symposium and a deeply factional student meeting. When it came to the scientific discussions – which took up most of the agenda – there were high-level presentations on the latest work in the field, but the meeting was also afflicted by the usual egomaniacs to be found in academic circles. Some of these types – no doubt familiar to many readers – were ably skewered by Rogers in his *Rolling Stone* article:

> An irrepressible gentleman from Switzerland monopolises a microphone for a baffling ten-minute dissertation on scientific ethics ... some well-intentioned American drones for an equivalent period about his role in the licensing of an obscure vaccine some years earlier which, it grows quickly clear, has very little to do with anything.[52]

Above all, Rogers noted, for the first two days 'it becomes immediately clear that the conference attendees would rather talk about almost anything than the issue at hand'.

There were various reasons for this odd situation. One was the way the agenda had been set up – Maxine Singer later admitted that 'there was too much traditional science at the beginning'.[53] But more fundamentally, it was a consequence of the outlook of the organising committee. As Baltimore emphasised from the opening session, they were not there to discuss ethics or moral considerations, the potential impact of the commercialisation of the new science, the possibility of gene therapy or the threat of using genetic engineering for biological warfare. These issues, which he accepted were very serious, were 'peripheral' to the central aim of the meeting, which was to 'design a strategy to maximise the benefits and minimise the hazards in the future'.[54] The fundamental question of what that research would be used for was not up for discussion.

Although few people realised it, the conference was sitting on an unexploded bomb. A few weeks earlier, a tense meeting had taken place in Berg's office in which he had learned that his university, Stanford, along with the University of California San Francisco (UCSF), had applied for a patent on the Cohen–Boyer cloning technique. If all went well, both universities – and Cohen and Boyer – stood to make a fortune. Berg was doubly infuriated. Not only was he opposed in principle to patenting scientific processes, he realised that were news of the patent application to get out there would inevitably be a huge row at the meeting and the conference organisers would be fatally compromised, for it would seem as though they were trying to get research restarted in order to make money.[55] Berg had no option but to keep quiet and hope that none of those inquisitive journalists discovered what was going on behind the scenes.

One other feature of the organisers' attempt to control the debate was their decision that there would be no votes – any positions had to be agreed unanimously. As Berg put it: 'If we come out of here split and unhappy, we will have failed in the mission before us.'[56] For the organising committee, the whole point of the meeting was to define the safety criteria that would permit experimentation on recombinant DNA to resume. There were a series of skirmishes around this issue, which crystallised into two positions. The organising committee's stance, based on the views of the working groups, was summarised in a thirty-five-page single-spaced document which was made available when delegates arrived. This set out broad biosecurity guidelines that would allow research to recommence, the details of which could be subsequently thrashed out by the NIH and equivalent national bodies. On the other hand, there was an undertow of opposition to any restrictions. Advocates of this view, primarily from the United States, simply wanted to move on and get back to work, many of them arguing that 'academic freedom' effectively gave them the right to do whatever they wanted. A few months later Brenner said that he was astounded by this attitude, which he identified as part of 'the social condition of American science'.[57]

Leading the charge against the organisers were the two Nobel Prize winners present – Joshua Lederberg and Jim Watson – who ended up dominating many of the discussions. Lederberg had always opposed any restrictions on scientific research, so his stance was expected. More surprising was the position of Watson, who was now eager to resume recombinant DNA research and was scornful of the decision to defer experimentation, despite having signed the Berg letter the previous year.[58] Watson has never properly explained why he changed his position – Berg has suggested that, as director of the Cold Spring Harbor Laboratory, Watson realised his institution could not afford the expensive biosecurity facilities that would be required. Given he now felt that the risks from recombinant DNA were actually quite low, he changed his view.[59] Whatever the case, Watson acted as a lightning rod for some of the conference attendees when he stood up at the beginning of the meeting and announced: 'I think we should lift the moratorium.'[60] Immediately challenged by Maxine Singer to clarify his new position and to explain what had changed over the past nine months, Watson could only burble.[61] He spent the rest of the meeting making snarky comments, a nay-sayer with little to contribute to moving things forward, for the simple reason that he did not want them to progress.

In contrast, Sydney Brenner's contribution to the meeting was both positive and decisive.[62] This was not only because of what he said, but also because of the way that he said it. For Berg, convincing Brenner to join the organising committee was 'the best decision we made'.[63] The *Washington Post* described Brenner as 'a captivating speaker who uses his mobile face, bushy eyebrows and body movements to punctuate his comments'. During the conference, Brenner again argued forcefully for 'disarming the bug' – creating bacterial strains, plasmids and viral vectors that would not be able to survive outside the laboratory. Over the course of the discussions, as different scientists thought about the possibilities, this option gradually grew more attractive and more realistic. Brenner was equally deft in dealing with the press. For example, he described a mutation in *E. coli* that would force it 'to actually explode and empty its contents when it gets out of the lab'; when faced with an incredulous reporter, he paused then said: 'Well, they generally don't make a loud noise.'[64]

For the *Washington Post*, the key role played by Brenner, apart from that of jester (according to *Rolling Stone* he sometimes sounded like a sleep-deprived leprechaun), was as the conscience of the meeting, reminding the delegates why they were there:

> The issue now is not to consider what might emerge in terms of ultimate human danger, in terms of danger to the planet. Rather the issue is how to proceed with scientific investigations consistent with the ability to do it with no risk to the investigator, no risk to the innocents within our institutions who clean and help in the labs, and no risk to the public, the innocents outside our institutions.

The alternative, he emphasised, would be externally imposed regulations and potential legal restrictions. One scientist later highlighted the significance of Brenner's intervention in the debate over animal viruses, which became extremely heated:

> I thought that the whole meeting would collapse at that point because there was this tremendous antagonistic atmosphere. It was only people like Sydney Brenner who cooled the whole thing.[65]

A couple of weeks after the meeting, Berg wrote to Brenner to express his 'admiration and deep appreciation for your magnificent contribution to the conference's outcome (success?). Your wisdom, your comments and ideas and certainly your humour rescued us on several occasions when we floundered.'[66] *

Although the research deferral advocated in the Berg letter appeared to have been followed around the world, there were clear signs that voluntary control was not reliable, as researcher Andrew Lewis explained to the meeting. Lewis had a collection of naturally

* I think one of the reasons Brenner was able to play such a decisive role in the meeting (apart from his natural ebullience and intelligence) was that he had experience of getting his way in the hurly-burly of political debate – as a youth in South Africa he was a far-left activist, studying Lenin and Trotsky. About a year before his death, I made a BBC World Service radio programme with Sydney about his life and work – *Sydney Brenner, A Revolutionary Biologist*, which can be heard here: https://bbc.co.uk/programmes/w3cstxnk

occurring recombinant SV40 strains that he distributed to laboratories that requested them. Concerned about possible misuse, he had begun to ask researchers to state how the strains would be used and to agree not to forward them to other groups. But several major laboratories – including Berg's and Watson's – had refused to abide by these simple requirements. This boded ill for any future voluntary limits on the use of recombinant reagents, argued Lewis.[67]

The final shift in the mood of the meeting came on the last evening and was the product not of Brenner's eloquence or humour, but of the far drier interventions from the three lawyers who had been invited to speak. They each talked about a fundamental threat to scientific research, not from regulations, nor from governments, but from the US legal system, in the shape of potential court cases and damages if an experiment were to go wrong. According to *Rolling Stone*, when law professor Roger Dworkin spoke the words 'multimillion dollar lawsuit' a chill settled over the restive meeting hall. This silence grew deeper when Dworkin pointed out that all the debates about the relative risks of various kinds of recombinant DNA research would be irrelevant under the terms of the US Occupational Safety and Health Act, which required the workplace to be free of hazard. 'Not relatively free, the statute says *free*', Dworkin emphasised. First violation of the law could involve a $10,000 fine and six months in jail, double that for a second offence.[68] As the *Washington Post* put it:

> It took a group of lawyers to bring the scientists back to real life that night and to shock them with some of the legal and financial implications of their experiments.[69]

It was money, not ethics, that scared the scientists. Duly chastened, everyone went to their beds.

Well, not quite everyone. Sensing that the wind might have turned in their favour, the organising committee sat up half the night producing a five-page summary of the conference discussions. As Paul Berg told me in 2021:

> When we had finished the statement – I'll never forget this – we were elated beyond anything I could imagine. We felt so secure

> that we had actually been able to put down in print what we thought was a reasonable summary – it wouldn't satisfy everybody, but it certainly seemed like a consensus statement. And then we walked around the conference grounds, under an absolutely bright full moon. Not a cloud in the sky. We were like on another planet.

The statement was presented to the meeting the next day – during the night it was typed onto stencils and feverishly duplicated in the small hours. The resultant document was then slipped under the doors of the sleeping delegates.[70] As Berg wrote to Brenner a few weeks later: 'that last night's effort will never be forgotten even if only for the warm feelings that developed between us as we worked against high odds.'[71]

The next morning, the attendees were given a thirty-minute period after breakfast to read and inwardly digest the statement – the meeting had to finish its business by noon, when the conference centre booking ended. A debate that would determine the future of a whole scientific field and potentially prevent disaster was being squeezed into a few hours, simply because time had been eaten up by the leisurely discussions at the beginning of the meeting.

By all accounts the next few hours were frantic, bad-tempered and hellish (Berg said to me 'We were a bunch of amateurs'). There were cranky, time-wasting niggles about formatting and grammar ('Is low risk in quotations one thing and low risk without quotations another?'), Lederberg complained repeatedly that the document was being railroaded through while Brenner (or perhaps it was Cohen) successfully got the meeting to agree to put each of the six sections of the document to separate votes. Although this eventually provided a path through the mess, it also led to more confusion and time-wasting. Meanwhile, the clock ticked on.

In those frenetic hours, one of the biggest arguments took place over one of the few points at which ethics did enter the debate. This was the issue that had begun the whole affair when Pollack first discussed with Berg whether some experiments, such as those on highly pathogenic microbes, should not be done under any circumstances. It would surely have been better for the meeting to address this point from the outset, instead of rushing through the issue at the end. Whatever the case, a clear statement to this effect was eventually included, with only five votes against.

The final declaration stated that recombinant DNA research could recommence so long as appropriate containment facilities and protocols were available (a rough description of these was included, together with a four-point scale of increasing risk* and the suggestion of using Brenner's disarmed bugs, which had barely been debated). Only a handful of attendees voted against. Exactly who the diehard opponents were differs from account to account (there was no formal headcount), but at least Cohen, Lederberg and Watson opposed the declaration, along with a young French researcher, Philippe Kourilsky, and perhaps a few others.[72] The document was soon published in a slightly edited form in *Science*, *Nature* and the *Proceedings of the National Academy of Sciences*, and a long report was submitted to the National Academy of Science.[73]

At one level, nothing much had changed. The same unmeasurable potential dangers still existed, no new facilities or 'safe' bacteria were available. The only experiments that were banned were those involving high-risk pathogens, but even here there was no way of stopping the research from taking place. The consequences of the potential commercial application of recombinant DNA went completely unremarked.

Although the gaps in the Asilomar declaration are evident, by codifying a set of rough criteria the meeting's attendees created a new situation, reinforcing their position as experts while at the same time putting power into the hands of the principal funder of US academic research in the area, the NIH. This organisation had no reach outside the United States, and indeed had no control over industrial research within the country, but there was now an international benchmark, devised by those most intimately involved, against

---

*This increasing scale of risk still holds; it was later summarised by pioneer biosafety officer Gwladys Caspar of Harvard as follows:

Level 1. Don't eat it
Level 2. Don't touch it
Level 3. Don't breathe it
Level 4. Don't do it here

which other initiatives could be measured. And, assuming those criteria were met, research could restart. The moratorium was over.

*

Asilomar was not the end of the recombinant DNA controversy, but it set the terms of the debate. Over the next few years, all over the world, but particularly in the United States, there were continued arguments and protests over what scientists should be allowed to do. By deliberately sidestepping moral questions and concentrating solely on what the organisers saw as the technical issue of relative risk, Asilomar had turned the debate into one of practicalities. This omitted fundamental questions that were particularly significant to the public, such as what felt right and what felt 'unnatural', the possibility of misuse through biological warfare or the manipulation of human genes, and above all the long-term aim of all this ingenuity beyond satisfying curiosity.

These questions continued to resonate in the following decades, in particular with regard to putting genetically modified food on the planet's plates and when it became possible to sequence and even alter the human genome.[74] Had these fears been openly addressed from the beginning, maybe things would have been different. Instead, the worries and nightmares, which had not been properly explored or exorcised at Asilomar, eventually surged to the fore as the public and politicians all around the world attempted to come to terms with the new power that scientists had discovered.

In 1977, the historian of science Jerry Ravetz was interviewed for a BBC TV *Horizon* documentary and put his finger on the central problem of how Asilomar had dealt with the debates around recombinant DNA:

> The scientists are not just clever chaps doing their own thing and risking a little leak from their labs. They are the midwives of a new technology and those people who are ready to exploit commercially and militarily their discoveries are certainly waiting, waiting to see what these clever scientists produce. And so if we just concentrate on the labs and preventing

leakings from them, then I think we are really missing the point, that a larger problem is coming very very soon. Namely the control on behalf of humanity and the natural environment of this technology which has the potential of transforming industry and if it goes wrong destroying a great deal of life on the Earth.[75]

– FIVE –

POLITICS

Asilomar closed with a combined sense of achievement and worry. Achievement, because despite the chaos of the meeting, an agreement had been hammered out. Worry, because many researchers remained alarmed about the risks to themselves, the population and the environment. This disquiet was not confined to safety issues – there were also fears that the Pandora's box opened at Asilomar would lead to legislative control over scientific research.

Initially, it seemed as if those concerns were well-founded. In the United States, there was a real political appetite for legislation to control the new technology, with more than a dozen bills introduced into Congress. There were also public meetings in several towns where universities wanted to set up high-level containment facilities. In the United Kingdom, regulation of recombinant DNA research not only involved trade union representatives but also saw the creation of statutory health and safety committees in universities and research establishments. In France, there were the inevitable open letters signed by intellectuals as researchers and technicians protested against techniques they felt were unsafe. These were met with equally inevitable technocratic disdain by the government of President Giscard d'Estaing which ploughed ahead anyway.

And yet five years later it all seemed to have been a bad dream.

By the end of the decade, all around the world, restrictions on recombinant DNA were reduced to a bare minimum. Genetic engineering seemed to have been domesticated. Introducing foreign DNA into various kinds of cells began to become routine, and in October 1980 Paul Berg's research received the ultimate seal of approval – the Nobel Prize. The pace of these changes was quite astonishing. As historian Susan Wright put it:

> One of the most remarkable aspects of the controversy about the hazards of recombinant DNA technology in the 1970s was the speed with which the whole issue faded away. Intensely debated in the period 1975–77, by 1979 the hazard question was almost a non-issue.[1]

This dramatic shift cannot be explained by a single factor. There was no disarmed bug that suddenly made the experiments safe, no widely available containment facilities, no new acceptance that the gains were worth the risks. Instead, there was a growing realisation that the existing approaches to containment were adequate, while a series of experimental results recalibrated the risks downwards. This all led to a rapid decline in the desire of US politicians in particular to control a technique that promised rapid innovation and potential profit at a time when the economy was slowly recovering from the worst recession in forty years.

*

The Asilomar meeting called on the NIH – the main US federal funder of molecular biological research – to introduce safety guidelines for research that it financed (the private sector could carry on doing what it wanted). A major difficulty faced by the NIH was that this involved estimating the risks associated with different kinds of experiments – but as Jim Watson had heckled at Asilomar: 'We can't even measure the fucking risks!'[2] The risks were, at best, known unknowns, and probably unknown unknowns.

The same fundamental difficulty applied to the search for practical solutions. DeWitt Stetten, an NIH administrator heavily involved in developing the guidelines, explained the messy business to a colleague:

> Conclusions which must be reached cannot be reached by scientific judgements alone. They are reached certainly in part intuitively, in part by analogy, in part by hunch, and in part by consensus. Certainly, none of these are the method of science.[3]

For one historian the regulations emerged from 'a patchwork of a priori logic, science and intuition'.[4] This meant there was ample room for opponents to conjure up hypothetical situations that might conceivably be encountered, without any requirement to demonstrate that these risks were precise or even real.[5]

Part of the problem was that the presence of new genes inside innocuous microbes might lead to new, unforeseen dangers, or it might not. This empirical question could never be fully resolved because it involved all possible combinations of all possible genes, viruses and bacteria. As a result, even the advocates of recombinant DNA research remained concerned that one such combination might inadvertently – or deliberately – create a new disease. At a meeting held at NIH headquarters in Bethesda in August 1976, one scientist repeatedly asked a colleague:

> What I want to know is, living in, say, Washington, can you make an epidemic in Washington? Can you make an organism so virulent that it will make an epidemic in Washington? ... Can you think of a circumstance in which you could make it spread? This is really the heart of the issue. Can it be done?

The response was blunt: 'I can't answer that.'[6]

Much of the political debate in the United States was driven by Senator Edward Kennedy, the Senate lead on health and biomedical affairs, who had a keen interest in public involvement in science. A few years earlier, in response to the horrific Tuskegee study, in which US scientists observed Black men with syphilis for decades and did not treat them, Kennedy helped set up a commission to oversee research on humans, with a majority of non-scientists on the board. When the senator compared this approach with the way that recombinant DNA had been discussed in 1974 and 1975, he found the Asilomar process wanting:

> It was commendable that scientists attempted to think through the social consequences of their work. It was commendable, but it was inadequate. It was inadequate because scientists alone decided to lift it. Yet the factors under consideration extend far beyond their technical competence. In fact they were making public policy. And they were making it in private.[7]

Reporting this statement, *Science* magazine noted acerbically: 'The Asilomar conferees may have been making policy without broad public participation, but they were hardly making it in private. Sixteen reporters were taking down every word.' Even if people were informed after the event, they were powerless. A tiny elite was taking decisions for the whole planet.

Opponents of recombinant DNA were also to be found within the scientific community. Shortly before the NIH guidelines appeared, Erwin Chargaff, a prominent and irascible 70-year-old biochemist, published a rambling and self-important letter in *Science* opposing the study of recombinant DNA. In typically curmudgeonly style, Chargaff unleashed apocalyptic predictions about 'freakish forms of life', warning of the coming of 'something much worse than virulence' when 'the little beasts escaped from the laboratory'.[8] His closing paragraph seemed to have been written by a Bible-bashing preacher:

> This world is given to us on loan. We come and we go; and after a time we leave earth and air and water to others who come after us. My generation, or perhaps the one preceding mine, has been the first to engage, under the leadership of the exact sciences, in a destructive colonial warfare against nature. The future will curse us for it.

Chargaff's religious tone was echoed by Bob Sinsheimer – along with Kornberg in 1967 he had synthesised DNA in a test tube and then persuaded a bacterial cell to integrate that DNA into its chromosome:

> We are becoming creators, inventors of novel forms that will live on long after their makers and will evolve according to their own fates. Before we displace the first Creator we should reflect whether we are qualified to do as well.[9]

The arguments deployed by the opponents of recombinant DNA were ably dissected by Carl Cohen, Professor of Philosophy at Ann Arbor, in particular the conjuring up of terrible consequences out of all proportion to the actual experiments involved: 'these heavy questions are so framed as to be either unanswerable, yet effective in surrounding the research with the aroma of catastrophe, or to be literally unanswerable. … They are devised as argumentative weapons shielded from rational response.'[10] Cohen's critique applied to most of the arguments about genetic modification that took place over subsequent decades, right down to the present day.

Some of the fears invoked by opponents of recombinant DNA were literally fictional. The historian Luis Campos has shown how Michael Crichton's first technothriller – *The Andromeda Strain*, published in 1969 and turned into a blockbuster film in 1971 – became a repeated reference point in many of the public discussions of the implications of recombinant DNA, despite the massive gap between the story and reality.[11] In Crichton's novel, a military satellite crashes to Earth, bringing with it a terrible infection – an extraterrestrial microbe, code-named Andromeda, which perpetually mutates, killing people and devouring matter. This plot was not particularly original (H. P. Lovecraft had written something similar in his 1927 short story 'The Colour Out of Space') and above all it had nothing to do with recombinant DNA.

Nevertheless, the term 'Andromeda Strain' soon mutated from Crichton's original alien epidemic to mean any out-of-control infection, including one created accidentally. It became a microbial synonym for 'Frankenstein'. Scientists could protest that *The Andromeda Strain* was science fiction, but a plague from space was exactly what NASA had been afraid of when the Apollo 11 astronauts returned from the Moon and were immediately quarantined. Recombinant DNA looked exactly like the realisation of science fiction's dreams and was fêted as such. And as fiction is fond of telling us, dreams can become nightmares.

Australian newspaper cartoon from 1976 explaining the controversy over recombinant DNA, showing how the issue spread all around the world.

*

After a great deal of argument, and three intensely debated drafts, the first version of the NIH guidelines was issued in June 1976.[12] Designed to be reviewed annually, the guidelines were based on the principles established at Asilomar – some experiments were thought of as too dangerous; other experiments could be done if there was no alternative and if appropriate containment facilities were available. The greater the danger, the greater the safeguards required.

But that did not mean the debates were over. In the United States there were many different levels of government that could potentially control recombinant DNA research. The federal administration, states and even cities all had relevant powers – between 1975 and 1979, nine US towns considered legislative control over recombinant DNA experimentation within their boundaries. The most extensive of these debates took place in the city of Cambridge, Massachusetts, home to Harvard University and MIT, the other side of the Charles River from Boston.[13]

At the beginning of 1976, Harvard University applied to the Cambridge city council for permission to build a P3-level containment facility in its Biological Laboratories – researcher Mark Ptashne had obtained $500,000 for the project from the NIH. P3 was not the highest level of protection, but still required the laboratory to have negative air pressure, so that any leak would not escape into the outside world, and all material had to be decontaminated inside the facility. A meeting of Harvard biologists to discuss the proposal ended up lasting six hours as there was substantial opposition from the wife and husband team of Ruth Hubbard and George Wald (Wald had won the Nobel Prize in 1967 for their joint work on the retina).

A month later, at the end of May 1976, there was a university-wide meeting to discuss the plans. This was attended by a Cambridge councillor, Barbara Ackermann, who just happened to have watched *The Andromeda Strain* on television the night before. Alarmed by what she heard at Harvard, Ackermann raised the issue with fellow councillors.[14] The mayor of Cambridge, Al Vellucci, a populist who had a frosty relationship with Harvard, seized upon the issue: 'They may come up with a disease that can't be cured', he said, 'even a monster. Is this the answer to Dr Frankenstein's dream?'[15] * Following Vellucci's intervention, Cambridge City Council voted unanimously to hold a series of public hearings on Harvard's proposed laboratory. This decision, which provoked dismay in scientific circles, mobilised both the radical group Science

* Science journalist John Lear sketched Vellucci thus: 'Beneath the mask of the loudmouthed absurdly gesticulating clown [there] was a shrewd and practical judge of situations.' Lear, J. (1978), *Recombinant DNA: The Untold Story* (New York, Crown), p. 155.

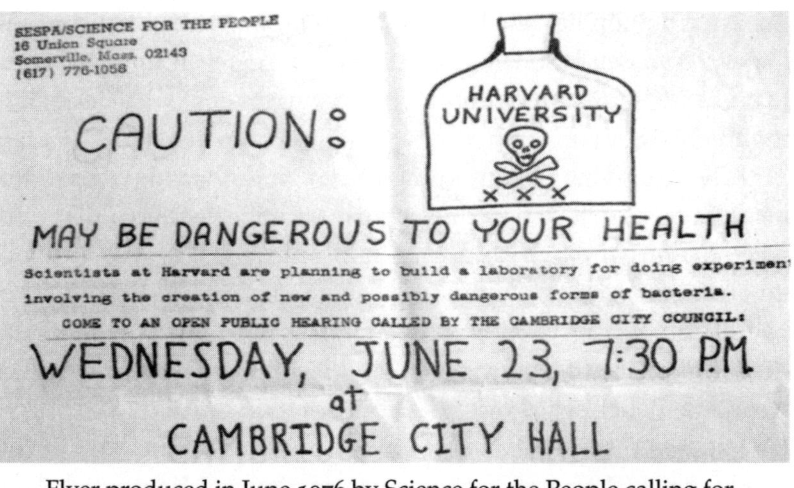

SESPA/SCIENCE FOR THE PEOPLE
16 Union Square
Somerville, Mass. 02143
(617) 776-1058
CAUTION:
HARVARD UNIVERSITY
MAY BE DANGEROUS TO YOUR HEALTH
Scientists at Harvard are planning to build a laboratory for doing experimen
involving the creation of new and possibly dangerous forms of bacteria.
COME TO AN OPEN PUBLIC HEARING CALLED BY THE CAMBRIDGE CITY COUNCIL:
WEDNESDAY, JUNE 23, 7:30 P.M.
at
CAMBRIDGE CITY HALL

Flyer produced in June 1976 by Science for the People calling for people to come to a Cambridge City Hall public hearing.

for the People and pro-recombinant DNA academics from Harvard and MIT. The conflict that had been rumbling for years was finally going to burst into the open.

The NIH sent Maxine Singer to attend the first Cambridge hearing, giving an indication of its concern at this outbreak of local democracy. The meeting had the air of a circus – the national media had got wind of the story and the chamber was packed with journalists and television cameras, while a high school choir sang 'This Land Is Your Land' from the gallery, from which was hung a banner reading 'No Recombination without Representation'.[16] Vellucci set the tone, admonishing the scientists who were about to give evidence:

> Refrain from using the alphabet. Most of us in this room, including myself, are lay people. We don't understand your alphabet, so you will spell it out for us so we will know exactly what you are talking about because we are here to listen.[17]

Less impressively, when introduced to Maxine Singer, Vellucci leered and asked for her telephone number.[18] After two of these hearings, in July 1976 the council set up an advisory group, the Cambridge

Experimentation Review Board (CERB), to investigate the issue and make recommendations to the council. In the meantime, there was to be a three-month moratorium on recombinant DNA work within the city limits.

The CERB was composed of seven Cambridge residents, including a nun, a community activist, an engineer and a postgraduate student who was studying philosophy.* The degree of public involvement seen at Cambridge was unique in the global debates over recombinant DNA, and in some respects was an exemplar of democratic scientific policymaking.[19] Meeting on two evenings each week, the panel heard over 100 hours of evidence and also visited laboratories at Harvard and MIT to see exactly what the experiments would entail. They even held a debate involving a clash of Nobels, as George Wald argued with MIT geneticist David Baltimore, who had won his prize a year earlier. The two men held similar political views and had both been active in the anti-war movement; on recombinant DNA, however, they were in complete disagreement.[20] The controversy cut across traditional political lines.

As in other debates over recombinant DNA, at times the gap between science and fiction became distinctly blurred. There was a great deal of alarm about an infestation of pharaoh ants in the Harvard building, which had escaped from a laboratory and were supposedly busy trailing radioactivity around, resisting all attempts at extirpation.[21] † Picking up on an idea in a recent scare-mongering article by scientist Liebe Cavalieri, there were lurid fears of the ants inadvertently carrying recombinant bacteria around the building and into the outside world.[22] This evoked Spider-Man fantasies of being bitten by a recombinant ant (pharaoh ants do not bite); it may also have revived memories of the recent movie *Phase IV*, in which ants in the desert begin to evolve into a hostile intelligence. The novelisation

---

* The student was Sheldon Krimsky, a physics graduate for whom the experience was literally life-changing. He subsequently spent much of his academic career studying the policy aspects of biology, writing a number of influential books, beginning with *Genetic Alchemy* (Cambridge, MA, MIT Press, 1982) which drew on his experience on the CERB.

† Harvard scientist Andrew Berry told me that the ants were finally eradicated at the end of the 1970s.

of the film blared on its cover: 'A race of super ants delivers an ultimatum to mankind – ADAPT OR DIE'.[23]

The members of the CERB were inevitably judgemental about the witnesses who came before them. They were particularly unimpressed by George Wald; as one CERB member recalled:

> Frankly, a lot of the opponents were rather foolish. I mean George Wald especially. Most of them appeared like little old ladies in tennis sneakers. ... I always hate to see an older man dress like a hippy-dippy.'

Another said:

> You're always going to get kooks that want to be heard, on almost any subject. But I think Dr Wald and his wife were kooks. He may have won a Nobel Prize, but, you know ...'[24]

Jim Watson went even further in his description of the critics – as well as calling them 'kooks', he also said they were 'incompetents' and 'shits'.[25]

In fact, none of the recombinant DNA sceptics had much useful to say – they made abstract claims about the importance of maintaining species barriers, used religious language (Wald extensively quoted the Book of Genesis in a 1976 article), or shifted the goalposts by raising hypothetical and often untestable dangers.[26] Sometimes they deliberately blurred the issue by raising irrelevant rhetorical points – in March 1977 at a fractious forum in which all of the leading players slugged it out, protestors provided an early, real-life example of Godwin's Law* by holding up a huge banner reading: 'We will create the perfect race – Adolf Hitler, 1933'.[27] (The year before, US novelist Ira Levin had published *The Boys from Brazil*, a best-selling thriller in which Nazi doctor Josef Mengele created clones of Hitler; a successful film starring Gregory Peck, Lawrence Olivier and James Mason appeared in 1978.)

---

*Stated by Mike Godwin in 1990, Godwin's law reads: 'As an online discussion grows longer, the probability of a comparison involving Nazis or Hitler approaches one.'

Campaigner Jeremy Rifkin, who had co-founded a group called the Peoples Business Commission that was resolutely hostile to recombinant DNA, also addressed a CERB meeting. Using a mixture of hyperbole and what turned out to be a vaguely accurate timeline, Rifkin predicted what was to come:

> With the discovery of recombinant DNA scientists have unlocked the mystery of life itself. It is now only a matter of time – five years, fifteen years, twenty-five years, thirty years – until the biologists, some of whom are in this room, will be able literally, through recombinant DNA research, to create new plants, new strains of animals, and even genetically alter the human beings on this earth.[28]

When it came to concrete proposals to ensure biosafety, the opponents put forward two options – either stopping the experiments or adding a layer of regulation. In the latter respect their proposals were remarkably timid. MIT biologist Jonathan King, a leading member of Science for the People, suggested nothing more radical than monitoring the health of researchers and the effluent of laboratories, appointing technicians to biohazards committees, restricting experimentation to trained researchers and introducing a system for reporting violations.[29] These were not exactly revolutionary proposals, and several of them were included in the CERB's final recommendations, which were delivered in January 1977 and formed the basis of the Cambridge Biosafety Ordinance, adopted a month later. This applied the NIH criteria to both private- and public-sector research in the city; it also included some additional measures, such as a biohazards committee with control over all recombinant DNA work in the city, and a monitoring system that could detect any escaped recombinant organisms.

Both sides of the Cambridge debate accepted the panel's verdict – although Wald still wanted a national ban on recombinant DNA, he described the CERB report as 'sober, sophisticated and thoughtful'.[30] For their part, Science for the People regretted that the CERB had not broadened its brief to cover more fundamental issues such as 'the likely specific uses of genetic engineering in class terms; the ecological or evolutionary dangers (in terms of infectious disease, soil ecology and other specific areas); and benefits and risks in broad social terms'. However, given that a central point of their campaign was to call for local control, they could not complain too much and they were left bemoaning 'the absence of a visible migration of popular opinion on the issues' which might have led to a different outcome.[31]

This was not a straightforward victory for the scientists, however. Ptashne's P3 laboratory had received the green light but fears still resonated in the popular imagination. In London, *The Times* headlined its front-page report of the affair '"Frankenstein" project given go-ahead in US'.[32]

The Cambridge debate had two ironic consequences. First, the subsequent relaxation of NIH regulations was so rapid that the facility was never actually used; by the time the building work was completed, Ptashne was able to do his experiments in ordinary laboratory space. In 1983 the containment facility was decommissioned and converted into offices at a cost of nearly $1 million to Harvard and the taxpayer.[33] Second, although the discussions had meant that there was a year in which research did not proceed, the region eventually became a focus for biotech development on the east coast as local academics and the city council embraced the technology.[34]

The conclusion of the Cambridge debate did not mean that Vellucci was done with recombinant DNA. In May 1977 he wrote to Philip Handler, the president of the National Academy of Sciences:

> In today's edition of the *Boston Herald American*, a Hearst Publication, there are two reports which concern me greatly. In Dover, MA, a 'strange, orange-eyed creature' was sighted and in Hollis, New Hampshire, a man and his two sons were confronted by a 'hairy, nine foot creature'.
>
> I would respectfully ask that your prestigious institution investigate these findings. I would hope as well that you might check to see whether or not these 'strange creatures' (should they in fact exist) are in any way connected to recombinant DNA experiments taking place in the New England area.[35]

I have been unable to find any trace of Handler's reply.

*

Meanwhile, following a series of terrifying events, the global public was increasingly concerned about the regulation of high-tech industries and rightly feared the consequences for both health and the environment should things go wrong. In July 1976 a massive explosion at a chemical plant in Seveso, north of Milan, led to a cloud of toxic dioxin gas covering the region, killing thousands of animals and polluting the soil. Two years later, in August 1978, the United Kingdom saw a second outbreak of smallpox. Janet Parker, a medical photographer at the University of Birmingham Medical School, contracted the disease from samples held in the laboratory of Professor Henry Bedson and died horribly a few weeks later. Bedson committed suicide shortly before Mrs Parker died. Exactly how she came into contact with the microbes remains unclear – the prosecution of the university under the Health and Safety at Work Act was eventually dismissed. As in London in 1973, a catastrophe was narrowly averted.

Around the same time, the United States experienced two awful environmental crises. At the beginning of August 1978, the critical state of a chemical dump at Love Canal in upstate New York led President Jimmy Carter to declare a federal health emergency; then, in March 1979, the nuclear power plant at Three Mile Island in Pennsylvania went into partial meltdown, releasing radioactivity into the environment and leading to the evacuation of around 140,000 people. Less than two weeks earlier, the thriller *The China Syndrome* had been released, the latest in a long line of Hollywood disaster movies, in which Jane Fonda played a journalist investigating how the nuclear industry tried to cover up a near meltdown. Truth and fiction were clearly not so far apart.

In contrast to these alarming events, many of the debates over recombinant DNA looked increasingly pedestrian. In the United Kingdom, where the discussions over the regulation of recombinant DNA research often took place behind closed doors under the Official Secrets Act, the Genetic Manipulation Advisory Group

(GMAG, pronounced 'geemag') was set up to regulate recombinant DNA research.[36] This committee was composed not only of scientists and lay members but also of four trade union members. In October 1978 the safety implications of genetic engineering were debated at a conference in London organised by the Association of Scientific, Technical and Managerial Staffs (ASTMS), a UK trade union with nearly half a million members (I was one of them).[37]

In my experience, there were two kinds of left-wing public meetings in the United Kingdom in the 1970s – those dominated by the far left, and those controlled by the dead hand of the trade unions. If the raucous 1970 BSSRS meeting had elements of the first kind, the staid ASTMS conference held eight years later was most definitely of the second. There was little debate; even supposed opponents seemed in agreement.[38] Sydney Brenner praised the role of the trade unions in GMAG and emphasised the difficulties of estimating risks 'in a field where we are dealing not with nuclear reactors, not with toxic substances, but with self-replicating biological entities', while MIT's Jonathan King, once an opponent of recombinant DNA, explained to the meeting that he now saw the technique as a major breakthrough and dismissed suggestions that recombinant DNA research might create an epidemic.[39]

To cement this slow shift in attitudes and to persuade legislators around the world of the need to relax regulation, the NIH's director, Donald Fredrickson, went on a series of globe-trotting trips as he attempted to convince scientists and governments around the world of the need to abandon regulation for all but the most dangerous experiments.[40] Fredrickson, like many other opponents of control over genetic experimentation, was driven by a growing conviction that the procedures were safe and by a profound agreement with the mood of deregulation that was becoming the new orthodoxy in the United States and Europe. In 1978 the European Commission drew up a highly controlling draft directive regulating recombinant DNA research, but faced with Fredrickson's arguments (or at least, that is how he tells the tale) the draft became ever less prescriptive until it was eventually adopted in an extremely neutered form in June 1982.[41]

At the same time, the various attempts to legislate in the United States, at federal or state level, all ran into the sand. Senator Kennedy lost interest, withdrawing his support for a key bill at the beginning of 1978.[42] Although the Senate Subcommittee on Science, Technology, and Space quizzed leading proponents and opponents of recombinant DNA, including Berg, King, Curtiss and Fredrickson, its March 1978 report said very little more than that the NIH regulations needed to be regularly revised, with greater public involvement.[43] The steam was going out of the debate.

Nevertheless, there were still problems. In France, a proposal to build a genetic engineering facility at the Institut Pasteur in the centre of Paris led to widespread opposition. One group of opponents, supported by 320 researchers and technicians, called for an immediate suspension of the experiments, control by all workers – 'not just by self-designated experts' – and for the public to be involved in the issue through open debates.[44] The administration of the Institut Pasteur condescendingly dismissed these arguments as based on 'obvious scientific errors and deliberate untruths' and emphasised that experiments at the Institut regularly involved all sorts of dangerous microbes (plague, rabies, cholera, polio and so on) without the slightest problem.[45] The idea that there might be anything to worry about got short shrift.* The French government eventually created its own version of the NIH criteria through a convoluted series of reports, resulting in a situation that was even less restrictive than that in the United States.[46]

During this period scientific societies increasingly argued against regulation of recombinant DNA.[47] The American Society of Microbiology mobilised its 25,000 members to lobby their political representatives to oppose restrictions on the new technology.[48] The European Molecular Biology Organisation (EMBO), now led by John Tooze, was opposed to regulation, arguing in December 1978 that there was a very small probability of a hazardous event and

---

*In June 1986, the Institut Pasteur admitted that there had been five unexplained cases of cancer among researchers who all worked in the same building. After nearly four years, a committee of inquiry published a report into all the deaths of people working at the Institut and found slight increases in the frequency of some cancers, but with such small numbers that it was hard to conclude either if there was an effect, or, if there was, what its cause was (*Le Monde*, 9 February 1990).

that the NIH guidelines should be rejected because they imposed unnecessary levels of physical containment.[49] A similar buccaneering role was taken by the Committee on Genetic Experimentation (COGENE), set up by the International Council of Scientific Unions, an international non-governmental organisation. COGENE and the Royal Society jointly expressed hostility to regulation in general and to GMAG in particular.[50] A month before Margaret Thatcher swept to power in the United Kingdom, opening the road to decades of deregulation in all sectors of the economy, the Foreign Secretary of the Royal Society adopted the prevailing rhetoric of the time to rail against the GMAG regulations as 'prison bars ... which prevent us from reacting rapidly to changing ideas evident in the rest of the world'.[51] Everywhere, regulation was seen as a fetter on innovation. As the US Senate Subcommittee put it:

> Recombinant DNA research appears to have considerable potential for utility in many fields and hence may result in new technologies with marked economic impact. ... Any significant US lag in genetic technology, for whatever reasons it may occur, could result in diminished power and prestige for this nation on the international scene. Such a possibility must not be overlooked as considerations for dealing with the recombinant DNA research issue going forward.[52]

*

The underlying explanation for this waning desire to regulate recombinant DNA research was partly that research had been safely carried out for three years or so under the kind of protocols outlined at Asilomar. Furthermore, many of the genuine concerns were being assuaged by data.

In the summer of 1977 Annie Chang and Stanley Cohen showed that naturally occurring *E. coli* restriction enzymes could cut plasmid DNA and that the fragment could then be integrated into the bacterial chromosome.[53] The result was indistinguishable from recombinant DNA produced in the test tube. This not only undermined those who argued that a supposed species barrier to DNA transfer was an essential feature of life, it also suggested that there might be little to fear from the process – nature had got there first. At around the same time, a small workshop of fifty researchers, held in Falmouth, Massachusetts, concluded that the widely used K12 strain of *E. coli* could not be transformed into a pathogenic agent.[54] This position was even accepted by Roy Curtiss, who had been one of the most critical scientists in the period before Asilomar and now agreed that existing biosecurity measures would suffice.[55]

Even the fundamental worry raised by the proposed Berg–Mertz experiment in 1971 turned out not to be so alarming. In December 1975, Sydney Brenner devised an experiment to see if mice infected with recombinant bacteria containing DNA from a virus that caused cancer in rodents would fall ill. After some debate and a two-year delay – partly caused by a legal challenge from a local resident – the experiment eventually went ahead in 1978 at the US germ warfare facility at Fort Detrick in Maryland. The results, which appeared in *Science* in March 1979, showed that none of the plasmids containing the virus produced disease.[56] Some critics continued to protest, but this experiment suggested that even if viral DNA contained in a recombinant bacterium did make its way inside the body, it was unlikely to cause disease.[57]

Finally, a completely unexpected discovery changed everything. In summer 1977, researchers at the Cold Spring Harbor annual symposium announced that some genes were split into pieces rather than being made up of a continuous sequence of nucleotides. This was not seen in bacteria, only in eukaryotes – organisms with a cell nucleus – and in some of their viruses. In turned out that in eukaryotic organisms the gene sequence was often interrupted by stretches of DNA that were not expressed and were initially thought to serve no purpose – these were called introns. Eukaryotic cells had evolved special machinery for dealing with this odd situation: when DNA is first translated into RNA, it has to be processed and all the irrelevant intron sequences are 'spliced' out. This eventually produces a mature messenger RNA (mRNA) molecule that can then be used by the cell to make a protein. Bacteria lack this cellular machinery – their DNA is turned directly into mRNA.

This had two implications for the genetic engineers. First, if you inserted a mammalian gene directly into a bacterial cell it was unlikely anything meaningful would happen. The bacterium would not be able to produce the relevant protein because it lacked the required biochemical processes to splice the initial mRNA transcipts. This suggested that some apparently worrying experiments would be pointless rather than dangerous. Second, it meant that rather than using bacteria to study mammalian DNA directly, researchers would have to obtain the mRNA produced by a particular gene, and then use the reverse transcriptase enzyme to turn the information in that mRNA molecule into cDNA, so that it could be introduced into a plasmid and thence into a bacterium. A new strategy and a new set of experiments would be necessary.

Faced with all these findings, the NIH gradually reduced the stringency of its guidelines, with a new version released in January 1980 that classified most experiments at the lowest level. By this stage the NIH was funding over 1,000 different projects using recombinant DNA to the tune of $131 million.[58] A year later, in 1981, the guidelines were made non-mandatory and by September of the same year, all special recombinant DNA guidelines were substantially relaxed, even allowing the cloning of previously banned toxin genes; only a handful of the most dangerous experiments remained forbidden.[59] In the United Kingdom, GMAG became less and less important as restrictions were weakened, and it was disbanded at the beginning of 1984. All over the world, the debate was over.

*

Opinions differed over whether the global panic over recombinant DNA had been worth it, or if it had all been a terrible mistake. For many of those most intensely involved, it seemed as though the debate had become a monster that had taken over their lives. In 1979, Maxine Singer wearily recalled the 1973 Gordon Conference which had publicly blown the whistle on the potential dangers of recombinant DNA:

> When I was elected to be co-chairman, I took that as an honour and looked forward to the Conference. In retrospect it turned

> out to be one of the worst things that ever happened to me, and my life changed from that day.[60]

A few months earlier, a similarly rueful Sydney Brenner explained quite how all-engrossing the controversy had become:

> I've been associated with this debate almost from its inception; and I can say that I have given a considerable amount of my time – perhaps four years of my life – to it, and certainly something like ten shelves in my office to the paper that has flowed out of these discussions.[61]

Inevitably, the opponents of regulation tended to have much more robust views. Jim Watson repeatedly claimed that it had all been a colossal waste of time, and decried his own role in launching Asilomar:

> All I can do today is say I was a jackass to sign the Berg letter. It was not a well thought document, but a silly emotional response which we as scientists should all be ashamed of.[62]

In contrast, Clifford Grobstein, who had been an opponent of recombinant DNA research, wrote:

> The public image of science was humanised and thereby enhanced. Asilomar is not interpreted generally as a brilliant conspiracy but as a conscientious effort by well-intentioned people who did not know what was best to do but did their best despite conflicting personal interests and inclinations.[63]

Probably the most considered view came from Paul Berg, over two decades after Asilomar finished its work. Berg argued that it had all been worth it, because of the science it allowed to be done with public support:

> So as I look back on the period, even though we were wrong – wrong is probably not the right word; certainly our assessment

of the potential risk was incorrect – by calling attention to it, I think the whole thing was better off in the long term. The science that has come out of it has just been absolutely mind-boggling. And so that's what in the end will justify it.[64]

– SIX –

BUSINESS

At the end of 1975, there was another telephone call to the Bay Area that changed the world. Unlike Bob Pollack's alarmed coast-to-coast call to Paul Berg in 1971, this one was local and very friendly. It was made by Bob Swanson, a 28-year-old unemployed MIT graduate. Shortly before losing his job with a small venture capital partnership called Kleiner & Perkins, Swanson had gone to lunch with his bosses as they schmoozed the heads of Cetus, one of their client companies. Cetus was based in Berkeley and had been set up four years earlier, mainly to sell equipment to molecular genetics laboratories, but had still not really found its way. There was the distinct feeling around that there was money to be made from molecular biology, but it was not clear how to do it.[1] Swanson's bosses thought they had glimpsed an opportunity: they were keen to explore the commercial possibilities of cloning and explained this over lunch, but the Cetus people were not interested.[2]

The idea stuck in Swanson's head, and after he found himself out of work he began casting around for ways of monetising recombinant DNA. If Cetus would not do it, he would. But there was a catch. Swanson had no scientific knowledge and no contacts with molecular biologists. His solution was to cold-call scientists on the Asilomar conference attendance list and to plead with them to work with him. He repeatedly got the brush-off until he got to Herb Boyer at the University of California San Francisco (UCSF), who was simply too affable to tell Swanson to get lost. That was the telephone call.

Emboldened by the fact that the polite Boyer did not immediately hang up, Swanson pressed his case, unaware that he was talking to the co-inventor of cloning. Bemused by the excited jabber coming down the line, Boyer reluctantly agreed to meet up for a twenty-minute chat a few days later and Swanson came to Boyer's laboratory on a Friday afternoon in early January 1976. In 2021, Boyer told me:

> We met in the lab, he was a young guy all dressed up in a three-piece suit. Of course, we looked like guys from off the street, in our jeans and whatever. My post-docs and students took one look at him and left the lab – they weren't quite sure what was going on.

Boyer and Swanson soon went off to a local bar, where the planned brief chat turned into a long discussion over several beers as Boyer found himself intrigued and excited by Swanson's proposition. Swanson later recalled:

> I can't be sure in retrospect, whether it was my persuasiveness, his enthusiasm, or the effect of the beers, but we agreed that night to establish a legal partnership to investigate the commercial feasibility of recombinant DNA technology.[3]

Each man put up $500 to create the partnership (Boyer had to borrow his).[4] The next task was to come up with a product that the new company could claim to be able to pioneer, to attract venture capital investors such as Kleiner & Perkins.

Nearly two years earlier, when Morrow, Cohen and Boyer integrated a frog gene into a bacterium, the Stanford press release had adopted the boastful style with which we are now so familiar, claiming that recombinant DNA might 'completely change the pharmaceutical industry's approach to making biological elements such as insulin and antibiotics'.[5] Insulin was the thing, Swanson now decided. This was a massive gamble – there would be immense technical challenges in getting the process to function and to churn out vast quantities of the desired protein that could then be purified, with no guarantee that it would work.

Insulin is a relatively small protein – only fifty-one amino acids long – and had been used by millions of people around the world for over half a century.[6] It was traditionally sourced from pig or calf pancreas – the animal version of insulin differed by a single amino acid compared to that in the human body and in around 10 per cent of patients this eventually caused an allergic reaction. Swanson realised that insulin was significant not only because it was so well-known and in demand, but also because drug approval should be straightforward and a recombinant product that was identical to the human version would be better – more natural – than the animal insulin currently on the market. As he recalled, the product he and Boyer were going to present to potential investors 'had to have an existing market. ... It had to be a significant market ... a product that could have a relatively short time through the regulatory hurdles. That was a key factor.'[7] For that dream to become a reality, they needed money.

In 1975, the United States saw a mere $10 million invested as venture capital – speculative, high-risk, short-term private investment in new businesses and processes.[8] At the time, the economy was stagnating and widespread speculation on biotechnology and computing had not yet begun. That would soon change – less than a decade later around $4.5 *billion* would be invested in venture capital projects, partly due to a legal change that allowed pension funds to invest their vast wealth in riskier companies.[9] In a bizarre coincidence, two of the sparks that began this explosion flashed into existence on 1 April 1976. On that day Swanson met with Kleiner & Perkins in San Francisco to pitch his recombinant insulin idea, while 60 km away down in Silicon Valley, Steve Jobs and Steve Wozniak set up Apple Computer Company.[10]

As Jobs and Wozniak took the first tiny step in the creation of what turned out to be a business behemoth, Swanson was already playing with the big boys, presenting his ex-bosses with a six-page business plan and asking for a cool $500,000. After postponing their decision until they met Boyer – he was the man with the scientific know-how upon which the whole project would depend – Kleiner &

Perkins eventually agreed to invest $100,000 in Swanson's six pages of paper. This was not such a crazy amount – a year earlier, the UK company ICI announced an investment of £40,000 in research at the University of Edinburgh to get bacteria to produce protein from a eukaryotic gene.[11] But Edinburgh, unlike Bob Swanson and Herb Boyer, had laboratories and employees. The American duo had an idea and that was it. On 7 April the new company was incorporated, with Perkins as chairman. Swanson was keen on calling it 'HerBob', but Boyer came up with the name Genentech and that was that.[12]

Boyer had not initially been interested in setting up a business – as he later recalled: 'I wasn't thinking about starting a company. I was just trying to think about how we could get these guys interested to take this and do things.'[13] But that did not mean he was not open to the possibility of making money. Eighteen months earlier, in November 1974, when Stanford University and UCSF had jointly applied for a patent on the Cohen–Boyer method of cloning, Boyer and Cohen were listed as co-inventors and potential beneficiaries of any licence revenues. This was the patent that had made Berg's blood boil in the run-up to Asilomar and which he had to keep quiet about.

The patent was drawn up at the initiative of Niels Reimer of Stanford's Directory of the Office of Technology Licensing; he had been inspired by newspaper articles on the frog recombinant DNA study earlier in the year, which, in turn, were largely based on the exaggerated Stanford press release that had mentioned the possibility of producing insulin. In other words, right from the outset there was a substantial element of hype underlying the whole business. There was also the potential for a great deal of rancour, which soon bubbled up. Potential patent co-applicants who had contributed to the development of the cloning technique, such as John Morrow, or any number of people from the Berg laboratory who had published articles in 1972 and 1973, were excluded from the patent.[14] Initially the application was criticised by the US Patent Office because of those prior publications, before eventually being accepted in 1980 (two other patents were approved in 1984 and 1988).[15] Differences between Cohen and researchers in Berg's biochemistry department grew and were amplified by issues around the patent, the intellectual claims to originality that it implied and above all by the financial windfall the patent eventually represented. Those conflicts ran deep and are still evident half a century later.

For whatever reason, Cohen was initially hesitant about the idea of patenting a basic scientific procedure and at first he signed over his share of any potential royalties to Stanford. In the 1980s he backtracked and began to receive personal royalties on what was becoming an immensely profitable patent.[16] The more boisterous Boyer took less convincing – from that first meeting in Honolulu in 1972 he had imagined practical applications – but his eventual share was entirely transferred to the University of California in order to fund research.

Berg later insisted that he had never thought about the possibility of making money from his science:

> 'Not me; I never heard anybody talk about it. ... I'd never heard anybody talk about potential commercial value.[17]

Berg was probably naive, but he was certainly not alone. That would soon change. With dollar signs lighting up in the eyes of university administrators and scientists, molecular biology's age of innocence was coming to an end and recombinant DNA was about to shake the world.

*

The new-born Genentech may have fixed its sights on creating insulin using recombinant DNA, but so too had several academic research groups – and they had the immense advantage of actually having scientists to do the work. Genentech was barely more than a rented office and some headed notepaper. The growing interest in insulin from molecular geneticists became clear at the annual Eli Lilly insulin symposium, held in Indianapolis in May 1976. Most insulin sold in the United States was produced by Lilly and the company hosted an annual meeting for leading insulin researchers to discuss the physiology of the hormone. At the 1976 event there was change in the air. As well as the usual crowd, including Bill Rutter's group from Boyer's department at UCSF, there were researchers with no track record in the field, such as Rutter's departmental head Howard Goodman and above all two molecular geneticists from Harvard – Wally Gilbert and Argidis Efstratiadis.

Faced with these molecular interlopers, the physiologists and medical researchers at the meeting were alarmed. As one of them recalled: 'what we had been working on for years suddenly became of interest to the molecular biologists. … These guys decided to leap-frog everything … if you're one of the people working in the field, you always feel a little bit put upon by it.'[18]

The presence of the Harvard duo was a consequence of Gilbert's recent interest in insulin. Efstratiadis had made his name working at Cold Spring Harbor, where he obtained mRNA for the globin blood protein from rabbit immature red blood cells. He used reverse transcriptase to turn the mRNA into a cDNA sequence, then used this to make globin in a test tube.[19] Now he was at Harvard, working with Gilbert. As Gilbert recalled:

> Somewhere very early on we switched to the idea of insulin, with the argument that it was a smaller protein, and actually a more useful one, to make in bacteria.[20]

From a technical point of view, working on globin had the advantage that it was relatively easy to obtain mRNA from blood cells. That would not be the case for insulin: the pancreatic cells that produce the hormone – the poetically named islets of Langerhans (a name I remember from my schooldays) – constitute only 1 per cent of the volume of the organ. Getting enough material from laboratory rats to obtain sufficient mRNA would mean killing around 500 animals and then carrying out hundreds of tedious and bloody dissections. At the Indianapolis conference, the Harvard duo discovered that they could get round this problem by using a strain of rat that suffered from pancreatic tumours – the pancreas in these animals grew extremely large, producing substantial quantities of mRNA.

At this point strategic differences emerged between the academic groups and the more business-focused Genentech researchers that were to have major consequences. Both Harvard and the Goodman–Rutter team at UCSF were intending to extract mRNA, use reverse transcriptase to make cDNA, insert cDNA into a plasmid and thence into bacteria and then, with luck, make the protein. Furthermore, they would do this twice – first with rat insulin mRNA as a proof of principle, and then they would use the rat cDNA as a way of obtaining human insulin mRNA more easily. Because of the close sequence similarity between the two versions of insulin, the rat cDNA could be expected to complement the human mRNA, allowing it to be isolated. This approach was not only slow, it also came up against the newly imposed NIH regulations, which restricted work on recombinant human DNA except in the strictest containment facilities. Gilbert's group soon hit another problem – the Cambridge row over recombinant DNA not only forbade them from pursuing key steps in their research programme, it also distracted them from laboratory work as they felt compelled to engage in debates and outreach activities.

None of this applied to Genentech – as a private company it was exempt from the NIH regulations. However, Genentech – then composed of just Boyer, Swanson and their investors – had no facilities or research staff of its own, so it had to subcontract work to NIH-funded researchers. Furthermore, in the feverish climate of 1976, with growing public unease at the supposed dangers of recombinant DNA, it was a canny move for private companies to voluntarily follow the NIH regulations. Much more significantly, Genentech decided not to use cDNA at all. Rather than wading through piles of rat corpses, it took a much more direct route: it would build an insulin gene from scratch, using the known amino acid sequence of the human version of the hormone.*

At the time, assembling a particular sequence of bases into a strand of DNA was a very time-consuming business. As Genentech set out on this road, Har Gobind Khorana's group at MIT was about to publish the results of a mammoth undertaking: they had synthesised the gene for a component of protein synthesis – a bacterial transfer RNA (tRNA) molecule only 199 base pairs long – and had

* In 2021, Herb Boyer told me that the germ of this idea had been planted by an off-the-cuff remark by Joshua Lederberg during the debates at Asilomar in February 1975.

shown it functioned when they transferred it into a bacterial cell. This work took Khorana's team *over nine years* to complete; it was eventually published in twelve back-to-back papers.[21] This was the kind of feat that the Genentech team was intending to emulate but under immense pressure to produce rapid results – a situation nothing like that normally experienced by academics. The insulin gene would be slightly smaller than Khorana's tRNA (153 base pairs), but this was still an immense challenge.

The Genentech gamble was not completely bonkers. Boyer was already working with Arthur Riggs and Keiichi Itakura at City of Hope National Medical Center, near Los Angeles, who had been able to synthesise 21 base pairs of the bacterial *lac* gene, clone it and express it.[22] This study involved topping and tailing the sequence with short stretches of synthetic DNA known as linkers that were targets for restriction enzymes, making it easier to cut and move DNA. The introduction of these linkers began the process of turning molecular biology into a kind of kit for manipulating molecules. As the City of Hope team put it:

> The combination of chemical synthesis technology with cloning technology will provide a powerful approach to many problems, including the chemical synthesis of long DNA sequences.[23]

Boyer suggested to Riggs that they should work together on cloning insulin using a synthetic gene, but this objective involved two major problems. First, the insulin gene would be much bigger than the *lac* gene that Riggs and his colleagues had worked on, and second, even if they succeeded in synthesising the gene, there was no guarantee that a human DNA sequence would function in a bacterial cell. All sorts of unknown control sequences might be required, and it was possible that the cell might simply destroy a protein it did not recognise (bacteria had never synthesised insulin in their three-billion-year history). Furthermore, this was the kind of experiment that the NIH guidelines were designed to prevent, or at least to make extremely difficult. But in a bizarre loophole, none of the guidelines applied to experiments with synthetic DNA. By adopting the synthetic gene approach, Swanson and Boyer were taking a huge gamble, but would potentially avoid a great deal of irritating bureaucracy. And after all, start-ups and venture capital projects were essentially one big gamble.

Riggs and Itakura were keen on insulin but suggested that they should first establish that it was possible to get a mammalian protein produced in bacteria. They had an appropriate project ready to roll – somatostatin, a small, recently discovered human brain hormone that was involved in promoting growth in mammals, including humans.[24] Swanson initially wanted to press on with insulin, rather than take what appeared to be a sideways step towards somatostatin with no financial reward, but the scientists were of one mind and he eventually bowed to the inevitable. The DNA sequence of somatostatin was unknown (DNA sequencing was barely in its infancy at this time), but its amino acid sequence had been established. Itakura would use the genetic code to establish a sequence of nucleotides that should, in principle, enable the cell to produce the protein. The genetic code is highly redundant – many of the twenty naturally occurring amino acids are coded by more than one three-base codon. This meant that the DNA sequence Itakura chose for his synthetic gene was almost certainly not the one actually found in the human genome.

Synthesising and cloning the somatostatin gene went straightforwardly, but there was no sign of protein synthesis in the recombinant cells. It turned out that the bacteria were simply degrading a protein they did not recognise. To get over the problem, the researchers fused the somatostatin gene to a bacterial gene, hoping that the cell would now accept the hybrid gene product because at least part of it was a typical bacterial protein. The team would snip off the unwanted bacterial protein segment after sufficient quantities of what was called a fusion protein had been synthesised. This also meant that they could claim they were being extremely careful with regard to biosecurity, as the human hormone was not directly produced in the bacteria. Although their use of a synthetic gene and private funding meant they did not need to do the experiment in a containment facility, they did so, for the sake of appearance.

Amazingly, this approach worked: a human gene could be expressed in a bacterium. The paper describing the procedures appeared in *Science* at the end of 1977 and was announced a few days before publication at a press conference in a Los Angeles hotel; the role of Genentech was barely mentioned.[25] This was the decisive scientific and technical breakthrough that made possible today's global biotech industry. The researchers recognised the significance of what they were doing, as Herb Heyneker recalled:

> It was an incredibly exciting time. I felt on top of the world. … It opened up so many avenues; you could envision so many things you could do all of a sudden.[26]

Riggs's memories were equally vivid, although he saw the experiment as having more fundamental implications:

> We did not take the natural gene for somatostatin, so we did not use a natural gene – from beginning to end, you might say. We designed it first on paper and then we made it. So in my mind, when it really worked, we had the final decisive proof that Watson and Crick were right, that the genetic code was right … you sit back and say, 'My God, that was all right! Science really works!'[27]

For Boyer, it was a key point in an incredibly rapid scientific journey – merely four years earlier he had been in a Hawaiian deli, brainstorming how to clone recombinant DNA. Now he had convinced a bacterium to produce a human protein.

Even before it was published, the work began to make waves. For a start, quite by chance, everyone was talking about somatostatin that autumn – its discoverers, Roger Guillemin and Andrew Schally, had just been awarded half of the 1977 Nobel Prize 'for their discoveries concerning the peptide hormone production of the brain'. Then, at the November 1977 Senate hearings into recombinant DNA, Berg highlighted the still-unpublished Genentech research on somatostatin in his evidence, generously describing it as 'extraordinary … astonishing … ingenious … elegant'.[28]

Berg also used the somatostatin work to emphasise to the Senate subcommittee the growing commercial potential of genetic engineering, although oddly he did not mention the involvement of Genentech:

> The ability to isolate pure genes puts us at the threshold of new forms of medicine, industry, and agriculture. Tailor-made organisms produced by recombinant DNA methods could provide valuable diagnostic reagents, probes for studying the operational status and efficiency of gene expression in health and disease, vaccines to immunise individuals and animals against the ravages of certain bacterial and viral infections and, possibly, even cancer.

The potential economies of scale represented by genetic engineering were enormous, Berg said. In a few weeks, the City of Hope and UCSF teams had produced 5 milligrams of somatostatin from 100 grams of bacteria, grown in a 7-litre vat. In comparison, the work by the Nobel Prize winners had also involved a few milligrams of somatostatin, but Guillemin and Schally each had to mash up millions of sheep brains to get the precious hormone.[29]

*

The somatostatin experiment proved that a human gene could be expressed in bacteria, opening the road to all kinds of genetic manipulation. Cells were increasingly seen as factories that could be made to do anything, so long as the right DNA was present within them. Life could be turned into an artificially programmed component of industrial production, but for industry to embrace that possibility, there had to be the prospect of profit. No matter how much of a breakthrough it was from a scientific point of view, somatostatin was never going to be a game-changer for the pharmaceutical industry. There were too few patients who would benefit from it, and anyway the hormone was already cheaply available through traditional manufacturing approaches. Insulin was where everyone expected the new technology would finally prove its worth, by providing a new approach to pharmaceutical production on a massive scale.

These high stakes produced a race to be the first to engineer the human insulin gene into bacteria. This competition pitted the cDNA groups, mainly those at Harvard and at UCSF, against each other and against the Genentech-funded scientists using the synthetic approach. The term race is not a retrospective interpretation – the members of the three principal groups were intensely aware of each other's work and of the potential threat to their hopes for success, fame and riches. One Genentech researcher later recalled: 'It was like each day it was almost a relief when you didn't hear word that someone else made an announcement that they'd finished the work.'[30] A science journalist embedded in Boyer's laboratory gave a graphic description of the feverish atmosphere:

> The lab seems in a perpetual frenzy. ... Carrying flasks and test tubes, people run between rooms like Keystone cops, their movements orchestrated by the jackhammer sound of timer bells, the dull flapping of dialyses and the hissing of autoclaves. There are periodic shouts of ecstasy and dismay.[31]

As Gilbert put it in one of the Cambridge hearings, the aim was to produce a unique drug – human insulin, without the different amino acid found in animal versions:

> We are making something which is, in a sense, beyond price. It is not that we are making cheaper medication. We are making something which we cannot get by other means.[32]

By spring 1977 Genentech had acquired a lot more funding and spent some of it on an office for Swanson on an industrial estate near San Francisco airport which eventually expanded into an adjoining warehouse.* Nevertheless, by the beginning of 1978, when Genentech was focusing on insulin, it still had 'no products, no salesmen, no full-time scientists, not even its own lab' as an early account of the period put it.[33]

In May 1977, the Goodman–Rutter group at UCSF were able

* The address was 460 Point San Bruno Boulevard – put that into Google maps to see what the area looks like now.

to clone rat insulin cDNA, but could not find any expression of the protein.[34] On the east coast, Gilbert's work led to interest from a European venture capital group as well as leading pharmaceutical companies and in March 1978 he helped set up a company called Biogen, along with other leading researchers such as Ken Murray of Edinburgh. With an eye on the main chance, the company was domiciled in the tax haven of the Dutch Antilles, while its central facility was in Geneva, avoiding those tiresome NIH regulations and with Swiss law making it easier for researchers to own patents on their work, without academic employers staking a claim.[35] Gilbert admitted to mixed motivations in setting up the company: 'wanting to do something socially useful, wanting to create an industrial structure, wanting to make something grow, wanting to make money'.[36]

By May 1978, Gilbert's group had expressed rat proinsulin in recombinant bacteria (insulin is not a direct gene product, but is synthesised in our cells via a molecule called proinsulin which the cell then modifies into mature insulin), but they still had to take the final, decisive step and repeat the procedure using human cDNA.[37] This was not really of much scientific interest – given the success with rat proinsulin cDNA it would be surprising if it did not work – but the potential practical and commercial impact was vast. Under the NIH guidelines, cloning experiments involving human cDNA had to be carried out under the highest containment level. There was only one such facility in the United States, at the US Army's Fort Detrick, and that was not available. The only solution was to go to Europe.

Rutter and Goodman sent their postdoctoral researcher Axel Ullrich to work in a Strasbourg facility, where over a period of months he cloned cDNA from human proinsulin mRNA obtained from pancreatic tumours.[38] However, he could not coax the resultant recombinant bacteria to produce any protein. Gilbert's group was able to get space in the Category IV facility at the UK Chemical Defence Establishment at Porton Down, and Gilbert, accompanied by three members of his group and with all their samples in their luggage, headed for darkest Wiltshire. There they encountered a culture clash as the staid civil servants and military staff, used to working nine to five, five days a week, struggled to cope with the

24/7 demands of a US academic research team engaged in a scientific race, pressured by the potential of immense financial rewards.

Meanwhile, the Genentech approach of simply synthesising a gene that would directly produce the insulin molecule itself was steaming ahead. Mature insulin is composed of a total of fifty-one amino acids and is made of two different molecular chains. The City of Hope group synthesised the DNA for each of the chains as fusion proteins and cloned them into separate plasmids, then made a hybrid plasmid that was introduced into the bacteria. These genes were doubly unnatural: there is no gene for mature insulin and the DNA sequences used to produce the two amino acid chains were those imagined by Itakura using his choices from the genetic code. This artificiality highlighted the new approach to life represented by genetic engineering.

The final step was to remove the unwanted bacterial amino acids from the fusion proteins and hope that the two chains would join in the appropriate shape – there was an element of chemical magic about this final step. Eventually, on 21 August 1978, it succeeded. In 2021, Herb Boyer chuckled when he told me of the mood in the Genentech team:

> They were exciting times. The excitement was there because we got into a competitive race with two other labs. Our guys were just ecstatic that we beat these two super-academic labs to the goal of making human insulin.

On 6 September the group held a press conference at City of Hope National Medical Center, announcing their success. For the stuffier members of the scientific community, the whole business had the whiff of a breach of protocol. At the time of the press conference a paper describing the result had been submitted to the *Proceedings of the National Academy of Sciences* but had not even been received by the journal office.[39] However, the publication route chosen by the City of Hope group meant that the final step was a mere formality – the paper had been submitted from a member of the National Academy of Sciences, which ensured that it would be published in a few weeks without peer review. And anyway, the results were the results.[40]

The article eventually appeared in January 1979, signed by researchers from Riggs and Itakura's group at City of Hope and by scientists from the grandiosely titled Division of Molecular Biology, Genentech, Inc.[41] (Boyer, who had not directly contributed to the research, did not sign the article, perhaps also to placate those at UCSF who were getting frustrated by his two salaries and potentially divided loyalties.)

On other side of the Atlantic, deep in the Wiltshire countryside, things were not going so well for the Gilbert group. As one of the Genentech team later crowed:

> On the very day when we were announcing success in insulin, [Gilbert] was, as he had for many days past, trudging through an airlock, dipping his shoes in formaldehyde on his way into the chamber in which he was obliged to conduct his experiments. While out at Genentech we were simply synthesising DNA and throwing it into bacteria, none of which even required compliance with the NIH guidelines.[42]

Things were even worse for Gilbert than the Genentech crew realised. Following weeks of work in the most trying conditions, after crossing thousands of miles of ocean and spending tens of thousands of dollars on the project, it turned out that the cDNA they were using was not in fact from human proinsulin mRNA. Somewhere along the way, some clones had got mixed up, and the team were inadvertently working on the rat proinsulin cDNA they had published earlier in the year. Even worse, when they dejectedly returned to Harvard, they discovered that all their human proinsulin cDNA clones had been lost in an incubator accident. Gilbert later described the trip as a total disaster.[43]

*

The insulin achievement propelled Genentech into a completely different world. Days after the result became known, Lilly paid the start-up a total of $500,000 to commercialise the synthetic gene approach to insulin production (Genentech would get 6 per cent royalties on all sales). There was a catch, of course – Lilly needed the scientific breakthrough to be turned into an industrial process, so they saddled the start-up with set of strict benchmarked criteria and very tight deadlines to scale up the method. These were met, and within two years of using a synthetic gene to produce insulin, Genentech had enough for a clinical trial. The results appeared in the *Lancet* in August 1980 and showed that, in healthy men, genetically synthesised human insulin seemed safe and effective and not statistically different in its effects from standard pig insulin.[44]

Two years later, the FDA approved the sale of what was now known as humulin, and Lilly announced an $80 million investment in industrial fermentation plants in Indianapolis and at Speke near Liverpool.[45] Ironically, the very fact that the effect of Lilly's product was identical to that of standard insulin, with the exception that allergic reactions were much less likely, led *Biotechnology Newswatch* to conclude that it was 'a giant step for genetic engineering, but only a marginal advance for medical science'.[46] It was also initially twice as expensive as traditional animal insulin, which hardly made it attractive. Nevertheless, all the 1970s fears about Andromeda Strain bugs eventually melted away and today all insulin is produced via various kinds of recombinant microbes.

Despite the promise that genetic engineering would lead to plummeting prices, in the United States the unit cost of insulin has risen ever since. In recent years, as the diabetes epidemic has increased use of the drug, US prices have spiralled, almost doubling from 2012 to 2016, and continue to soar, even though the patents have long since expired. Drug prices are not just about production costs.

The involvement of private companies in research, and concern that the pressure of competition and potential profit might drive scientists to make dangerous short-cuts, led to close questioning by the Senate Subcommittee that discussed recombinant DNA in November 1977. The senators particularly focused on what appeared to be a series of scandals involving systematic disregard for NIH guidelines at UCSF. Earlier in the year the journalist embedded in the Boyer laboratory had revealed cavalier behaviour by researchers and students – eating in the laboratory, mouth pipetting, experiments not done at the appropriate containment level – and included a phrase seized upon by the senatorial inquisitors: 'Among the young graduate students and postdoctorates it seemed almost chic not to know the NIH rules.'[47] Worse, just before the hearings, a three-page article in *Science* revealed that the Goodman–Rutter group at UCSF had breached NIH guidelines by using an uncertified plasmid.[48]

In many respects, this supposed scandal was much ado about nothing. Boyer had been told informally that the NIH had approved one of his new, safer plasmids, so the Goodman–Rutter group went ahead and used it. However, it later transpired that there was an additional, formal step of NIH certification that the plasmid had not yet received, so its use was technically in breach of the guidelines. Once this became apparent, Goodman ordered work to halt and the clones were destroyed. The experiment was repeated with another type of plasmid and was published in June.[49] The NIH was not told of any of this, and the forbidden plasmid was finally certified in July 1977 (it is still used by researchers).

In spiky exchanges about the role of Genentech (which had nothing to do with the affair), the senators grilled Rutter and Boyer over whether competition and the profit motive were behind the breach of the NIH guidelines. It seemed as though the scientists – perhaps with the connivance of the NIH – were prepared to sidestep the guidelines in order to make money.[50] In the end, the affair was nothing more than a minor cock-up that had been corrected, and the senators limited themselves to some stern finger-wagging.

Bizarrely, that was not the end of the matter. In 1990, the University of California sued Eli Lilly in a breach-of-patent case which to everyone's surprise involved a return to the forbidden plasmid. Wally Gilbert gave evidence for Lilly, suggesting that the UCSF team had gained an advantage by breaching the NIH guidelines through their use of the uncertified vector. Despite the counterclaims of the UCSF researchers, the rat proinsulin cDNA sequence data published in *Science* in June 1977 – the basis of the patent – were indeed from the uncertified plasmid. In response, the UCSF researchers could only emphasise that they were not responsible for errors in 'secretarial transcription' and insisted that they had not used Boyer's plasmid. The judge ruled that the UCSF team had engaged in 'inequitable conduct' and dismissed the University of California claim. When the case went to appeal in 1997, the Court of Appeals set aside the argument over the plasmid as legally irrelevant but said nothing about its veracity. It also upheld the lower court's dismissal of the claim. The university came away with nothing to show for its pains but a $12 million legal bill, while *Science* was unable to verify the origin of the data it had published two decades earlier.[51]

Another notorious piece of bad behaviour that occurred at this time also ended up as part of a tangled court case decades later. In August 1978 the University of California filed a broad patent relating to the production of protein in bacteria, but the inventors included Goodman and Rutter, neither of whom had done any work on the project. The postdoctoral researchers responsible for the breakthrough – Axel Ullrich, Peter Seeburg and John Shine – were all left off. They were pretty miffed and relations in the department began to sour. A key flashpoint focused on another project Seeburg had been working on in the evenings with his friend and colleague John Baxter – an attempt to clone growth hormone.[52]

Seeburg, along with Goodman and other researchers at UCSF, had been working on various aspects of the physiology of growth hormone, and at the end of 1977 had published a cDNA sequence of rat growth hormone.[53] Work was progressing on extracting mRNA from human pituitary tumours when the row over the patent broke. Genentech, well aware of the tensions in the department, invited Seeburg and Ullrich to come and work for them. The start-up had recently signed a contract with the Swedish company KabiVitrum, which was the leading global supplier of human growth hormone. The hormone was too large to be produced using a synthetic gene, so they would have to adopt the cDNA approach; Genentech asked UCSF for aliquots of the cDNAs that Seeburg had made but was sent away with a flea in its ear.

Determined not to be deprived of his work, Seeburg organised a daring raid on the freezers of his ex-employers at UCSF. At about 11 o'clock in the evening on New Year's Eve, accompanied by Ullrich, who was about to join Genentech, Seeburg went to the UCSF laboratory and removed some of the growth hormone cDNAs he had created, taking them straight to the Genentech laboratory. As they stole into the Genentech building, a police car pulled up. One of the officers came up to them and asked what they were doing. 'We're scientists', insisted the pair. 'You don't look like scientists', replied the cop, before letting them go.[54]

Seeburg was quite open about what he had done – two years later he told the *New York Times*:

> I had largely started the growth hormone project and worked on it since 1975. Why shouldn't I take material which I had acquired? I didn't take anything exclusively. Whatever I took I left some behind.[55]

In the summer of 1979 Genentech and UCSF engaged in a race of press conferences and papers as each tried to be first to announce the synthesis of recombinant human growth hormone. In terms of publication date, the UCSF team got there first, but the paper from Genentech and City of Hope involved a semisynthetic approach that was more suitable for manufacturing.[56] Genentech had won the industrial race and the profits that would potentially go with it – although the company encouraged its staff to publish in scientific journals, making money was, after all, its raison d'être.

However, there was to be no immediate US approval for the Genentech recombinant human growth hormone. It showed no benefits over the natural product, which was derived from the pituitary gland at the base of the brain in cadavers, and there would be no problems with supply so long as people kept dying. This situation changed radically with the global outbreak of Creutzfeldt–Jakob Disease (CJD). Although normally an inherited condition, this neurodegenerative disease can be transmitted through the introduction of nervous tissue or its components into the body, either by ingestion or injection. In 1985 it was discovered that four adults who had received human growth hormone injections as children had tragically died of CJD, presumably transmitted with their treatment. Under massive pressure, regulators rapidly allowed Genentech's recombinant growth hormone to be prescribed. Within two decades sales of the drug had passed $2 billion.[57]

Nevertheless, there was a shadow hanging over Genentech's success – the origin of the cDNA they used to clone the growth hormone gene remained murky. At the time, they claimed they recreated cDNA from mRNA they legitimately obtained from UCSF; they insisted they had not used the samples that Seeburg snatched on New Year's Eve. Not everyone believed this story. In 1990, the University of California sued Genentech, alleging theft of their cDNA samples, thus beginning another long court case that finally ended in 1999. During his evidence, Seeburg sensationally claimed that the *Nature* paper contained 'technical inaccuracies' with regard to the source of the cDNA, and that he had in fact used the cDNA he had taken on the infamous raid.

The origin of the cDNA did not affect the scientific result, but it had massive implications for the outcome of the court case. Seeburg's co-authors disputed his claim in ferocious exchanges in the letters pages of *Science*, but in the end, Genentech – while maintaining its innocence – agreed to pay $200 million to settle the case out of court.[58] This sum – which the markets thought was a good deal – included $50 million to be paid to UCSF towards the construction of a flagship research building on their new campus (the building is called Genentech Hall), $65 million to the university, and an astonishing $85 million to be split equally between the five people at UCSF who were working on the project. In a twist worthy of fiction, one of those beneficiaries was Seeburg, who, by effectively testifying against his past self, had made a fortune.[59] Seeburg died in 2016; a devoted Jimi Hendrix fan, his funeral service played out with 'Voodoo Child'.[60]

*

The influx of massive sums of money into genetic engineering changed many things, not least the lives of the people involved. Genentech was the first and most successful of the genetic engineering start-ups – by the beginning of 1980 it had agreements with three major pharmaceutical companies to develop completely new products. It had signed up with Hoffman-La Roche to find a way of producing a supposed anti-cancer drug called interferon about which there was a great deal of excitement.[61] It had an agreement with Monsanto to produce animal growth hormone for use in agriculture, while the French vaccine giant Institut Mérieux had signed up for the production of a vaccine against hepatitis B. It was a venture capitalist's dream.

In the United States there was growing awareness of the possibilities of making large sums of money out of recombinant DNA through venture capital investment or, to put it more crudely, speculation. Even those in the know were surprised by the level of interest. In autumn 1979, Nelson Schneider, a pharmaceutical analyst with the stockbroker E. F. Hutton, organised a meeting on the prospects for investing in biotech. He circulated a report to his clients and then invited Bob Swanson and representatives of three other biotech start-ups to speak at an event he expected to attract no more than a few dozen people. To Schneider's amazement, around 500 excited investors, investment bankers and analysts turned up.[62] The smell of money was in the air.

As interest heated up, Swanson and Perkins, respectively CEO and chairman of Genentech, decided to sell the company. Better to cash in as the market rose, rather than find yourself the other side of a burst bubble and miss out. Plus, getting into bed with a big company would guarantee an investment stream. But Genentech's products were too odd, and its recombinant DNA business plan too uncertain, to interest either of the two major companies they approached – US pharmaceutical giants Johnson & Johnson and Eli Lilly. Especially at an asking price of $80 million.

With public and institutional fears about recombinant DNA evaporating and an increasing appetite for speculation as the US stock market rose, the Genentech founders abandoned the idea of selling to the highest bidder and instead made plans to go public. But there was a problem – although the company was in profit, it was not yet making much money. In the first half of 1980 it earned a mere $81,000 on revenues of $3.5 million. To attract investors, the company had to emphasise the potential riches that could flow from its demonstrated mastery of the new technology.

After a gruelling series of roadshow presentations by Boyer and Swanson to potential investors in the United States and Europe, the initial public offer (IPO) date was set for 14 October 1980, with 1.1 million shares offered at a price of $35 each. Within twenty minutes of Wall Street opening, the value had soared to $89 a share, as investors scrambled to get a piece of the action. By close of trading on that first day the share price was down to $71, but that was still more than double the initial value – the company was now worth $532 million. According to the *Wall Street Journal* this was 'one of the most spectacular debuts in memory'.[63] Boyer, who was on an academic salary of $50,000, went straight out and bought himself a Porsche. With 925,000 shares apiece, he and Swanson were each worth over $65 million. Kleiner & Perkins, which had bought 938,000 shares back in 1975, had made a profit of around $64 million.

Amongst all this obscene wealth, spare a thought for poor Axel Ullrich, who a few years earlier had sold 800 shares for $8,000 to buy a car. 'At some point, those eight hundred shares were worth more than a million dollars', he recalled, exaggerating slightly, '… Oh god!'[64]

The Genentech sale, which occurred as the US economy was again mired in recession, heralded a new period on the US stock market. Investors and speculators flocked to make a fast buck through the twin novelties of personal computing and genetic engineering and the stock market soared, buoyed up by the election of Ronald Reagan in November and the prospect of good times to come, for the rich. For the president of the New York Stock Exchange, 1980 was 'probably the most profitable year in the history of Wall Street'.[65] Two months after the Genentech IPO, the far more reliable prospect of the Apple Computing Company – already turning a tidy profit – followed suit. It was the most heavily oversubscribed launch in history and by the end of the day Apple was worth $1.79 billion – over three times as much as Genentech (at the time of writing, Apple is valued at $3 *trillion*).[66] Although Boyer might not have had Jobs's style, he was still a golden boy. In March 1981 his jovial, moustachioed face appeared on the cover of *Time* magazine. Bob Swanson's telephone call had indeed changed the world.

Swanson died in 1999, aged only fifty-two. He had stepped down from the day-to-day management of Genentech in 1990 when Hoffman-La Roche bought up 60 per cent of the company in a takeover worth $2.1 billion. In 2009 the Swiss giant bought the remaining 40 per cent of stock for an astonishing $47 billion.

*

On 14 October 1980, as the champagne corks were popping on Wall Street to celebrate the Genentech IPO, yet another telephone call was made to the Bay Area. Once again it was to Paul Berg, but this time it was from Stockholm. The Nobel Prize committee announced that the prize in chemistry had been awarded to three giants of the study of DNA – Wally Gilbert and the UK's Fred Sanger shared half of the prize for their rival methods of sequencing nucleic acids (this was Sanger's second Nobel), while Berg won the other half 'for his fundamental studies of the biochemistry of nucleic acids, with particular regard to recombinant DNA'. The *San Francisco Examiner* ran the story on its front page, but the main headline was 'Genentech jolts Wall Street'. The two strands of glory emerging from recombinant DNA, one academic, the other financial, sat together in the space of a few column inches.

The Nobel citation gave no recognition to Boyer and Cohen, the two men who had discovered the procedure that made recombinant DNA a practical proposition for research and industry. For over forty years, the Nobel Prize Committee has considered that Boyer and Cohen's transformative application of the technique did not deserve the ultimate accolade. Both researchers have made their views about this clear. In 1995, Cohen expressed his bemusement at Berg alone being rewarded, and recalled that many colleagues shared this opinion:

> And you know, the Nobel Prize issue bothered me for a long while, to be honest about it, but it's something I've gotten over. The fact is, not winning the Nobel Prize hasn't affected my ability to carry out my research, which is really what a scientific career is all about in the most important sense.[67]

In 2009, Boyer talked to Jane Gitschier, a one-time colleague at Genentech, about his life in science. At the end of the interview, she probed his feelings about the Nobel Prize. His response was dignified, but tinged with understandable sadness:

Stocks up 2.30 Page C1

116th Year No. 107

San Francisco Examiner

Tuesday, October 14, 1980 20¢

★★★★ Final edition Complete stocks

Genentech jolts Wall Street

Today

Topic A

THE MOST spectacular stock offering in at least a decade hit Wall Street. Shares of Genentech Inc., a South San Francisco genetic-engineering firm, were sold to the public and immediately more than doubled in price. Page A1.

City/State

PROFESSOR PAUL Berg of Stanford University was among five scientists who won Nobel prizes in physics and chemistry for their work in genetic engineering and research of the origins of the universe. Page A1.

FOR THE fourth time in a month, San Francisco police believe that a Municipal Railway bus may have been the target of gunfire. Page A2.

ANTI-HIGH-RISE ACTIVISTS reacted with cynicism upon learning that new five skyscraper projects are on the San Francisco Planning Department docket. Page B1.

SUPERVISOR QUENTIN Kopp, suspicious of the San Francisco Redevelopment Agency's selection of the Yerba Buena theme park developer, said he will subpoena to a special investigation of the selection process. Page B1.

A BENEFIT for the "No on Proposition N" campaign brought out several supervisorial candidates, including David Scott, who presented himself as a liberal alternative in last year's mayoral race. Page B1.

ALAMEDA COUNTY Supervisor John George was arrested and briefly jailed in Oakland on charges of assault with a deadly weapon and battery. Page B4.

Nation

THE SUPREME Court awarded a victory today to DES daughters by refusing to review a California decision allowing lawsuits by women who contracted cancer because their mothers took the drug. Page A1.

JOHN ANDERSON, the independent presidential candidate, presented a five-year budget plan for the United States. However, the plan did not include estimates on how he would control inflation if elected president. Page A6.

A SPECIAL prosecutor has been appointed to investigate Tim Kraft, President Carter's former campaign manager, who has been accused of used cocaine. Page A7.

World

AMERICANS JAILED in Cuban prisons will be released "within hours," sources in Washington said. There is speculation that Fidel Castro's decision may have been motivated by a desire to help President Carter win re-election. Page A1.

TURKISH ARMY sharpshooters cut through the door of a hijacked jetliner and rescued more than 100 hostages. The five hijackers, who smuggled guns aboard in hollowed-out Korans, were captured. Page A2.

Secret decoder

Examiner: Bill Tate

FOR PAUL BERG, CO-WINNER OF NOBEL PRIZE, IT'S A DAY OF CONGRATULATIONS
Stanford biochemist developed tools for decoding the secrets of life

Stanford's Berg shares Nobel Prize in chemistry

Examiner News Services

STOCKHOLM, Sweden — Professor Paul Berg of Stanford University was among four Americans and a Briton who won the 1980 Nobel prizes in physics and chemistry today for their work in genetic engineering and research of the origins of the universe.

It was the second Nobel in chemistry for British Professor Frederick Sanger of Cambridge Sanger and Harvard University Professor Walter Gilbert.

The three were honored for their biochemical studies of nucleic acids, the master blueprints of life.

This year's Nobel prizes carry a stipend of 880,000 Swedish kronor — equivalent to $212,000.

Seven of the nine Nobel prizes awarded so far have gone to Americans. Still to be announced is

Developer of interferon soars from $35 to $71¼

The most spectacular new stock offering in at least a decade hit Wall Street today. Shares of Genentech Inc., a South San Francisco genetic-engineering firm, were sold to the public and immediately more than doubled in price.

One million shares of the company were sold at $35 each. Shortly after they began trading in the over-the-counter market, they were quoted at more than $80 bid. The price rose to $88 and then closed at $71¼.

Genentech was founded four years ago. It's involved in such genetic engineering as development of interferon, which has received wide publicity as a potential agent against cancer and other diseases.

There is talk of wide applications in such diverse fields as energy and agriculture for biochemical products produced through the manipulation of deoxyribonucleic acids — DNA, for short — which are found in the nuclei of cells.

The Bay Area has several of the world's leading companies and research centers in the young science of genetic engineering, although none have indicated any willingness to offer stock to the public. Cetus Corp. in Berkeley is probably the best known rival of Genentech, with research in a variety of industrial applications as well as the anti-viral agent Interferon.

But another company — virtually unknown even to analysts who follow the fledgling industry — is quietly working on other applications for gene-altering techniques. International Plant Research Institute in San Carlos is a leader in bringing the new technology to green plants.

The company was founded four years ago by Martin Apple, a University of California at Berkeley researcher. It's working on genetically altered vegetable and grain crops that can grow in salt water. Other projects include successful attempts to create new forms of plants that can tolerate extremes of heat and cold or are more resistant to diseases.

The University of California at Davis has been an early leader in academic research that Apple's company is now trying to commercialize. One of its first breakthroughs was in development corn and other seed crops that can produce their own nitrogen, thus eliminating the costly use of chemical fertilizers.

Genentech had revenues of $3.46 million and earnings amounting to just one cent a share in the first six months of this year. But enthusiastic speculation abounds in the financial community over its prospects for future earnings.

High court backs DES 'daughters'

Court clears way for public to see Abscam tapes Page A8

WASHINGTON (UPI) — The Supreme Court handed a victory to so-called "DES" daughters today by refusing to review a California ruling allowing lawsuits by women who contracted cancer because their mothers took a drug to prevent miscarriages.

The justices refused to hear an appeal by drug manufacturers of a March 20 ruling by the California Supreme Court. It allows the DES daughters to sue drug companies without naming which manufacturer's product caused the injury. Thus the ruling divides liability for damages among the drug's manufacturers, according to their share in the market in DES — diethylstilbesterol.

A manufacturer may escape liability only by showing it could not have made the DES in question.

Between 1941 and 1971, about 200 drug companies made and sold DES, a synthetic compound of the female hormone estrogen. The pharmaceutical manufacturers included the biggest names in the business — E.R. Squibb and Sons Inc., Upjohn Co., Rexall Drug Co., Eli Lilly and Co. and Abbott Laboratories.

In 1947, the Food and Drug Administration authorized the marketing of DES on an experimental basis to prevent miscarriages and up to 3 million women used the drug. Later evidence showed DES can cause vaginal and cervical cancer in daughters of women who took the drug while

The front page of the *San Francisco Examiner*, 14 October 1980.

It is not for me to decide whether I should or should not win a Nobel Prize. I've received many prizes and honours and I am indeed grateful for the recognition. You can imagine from what I've said about my boyhood, that I never would have expected to do what I've done. … Disappointed at times? Yeah. But I've been through quite a few periods in my life where I've had strong emotional reactions to one thing or another. … All in all, these experiences can be of great value to your outlook on life. I have been rewarded, and I am so lucky, Jane. And I'm so grateful.[68]

– SEVEN –

BIO-RICHES

1980 was a pivotal year for genetic engineering. Not only did Paul Berg win the Nobel Prize and Genentech launch on the stock market, but that year was also marked by two decisive legal changes in the United States which shaped how science was carried out and applied across the planet. The focus of both changes was the right to own and exploit discoveries through patents, including on forms of life.

Part of what made Genentech and other companies so attractive to investors was the prospect of their patent portfolios being lucratively licensed to other companies. But before that could happen, a legal argument had to be settled. Under long-standing US law, as drafted by Thomas Jefferson, patents could be granted to 'any new and useful art, machine, manufacture, or composition of matter, or any new or useful improvement thereof' (in 1952 'art' was replaced by 'process').[1] By implication it was not possible to patent living things or their components, because at best they had been discovered, not created, and were therefore not new.

That did not mean that natural products could not be patented. In 1900 a patent was granted for the hormone adrenaline, because it had been extracted and isolated from the body.[2] And in a major exception, patents on plants were explicitly permitted – clearly not all life forms were equal in the eyes of US legislators. In the earliest days of molecular biology, patents had been filed for derivatives of the RNA nucleotide uracil, and for a process for synthesising nucleic acids as well as the viral RNA that was thereby produced.[3] None of these patents made a penny, because they had no application outside of an academic setting. With the notable exception of plants, the general view of the US Patent Office was that living things, and their fundamental components, could not be patented.

This posed a problem for Ananda Chakrabarty, a biochemist who worked for General Electric.[4] While a postdoctoral researcher at the University of Illinois, Chakrabarty became interested in the ability of some strains of *Pseudomonas* bacteria to degrade hydrocarbons, such as oil. After moving to General Electric, Chakrabarty pursued this work in his own time and discovered that plasmids were responsible for this ability. By crossing different strains of *Pseudomonas* and using X-rays to fuse their plasmids, Chakrabarty produced two strains of bacteria that could digest crude oil. With the encouragement of his employers, in June 1972 Chakrabarty filed a patent both for the process he had used and for the two strains he had created. He claimed rights over a life form.

The Patent Office rejected the claim, arguing that it would 'open the flood gates to patentability for all newly produced microorganisms as well as for all newly developed multi-cellular animals such as … chickens and cattle'.[5] Chakrabarty appealed and in 1979 the Court of Customs and Patent Appeals ruled in his favour. The flurry of genetic engineering patents that had been filed following Cohen and Boyer – there were a total of 114 patent applications pending on living organisms – required clarification of their legal status, so the Patent Office decided to take the question to the Supreme Court, which would decide the matter once and for all.[6]

The case was heard in March 1980 under the title *Diamond v. Chakrabarty* (Sidney Diamond was the Commissioner of Patents at the time). It was accompanied by what are known in the US legal system as *amicus curiae* briefs – submissions from interested parties. Genentech and the Pharmaceutical Manufacturers Association, along with various other groups, argued that the patents should be awarded. Genentech tried to sway the court by characterising genes as mere tools that could be wielded by humans at will. According to their argument, the plasmids in question were 'new but dead'. They had a purely functional role, like a part in a motor vehicle – 'properly installed, they permit the bacterial engine to cough into useful life'.[7]

There was a lone submission supporting the Patent Office's position, from the Peoples Business Commission (PBC), which had intervened in the Cambridge debates. Their contribution was not so much a legal argument as a general critique of genetic engineering, full of hostility to the principle of patenting organisms.[8] The PBC claimed that the patenting of plant varieties had led to a decline in crop genetic diversity and an increase in monoculture which left plants vulnerable to disease. If the Chakrabarty patent were accorded, animal patents would inevitably follow, with the same catastrophic results in terms of disease and loss of genetic variability, they argued. While accepting that genetic engineering represented a biological revolution that was making 'astounding and awesome strides', the PBC predicted that using this power would have alarming consequences that the Supreme Court could prevent:

> Genetic engineering will, within the lifetime of many of us, give some individuals or institutions the final and awesome power to irreversibly violate three billion years of evolutionary wisdom through the creation of novel life forms, or the genetic alteration of living entities now existent.

Accurately foreseeing the future, if somewhat telescoping events, the submission imagined the asexual reproduction of animals through cloning and the advent of 'genetic surgery' to alter the heredity of complex organisms. When it came to what the PBC claimed was 'the essence of the matter', the submission began to go off the rails, describing 'the manufacturing of mammals, including human beings, to specification' and 'the creation of super-intelligent beings'.

The Supreme Court was not swayed by this rhetoric and in a narrow decision ruled by five to four that the two strains of bacteria that were the actual subject of the case were the product of Chakrabarty's ingenuity and were therefore patentable, even though microbes had been explicitly excluded from a 1970 statute

Cartoon by Herb Block that appeared in the *Washington Post* in June 1980 after Ananda Chakrabarty won the right to patent a life form.

that regulated the patenting of plants.[9] Even the four minority judges dismissed the PBC's brief – both sides of the Court considered that the political implications of genetic engineering, and potential future laws regarding the patenting of life forms, were not their affair. In rejecting the government's argument, the majority ruling referred to the PBC's concerns regarding the grave risks that might be incurred if the patent were to be supported: 'The briefs present a gruesome parade of horribles', said the judges. However, the Court considered that the horribles had nothing to do with the case, and these profound issues should be dealt with by Congress.[10]

The case was over, and life could be patented, in the United States at least. General Electric, alarmed at potential public opposition and the complex regulatory hurdles it would have to leap before deploying the bacteria, did not develop Chakrabarty's invention any further.[11] On 2 December 1980, riding on the back of the Supreme Court decision, the US Patent Office granted the Cohen and Boyer cloning patent that had been submitted a little over six years earlier.[12] Together with all the other patents relating to genetic engineering, it had been on hold while the Chakrabarty case trundled through the legal system. Once granted, the Cohen–Boyer patent lasted seventeen years (in 2011 US patents were extended to twenty years, matching the situation in Europe) and was licensed to 468 companies, initially at a cost of $10,000 at sign-up and $10,000 per annum (there was no charge to academic researchers). In total, 2,422 products were developed by these licensees, generating over $35 billion in sales.[13] The total revenue from the patent was around $255 million; each university received $90 million, with Boyer and Cohen sharing the rest.

*

The financial impact of *Diamond v. Chakrabarty* was reinforced by a second key US legal event of 1980: a law that was enacted by President Jimmy Carter shortly before handing Ronald Reagan the keys to the White House. Earlier in the year, Democrat senator Birch Bayh and his Republican colleague Bob Dole had sponsored a brief amendment to the Patent and Trademark Act with the aim of forcing the US government and recipients of governmental research funds, such as NIH grant-holders, to take out patents on their discoveries and to license those patents. At the time, the US government held nearly 30,000 patents, the vast majority of which were unlicensed and unexploited. For Bayh and Dole, innovation and economic growth could be encouraged by decentralising control over the exploitation of discoveries that had been made through federally funded research.

This claim combined suspicion of the supposed stifling effect of government on innovation and economic growth – a key component of the neoliberal outlook that dominated the world for decades, beginning with the Reagan years – with a fundamental trope of US culture. A key part of the mythology of the American Dream was – and still is – that if the little guy (or gal) comes up with a brilliant idea, they can acquire great wealth through the operation of the market, as long as the invention is protected by patent law. As the nineteenth-century American writer Ralph Waldo Emerson is supposed to have said: 'Build a better mousetrap, and the world will beat a path to your door.'*

American universities were certainly not little guys, but what became known as the Bayh–Dole Act encouraged them to focus their attention on the possibilities of getting rich quick through patents and involvement with business, in particular through the application of genetic engineering. At one level, this was nothing new – US universities had worked closely with industry throughout the first part of the twentieth century.[14] The widespread notion of pure research, with no interest in application, especially in biology, is partly a consequence of the relatively well-funded global research environment of the 1950s and 1960s.[15] Nevertheless, some universities had haughtily declined to patent their findings – from the 1920s Harvard refused to profit from health-related discoveries by its staff.[16] The poster boy for such altruism was Jonas Salk of the University of Pittsburgh. In the 1950s Salk successfully pioneered a polio vaccine, backed by the National Foundation for Infantile Paralysis and its fundraising scheme, the March of Dimes. Famously, neither Salk nor the Foundation claimed any rights over his discovery. When the television presenter Ed Murrow asked who owned the vaccine, Salk replied: 'There is no patent. Could you patent the sun?'

By the 1970s the world had changed radically, and universities and researchers had begun to revel in the seductive new power of hyper-profitable links with industry.[17] In 1974, the chemical giant Monsanto agreed to give Harvard $23 million over a twelve-year period in return for rights to a tumour factor that no one was sure actually existed. The company was quite open about its calculation: 'We realised we couldn't build a biology department by hiring people away from the medical schools, so we thought about collaborating.' The deal was pretty good for Monsanto – those tax-deductible millions were a drop in the ocean of its global sales of $3.5 billion.[18] A year later Harvard gave its staff the right to exploit discoveries in health

*What Emerson actually wrote was somewhat wordier and made no mention of mousetraps: 'If a man has good corn or wood, or boards, or pigs, to sell, or can make better chairs or knives, crucibles or church organs, than anybody else, you will find a broad hard-beaten road to his house, though it be in the woods.'

areas, and in 1977 it set up a patent office. For many years Stanford had a more entrepreneurial attitude which was partly responsible for the growth of the technology companies in Silicon Valley – hence its push to get Cohen and Boyer to patent their discovery.[19]

The combined effect of the granting of genetic patents and the passing of the Bayh–Dole Act was that business's existing interest in genetic engineering was rapidly amplified and many big companies struck major agreements with US universities in order to exploit the new technology. In 1982 the German pharmaceutical company Hoechst effectively set up its own department of molecular biology at Massachusetts General Hospital, employing over fifty staff. Harvard, having got over itself, struck a five-year deal with DuPont worth $6 million which involved creating a new genetics department.[20] In the same year, Monsanto agreed to hand over $23.5 million to Washington University, the sum to be allocated to research that was 'directly applicable to human diseases'. Any patents from the work would be held by the university, but with an exclusive licence to Monsanto. Howard Schneiderman, a vice-president at the company, unctuously described the agreement as 'the culmination of a long love affair between two institutions'.[21]

For many scientists, such organisational arrangements were less exciting than the prospect of making an individual discovery and an individual fortune. Genentech became the exemplar of the genetic engineering gold rush as scores of new start-ups were created, none of which had either the scientific or financial impact of Boyer and Swanson's creation.[22] In 1981, Wally Gilbert claimed that half of his colleagues were involved in companies, while the following year *Science* reported that most of the United States' leading biologists were affiliated with a biotechnology company.[23] As the magazine noted: 'Scientists who 10 years ago would have snubbed their academic noses at industrial money now eagerly seek it out.'[24] According to Donald Kennedy, President of Stanford: 'The tremendous wave of activity sweeping over the field as molecular biology is privatised is creating unique opportunities for certain professors to secure financial gain.'[25]

Some of the scientists had far-reaching ambitions that were extremely naive. In 1980, Gilbert boasted: 'Industry likes academics

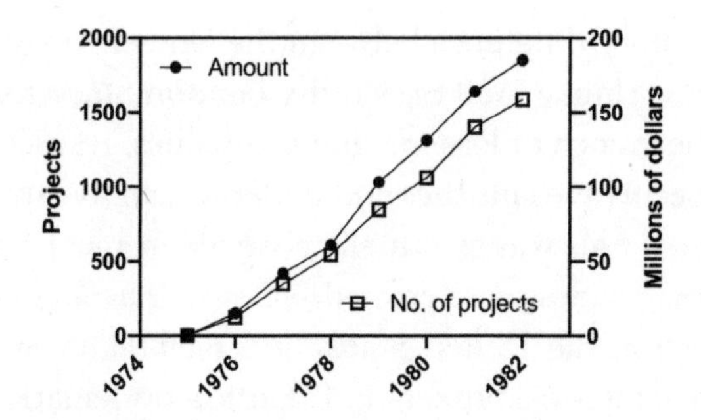

Increases in NIH funding for genetic engineering between 1975 and 1982. Data from Wright (1986), *Social Studies of Science* 16:593–620.

to be consultants. What we are seeing here is an attempt by academics to control industrial development!'[26] To prove his point, Gilbert left Harvard to become CEO of Biogen. He was ousted four years later when institutional stockholders insisted on a more professional management style.

The combination of the new science that could be done with genetic engineering and the financial rewards that it promised was a heady mix for all concerned. Between 1980 and 1982, NIH increased its support for recombinant DNA research by around 50 per cent.[27] This was matched by growing interest from the private sector: nine genetic engineering start-ups were created in the United States in 1979; there were eighteen in 1980 and thirty-three in 1981.[28] Even more companies were set up under the general category of biotechnology – some were focused on vaccine development, others on creating new diagnostic molecules.[29] Whatever the precise techniques they employed or the products they aimed to produce, they all hoped to bask in the scientific glory and potential riches represented by genetic engineering.

For investors, the key challenge was to pick a winner. In February 1981, Margaret Thatcher visited the laboratories of Genex, a start-up based near Washington DC that hoped to produce the artificial sweetener aspartame through genetic engineering. When the company's venture capital financing was explained to her, she replied enthusiastically: 'It's as exciting as betting on horses!'[30] Thatcher's apparent unfamiliarity with the system flowed partly from a regulatory difference between the United Kingdom and the United States. Unlike Wall Street, the London Stock Exchange did not allow the launch of loss-making companies, restricting the possibility of speculative launches and undermining the involvement of venture capital (this was eventually changed in 1993).[31]

Trade magazines were created – *Genetic Engineering News* and *Bio/Technology* in the United States and *Biofutur* in France. In the United States, the stockbrokers E. F. Hutton trademarked the word 'biotechnology' for their newsletter (the term had been coined by Karl Ereky in Hungary in 1917, eventually becoming synonymous with industrial fermentation).[32] The whiff of money soon permeated the pages of the scientific journals too. In 1980, *Nature* licked its lips at the prospect of future wealth, advising the readers of a ten-page survey of genetic engineering applications that 'greater profits can be found in a combination of closely-guarded research and rapid commercialisation of the results'.[33] Two years later, the journal launched a monthly update on the performance of biotechnology stocks, crassly entitled the '*Nature* Guide to Bio-riches'.[34] This nonsense initially took up nearly a whole page of the journal and included an index of biotechnology shares. It was eventually discontinued in January 1984 as a sad graph showed the boastful biotech index declining while the Dow Jones soared.[35]

Scientists were discovering that stock markets are fickle beasts, stimulated by rapid profits and easily swayed by fashion. Venture capital fund manager Kathy Behrens almost yawned when she explained what was happening:

> It's a pretty routine scenario. … First, the breakthrough technology. Everyone gets excited. The money starts rolling in. Not too many companies around. Everybody and his brother starts jumping into the pool with 'me-too' companies. Competition becomes intense. Company evaluations either decline or level off. The money gusher begins to dry up.[36]

The business atmosphere of the 1980s could be summed up by the misquoted words of trader Gordon Gekko in the 1987 film *Wall Street*: 'Greed is good.'[37] But by the end of the decade that attitude was beginning to rankle in certain quarters and even *The Economist* was bewailing 'the money-guzzling genius of biotechnology'. By that point, $10 billion had been invested in the field and 150 companies had gone public, but although they raised an additional $4 billion, only Genentech was making a sustained profit and that was 'disappointingly small', sniffled the bosses' magazine.[38]

The financial excitement in the United States was above all a function of the uniquely American layer of venture capital speculators that appeared after the Second World War and had no true equivalent in other countries. In the absence of such intermediaries and with different stock market rules, in Europe, the United Kingdom, Japan and India, the discoveries of genetic engineering were exploited by the established pharmaceutical and chemical companies, with the state playing a key role in funding much of the research.[39]

In Japan, MITI – the Ministry for International Trade and Industry – had long encouraged industrial development through investment and policy choices. MITI initially focused on electronics, new materials and biotechnology, but by 1983 it was emphasising biotechnology in particular. The following year one of the first biotechnology products became wildly popular in Japan; this was a lipstick called Lady 80 Bio that contained shikonin, a traditional plant-based dye now produced in plant cell cultures (but not by genetic engineering). More significantly, MITI promoted the grandly titled Next Generation Basic Technology (NGBT) project, which coordinated and developed the activities of fourteen companies, including the development of recombinant DNA technology. This was funded to the tune of ¥26 billion (around $100 million) over ten years.[40]

The United Kingdom was much less systematic in its support of the new industry. Stung by criticism that it had failed to patent the 1975 discovery of monoclonal antibodies by César Milstein and Georges Köhler in Cambridge (these did indeed turn out to be hugely significant*), in 1979 the dying Labour government relaxed

---

*Even more strikingly, no patents were taken out on Fred Sanger's incredibly influential, double Nobel Prize-winning techniques for sequencing proteins (1950s) and DNA (1970s).

regulations on genetic engineering and set up a Joint Working Party under former ICI research head Alfred Spinks 'to recommend action by Government or other bodies to facilitate British industrial development in biotechnology'.[41] As a civil servant recalled, 'the impetus for it really came from all these investments in the small companies in the States and a feeling that there was a race we are losing in some way'.[42]

The Spinks Report appeared in March 1980, ten months after Margaret Thatcher entered Downing Street, and its call for intensive government support for the new technology did not sit easily with the Thatcher government's increasingly neoliberal appetites. Nevertheless, by the end of the year the government had set up a company, Celltech, which initially had preferential rights to exploit any discoveries made through Medical Research Council funding. In 1983 the Agricultural Genetic Company was created, with similar links to the Agriculture and Food Research Council. Although UK government investment in biotechnology reached £30 million in 1981, this was only around one-third of the money spent by NIH alone that year.[43]

1982 saw an increase in research council funding and the creation of a Biotechnology Unit in the Department of Trade and Industry with an annual budget of £15 million – roughly equivalent to that of the NGBT in Japan.[44] However, this budget was not sustained over the medium term because the government was in thrall to the myth that only the private sector drove innovation. In reality, all of the discoveries underlying the genetic engineering revolution had been made in US universities via federally funded research. That state-subsidised research was then exploited by the private sector – just as had occurred and would continue to occur in electronics and computing. In the United Kingdom, this did not happen because the government failed to fund the necessary basic research, focusing instead on cutting government expenditure, including in the universities. As one historian has put it: 'the government's commitment to promoting biotechnology was subverted by the very construction of the neoliberal strategy adopted to promote it.'[45]

The involvement of increasing numbers of scientists in commercially sensitive projects began to cause ripples in academic life, with many researchers noting the corrosive effects of the new situation.

In 1977, at the very beginning of Genentech's involvement with UCSF, microbiologist David Martin told *Science:* 'Capitalism sticking its nose into the lab has tainted interpersonal relations – there are a number of people who feel rather strongly that there should be no commercialisation of human insulin.'[46] A postdoctoral researcher in Boyer's laboratory described how business changed the relationships between those who were working on Genentech projects and the rest of the researchers: 'You knew they weren't free to talk about their work. They were your friends, so you didn't want to make them feel awkward. So you sort of censored yourself.'[47]

Paul Berg also noticed the change: 'No longer do you have this free flow of ideas. You go to scientific meetings and people whisper to each other about their companies' products. It's like a secret society.'[48] One of the factors underlying this atmosphere was that although in the United States patent applications could be made within a year of a discovery being announced, elsewhere in the world any prior disclosure of a finding – including in a conference presentation – invalidated a patent claim (as a result, the Cohen–Boyer patent, based on their 1973 paper, was valid only in the United States).

These cagey attitudes soon permeated fields where there were no immediate commercial stakes. I recall 1980s conference presentations in *Drosophila* neurobiology where researchers would flash up a laboriously obtained DNA sequence for a few seconds while some in the audience furiously tried to scribble it down. There were suspicions – or maybe boasts – that some people deliberately included errors in the sequences they briefly showed in order to deceive competitors.

More significantly, scientists' financial entanglements began to produce potential conflicts of interest in the research they were doing. A study of 789 scientific articles published in 1992 in fourteen leading journals found that over one-third of the articles met the criteria for a conflict of interest.[49] Not one of the more than 250 papers disclosed the authors' financial interests in the research that had been published. Worse, the journals seemed not to care. In response to the study, *Nature* published a snooty editorial entitled 'Avoid Financial "Correctness"' in which the journal dismissed any concerns in its usual patrician style:

> It would be reasonable to assume, nowadays, that virtually every good paper with a conceivable biotechnological relevance emerging from the west and east coasts of the United States, as well as many European laboratories, has at least one author with a financial interest – but what of it? ... The work published makes no claim that the undeclared interests led to any fraud, deception or bias in presentation, and until there is evidence that there are serious risks of such malpractice, this journal will persist in its stubborn belief that research as we publish it is indeed research, not business.[50]*

In 1983, just as the era of bio-riches was kicking off, this complex intertwining of science and business was explored by UK historian and writer Edward Yoxen in his television series *The Gene Business*. Never mind potential compromises of scientific integrity – Yoxen thought genetic engineering was laying the basis for a new phase of capitalism:

> As technology controlled by capital, [genetic engineering] is a specific mode of the appropriation of living nature – literally capitalising life. It constitutes a new set of social relations between fundamental research scientists, industrial corporations and state agencies. ... Biotechnology – formed from the institutions, skills and concepts of a directed molecular biology – sustains the circulation and accumulation of capital by offering a new set of products, processes and markets. Life becomes a productive force, playing a growing role in reconstituting the social relations of a new capitalist order.[51]

Dramatic stuff, but shorn of its Marxist vocabulary this vision would have been embraced by many venture capitalists and pro-business scientists. Where Yoxen saw an alarming negative, they saw a very

---

* *Nature* and its sister journals now require a full declaration of authors' financial competing interests, including stocks and shares, consultancy fees and so on, 'in the interests of transparency and to help readers form their own judgements of potential bias'. Better late than never.

attractive positive. Four decades later, our modern world proves both those views wrong. For good or ill, commercialised biotechnology did not lead to a novel form of capitalism, but neither did it transform the economy.

*

For many researchers, the holy grail of genetic engineering was the ability to transform mammalian cells. In 1979, scientists at Columbia University in New York, led by Richard Axel (he would share the Nobel Prize in 2004 for his work on the molecular basis of the sense of smell), described a procedure that introduced DNA from two or more sources into mouse cells.[52] This was effectively a mammalian version of Cohen and Boyer's method for creating recombinant bacteria – one gene would produce an innocuous marker protein indicating that the process had worked, while the other would alter the activity of the organism in some way. Often known as the Wigler method after Axel's PhD student Michael Wigler, this technique was not only of decisive significance for the development of genetic engineering in mammals (it is still used today), it was also a massive money-spinner. Over a period spanning 1983–2002, Columbia was granted five patents for this invention; during the lifetime of these patents the university accrued more than $600 million, while the three inventors and a fourth, unnamed, person shared a cool $110 million.[53]

Even more significant, in terms of scientific progress, was the development of gene targeting, which enabled researchers to alter genes at will in mammals. On rare occasions, a cell will respond to the presence of a piece of DNA corresponding to one of its genes by using it as a template to alter the original, in a process known as homologous recombination. In the mid-1980s Mario Capecchi and Oliver Smithies independently realised that if they could use this process in mouse embryonic stem cells (these cells had been identified by Martin Evans) they would be able to create a mouse with the desired genetic change in every one of its cells. As Smithies' group put it in 1987, it was a way 'of obtaining specific, predetermined germline changes in experimental animals.'[54]

However, this method was extremely time-consuming and had very low efficiency. According to Smithies, the initial version he published was 'impossibly laborious', with a success rate of about one in a million cells.[55] Nevertheless, the groups of both Smithies and Capecchi were able to successfully target and transform the *HPRT* gene (this gene was chosen because it is on the mouse X chromosome, so male mice, which only have one X chromosome, only have one copy of the gene that would need to be changed). This technique was soon widely adopted and saw the creation of thousands of different kinds of 'knock-out' mice expressing – or not – a particular gene. This transformed our ability to control genes in mammals, leading to huge advances in our understanding of mammalian biology, in particular in the fields of embryonic development, physiology and neuroscience.* It also led to Capecchi, Smithies and Evans sharing the Nobel Prize in 2007.

This breakthrough was clearly in the air: in the middle of the 1980s, six groups, based in the United States, the United Kingdom, Switzerland and Germany, independently reported the creation of recombinant mice that transmitted a genetic change to their offspring.[56] It was in one of these papers that Jon Gordon and Frank Ruddle of Yale University coined the term 'transgenic', one of the many words that have been used to describe genetically modified organisms.[57]

In 1984, a series of chance events enabled two of those pioneer researchers, Richard Palmiter of the University of Washington and Ralph Brinster of the University of Pennsylvania, to produce a mouse containing SV40 viral genes. The animals developed characteristic brain tumours, allowing researchers to study the physiology

* A key factor in the rapid spread of genetic engineering in this period was the 1981 publication of *Molecular Cloning: A Laboratory Manual.* Written by Tom Maniatis, Joe Sambrook and Edward Fritsch using protocols they had produced for their Cold Spring Harbor Laboratory training course, the book – popularly known as 'Maniatis' or simply 'the Bible'– soon appeared on laboratory benches all over the world, its spiral-bound pages becoming stained and wrinkled with use and abuse. Still in print and now in its fourth, multivolume edition, the book has sold hundreds of thousands of copies. Without it, genetic engineering would not look the same. Creager, A. (2020), *BJHS Themes* 5:225–43.

and genetics of cancer in precise detail.[58] Out of naivety or principle, they did not patent the procedure or the mouse. At around the same time, Harvard researchers Philip Leder and Tim Stewart, funded by DuPont, developed a mouse that expressed the cancer-inducing oncogene *myc*. Unlike Palmiter and Brinster, the Harvard group and DuPont patented their mouse and even gave it a trademarked name (OncoMouse™).* The patents broadly claimed not just that particular oncogene mouse, but any transgenic non-human mammal that contained an oncogenic sequence.[59] In other words, the transgenic animal produced earlier by Palmiter and Brinster also belonged to Harvard and DuPont, as would any future transgenic oncogene rat, cat or wombat![60] Inevitably, there were long legal battles and a huge moral debate erupted about the legitimacy of patenting animals, with protests around the world and pickets of patent offices.[61] Eventually, the ambitious claim to all oncogenic mammals was struck down, but the specific claims with regards to oncogenic mice were accepted in the United States and the European Union (but not in Canada, which rejected the application).[62] DuPont's investment in Harvard science had come good, for their shareholders at least.

In the 1980s more than 1,000 patents on genes or genetic sequences were filed in the United States.[63] Exactly as predicted by the PBC contribution to *Diamond v. Chakrabarty*, the US Patent Office soon began approving patents on multicellular life forms. Among the first to be granted was for triploid Pacific oysters that could be harvested and sold all the year round; the DuPont OncoMouse™ soon followed.[64] What had long been seen as either immoral or unconstitutional became widely accepted. Life itself could be owned, and everyone thought it was normal.

The enthusiastic patenting of genes and genetic variants really took off at the beginning of the 1990s with the creation of a series of projects aimed at sequencing the human genome – the National Human Genome Research Institute in the United States (NHGRI

*In 2000, US artist Brian Crockett created *Ecce Homo*, a metre-high resin and marble sculpture of OncoMouse™. With its paws held out in Christ-like supplication, Crockett's giant hairless mouse suggested the rodent was the medical and spiritual saviour of humanity.

– led by Jim Watson) and the international Human Genome Organisation, together with parallel organisations in France, Japan and the United Kingdom. There were intense debates over the best strategy to adopt. NIH researcher Craig Venter was in favour of taking mRNA sequences from expressed genes in given tissues (he was particularly interested in the brain) and turning them into cDNA. He would then sequence around 400 base pairs of cDNA, rather than the whole gene, which was challenging for technical reasons. His argument was that this 'shotgun sequencing' of what were called expressed sequence tags (ESTs) meant that researchers would soon be able to discover at least some of the genes involved in key genetic diseases.[65]

This approach was immensely productive – Venter later estimated that his laboratory was able to discover up to sixty new human genes a day, at a time when fewer than 2,000 were known.[66] It also raised the prospect of patenting those ESTs, even though in virtually all cases they had no known function. Between 1991 and 1992, Venter took out patents on over 7,000 ESTs with the support of the NIH. Apart from the dubious morality involved, this Wild West view of patenting – register the sequence, whatever function it might turn out to have – threatened to cause chaos by leading to multiple patents being awarded for the same gene.

When Venter casually announced his intention to patent ESTs at a Congressional hearing in summer 1991, a huge row erupted over the principle of patenting human genes.[67] Watson described the idea as sheer lunacy and began agitating against the proposal, against Venter, and against Bernardine Healy, the head of the NIH who supported the idea. Faced with governmental and institutional opposition, Watson realised that his position as head of NHGRI was untenable and in April 1992 he resigned.[68] After Bill Clinton became president at the beginning of 1993, Healy was replaced by Harold Varmus; in the changed political climate, the NIH now adopted Watson's position and abandoned all its claims to human genetic sequences.* But all that happened was that the private sector moved

---

* In 2017, Watson told me that opposing the patenting of human genes was the most important thing he had done in his life – far more significant than his co-discovery of the structure of DNA. He explained that the discovery of

in – for example, in 1997 biopharmaceutical company Incyte applied for patents on 1.2 million sequences.[69] As the long-time observer of genetic engineering, Sheldon Krimsky, put it in 1999: 'the human genome was under colonisation'.[70] Venter himself set up Celera Genomics, which contributed to the human genome sequencing project with the aim of commercial exploitation. The result of this surge of activity – apart from untold wealth for the patent lawyers – was that by 2011 around 50,000 US patents on human DNA had been granted, nearly 80 per cent of them held by for-profit organisations.

Everything changed in 2013, when the Supreme Court struck down over 4,300 of those US patents and undermined tens of thousands of others, in the culmination of a legal battle that had spanned more than a decade. In 1994 and 1995, Myriad Genetics (co-founded by Wally Gilbert) patented two genes that are involved in some forms of breast cancer – *BRCA1* and *BRCA2* – with the aim of developing a diagnostic test. Myriad then warned rival companies working on these genes that they were infringing the patent but refused to grant a licence. A number of interested parties, including campaigning organisations and patients' rights groups, took Myriad to court.

In 2013 the case finally arrived at the Supreme Court, where the judges unanimously rejected Myriad's case. The Court overturned more than thirty years of patent activity by ruling that 'a naturally occurring DNA segment is a product of nature and not patent eligible merely because it has been isolated'.[71] The Myriad decision was a pleasant surprise to many (but not Myriad and its shareholders). NIH director Francis Collins tweeted his glee at the decision: 'SC: "A naturally occurring DNA segment is a product of nature and is not patent eligible merely because it has been isolated" Woo Hoo!!!'[72]

The situation was not quite as clear as Collins made out, however, for the Court also ruled that cDNA could be patented where it was not identical to the genomic DNA, because it does not occur naturally (Venter's EST patents would therefore have stood,

---

the double helix was inevitable and would eventually have been found by somebody, but his arguments had been pivotal in NIH deciding not to patent human genes. When I asked him why he had felt so strongly about the issue, his reply was simple: 'I'm a socialist', he said. I raised an eyebrow and he explained – 'I believe in universal education and universal healthcare.'

Francis S. Collins
@NIHDirector

SC: "A naturally occurring DNA segment is a product of nature and is not patent eligible merely because it has been isolated" Woo Hoo!!!

4:01 PM · Jun 13, 2013 · Twitter for iPhone

Tweet by Francis Collins, the director of the NIH, responding jubilantly to the US Supreme Court decision striking down many gene patents in June 2013.

had the NIH not already rescinded them). The current situation in the United States is therefore that genes cannot be patented simply from a genomic sequence.

In Europe (and this includes the United Kingdom, even after Brexit), things are slightly different. The European Biotechnology Directive, first adopted by the European Commission in 1988, allows the patenting of gene sequences, from a human or any other organism, if they have been isolated from the body and there is a demonstrated function. The Directive was eventually passed by the European Parliament in July 1998 following a ten-year battle that involved intensive lobbying by pharmaceutical companies (SmithKline Beecham alone had a budget of €30 million devoted to campaigning on the question).[73]

*

It is claimed that patents both reward and encourage innovation by providing income to the discoverers and producing a regulated environment in which entrepreneurs can take advantage of a new invention. However, in 1998 legal scholars examined the evidence and concluded that, on the contrary, patents may stifle innovation.[74] For example, over 800 US patents have made claims about the adrenergic receptor, which is found in our bodies and is the target of natural substances like adrenaline and drugs such as beta blockers. Any company wishing to create a product relating to this receptor might have to negotiate dozens of different patents. Even worse, many companies have imposed licence agreements that give the patent holders rights over subsequent inventions made using the licence. For example, the OncoMouse™ patent required licensees to consult DuPont before commercialising any discoveries made through their use of the mouse. The terms of the licenses gave DuPont a say in any future negotiations relating to novel commercial products.[75]

The commercialisation of the most significant breakthrough in molecular biology to occur in the 1980s – Kary Mullis's invention of the polymerase chain reaction (PCR) – was even more far-reaching. This incredibly efficient way of amplifying specific bits of DNA transformed science and medicine, rapidly becoming an essential component of molecular biology and diagnostics and entering into everyday vocabulary during the COVID-19 pandemic.[76] Mullis worked for the pioneer biotech company Cetus, which immediately patented the procedure, estimating that by 1998 the market for PCR-based diagnostics alone would be worth around $1.5 billion.* But Cetus also claimed that they would own part of the rights to any subsequent commercial application of the PCR process – and that included work done in an academic setting. As Robert Fildes, president of Cetus, put it in 1988: 'Our attitude to the universities is: use PCR, but don't give the results to the pharmaceutical houses without giving us a piece of the pie.'[77] In 1991 Hoffman-La Roche acquired the rights to PCR for $300 million and removed those highly restrictive downstream claims.[78]

Academic argument about the impact of patents has continued over the last two decades, but it is evident that while the kind of all-encompassing claims made by DuPont and Cetus may have increased the income of the patent owners, they did not encourage innovation. Reviewing the effect of DuPont's patent on all oncogenic mice in 2007, transgenic mouse pioneer Richard Palmiter was savage. The whole point of the mice should have been to increase patient

* The initial costs of using the *taq* polymerase enzyme patented by Cetus were too great for many smaller laboratories and some researchers sought to bypass the official channels. I recall being surreptitiously handed some bootleg *taq* by a fellow researcher when we met in a Parisian bar.

survival by leading to the development of new drugs. Instead, DuPont had adopted what Palmiter termed the 'socially controversial if not uncommon business practice' of using patents simply to generate revenue:

> DuPont ... has not obviously sought to generate or import oncomice so as to use them in its internal drug development programmes. Rather, DuPont has focused on sublicencing the Harvard patents, with hefty fees and restrictions ... DuPont's business plan ... has complicated and arguably slowed commercial applications of oncomice to the development of new cancer drugs.[79]

The same counterproductive, stifling effects appear to have occurred with gene patents. A comparison of genetic variants patented by Craig Venter's Celera Genomics with sequences that were publicly released by the Human Genome Project revealed that the patented genes generated up to 30 per cent less research and product development.[80]

Similar debate surrounds the impact of the Bayh–Dole act, which has been copied around the world. For example, Japan, Taiwan, Brazil, South Africa and the Philippines all adopted similar policies, while Germany has moved away from ownership of discoveries by individual researchers towards institutional ownership. Although the law might seem to have had a straightforwardly positive effect on innovation and commercialisation, many observers do not see things that way. In 2007 a vice-president of Hewlett-Packard claimed that the act 'fuelled mistrust, escalated frustration, and created a misplaced goal of revenue generation, which has moved universities and industry farther apart than they've ever been'.[81]

At the beginning of the 2020s we live in a very different world, with so many DNA patents ruled invalid by the courts, and with two novel contradictory factors shaping science and medicine. There is a growing preparedness to share data, as seen by the open science movement and the use of preprints which rapidly dominated scientific investigations of COVID-19. On the other hand, we are increasingly aware of privacy and individual rights; researchers recognise that large-scale genomic studies might have to deal with millions of individual claims over the use of personal genetic data in that research.[82] Furthermore, indigenous groups realise that their DNA might be used to develop products and are understandably hostile to the exploitation of their most personal data. Finally, despite initial claims from many universities racing to develop a COVID-19 vaccine that any successful product would be made available cheaply, or free, to developing countries, virtually no university or research institute would now allow its employees to refuse to patent a significant vaccine.* Scientists and university administrators have eaten from the tree of wealth, and things will never be the same again.

*At the very end of 2021 an honourable exception appeared. Texas Children's Hospital and Baylor College of Medicine announced they were making their COVID-19 vaccine freely available around the world, starting with India. Mind you, this was only possible because a number of private donors, including philanthropic foundations, made substantial donations to the developers. In May 2022, leading members of the World Trade Organization proposed waiving patents on COVID-19 vaccines, in order to encourage the production of existing vaccines around the world. The pharmaceutical companies have been lobbying against the proposal, and as this book goes to press, no agreement has been reached.

– EIGHT –

FRANKENFOOD

One of the main ways that the general public is aware of genetic engineering is through one of the field's great commercial successes: GM crops. This has also been a story of brilliant scientific success – in the space of a decade the creation of transgenic plants went from riddle, through cutting-edge science to reliable technique. The scientists involved have written memoirs of this period, but it has yet to be investigated by historians – it does not feature in accounts of the history of molecular biology, even though it was one of the most significant developments of the 1980s.[1] As well as resolving a tricky scientific problem, it opened the road to many other methods for modifying plants and led directly to the controversial transformation of parts of global agriculture.

In the 1960s, Belgian researchers led by Marc Van Montagu and Jeff Schell became interested in the origin of plant galls, a form of tumour. A particular bacterium, *Agrobacterium*, turned out to play a key role and in 1974 the Belgian group reported the presence of a giant plasmid in the cells of this species. They immediately suspected that the plasmid might play a role in transporting DNA from the bacterium to the plant, where it produced uncontrolled growth – a gall. They also realised that this might be a route that could be used to create recombinant plants. Van Montagu told me in 2021 that this insight led to a period of 'happy science', as he and his colleagues amicably competed with two other research groups to turn this intuition into a technology.

In 1977, Mary-Dell Chilton and her team at the University of Washington showed that, as Van Montagu and Schell suspected, galls were caused by the introduction of a sequence from a tumour-inducing plasmid into the DNA of the plant host.[2] This remarkable discovery was a naturally occurring example of the transfer of genetic material between different kingdoms of life and was part of a wave of discoveries showing the existence of recombinant DNA in nature, alleviating fears about the technology. Six years later, at the beginning of 1983, all three groups working on plant genetic engineering announced they had been able to remove the tumour-inducing sequences from the plasmid and replace them with a chosen piece of DNA, which was then integrated into the plant genome. Plants could be genetically engineered.

These research groups were all funded by the US company whose name is indelibly associated with the power and the problems of GM crops – Monsanto. In Belgium, Van Montagu and Schell were supported by Monsanto, while Mary-Dell Chilton received Monsanto funding before leaving academia in 1983 to create a genetic engineering facility for CIBA-Geigy (now Novartis).[3] The third team came from Monsanto itself – in 1980 the company had set up a genetic engineering group under Ernie Jaworski; Chilton later described him as 'an early visionary'.[4]

One of the key members of that third team, Robb Fraley, played a fundamental role in the development of genetic engineering in plants. In 2021 he told me of the euphoric atmosphere in the laboratory when a key experiment produced the desired results:

> It was an exciting time. I still remember, in the Fall of 1982, Rob Horsch, who was our tissue culture expert, running down the hallway in the research labs in Monsanto, just screaming at the top of his voice, 'It worked! It worked! It worked!'[5]

The breakthrough made by these groups was announced at a Miami symposium held in January 1983, with publications appearing over the subsequent year, beginning with an article from the Belgian team. [6] Although the Monsanto researchers were the last to publish – and therefore 'lost' the race according to the strict criterion of priority – media reports of the January 1983 symposium presented the discovery as theirs. *The Times* focused on the role of Monsanto and the ultimate goal of improving plant productivity, while the *Wall Street Journal* heralded the feat as the work of Monsanto scientists.[7] Inevitably, given the patent gold rush of the time, all three groups filed patent claims. This kept the lawyers busy for over two decades – the sorry mess was finally resolved in 2005 when it was agreed to share the licences.[8]

The relative flexibility of plant biology meant that within a few years there were several methods for making recombinant plants, some of them almost trivial in their simplicity. For example, in 1985 Monsanto researchers showed that if you punched out a small piece of leaf and left it brewing in a soup of engineered *Agrobacterium*, the recombinant plasmids would find their way into the tissue where they would integrate into the genome. Plants being plants, in principle it was then possible to get these discs to produce roots and – hey presto! – you had a transgenic plant.[9] A year later, researchers in Basel resolved a long-running debate by showing that DNA could also be taken up and transmitted to offspring by using what are called protoplasts – plant cells that have had their cell wall removed. The title of their article said it all: 'Direct Gene Transfer to Plants'.[10]

All these studies were on dicotyledons, a group that includes a large number of agriculturally significant plants, such as cotton, soybeans and sunflowers. But many key crops – rice, maize and other cereals – are monocotyledons and for various reasons this group proved recalcitrant to transformation. Although monocotyledon protoplasts would take up foreign DNA if pores were created in the cell by giving it an electric shock, it was difficult to get the transformed cells to grow. By 1987 this was widely seen as a key limitation of the potential of genetic engineering in plants. *The Guardian* noted that 'all attempts to bring these techniques to the improvement or hardening of the world's major species of cereals – such as rice, maize or wheat – have failed'.[11] But even as the article was printed, this enigma was being resolved in the most bizarre fashion imaginable.

In 1987, a group of Stanford researchers announced the development of a technique that could only have been invented in the United States – the gene gun.[12] This device did exactly what you might expect: it shot tiny DNA-covered particles into plant tissue, leading the foreign DNA to integrate itself into the plant's genome.[13] Initially made out of an airgun and used on onion cells (onions are monocotyledons), the steampunk-sounding gene gun soon employed gunpowder. Co-inventor John Sanford recalled that developing the gun was 'really very much fun', with loud bangs and debris filling the air. All their colleagues thought they were crazy, and what with the blasted onions and the gunpowder, the laboratory soon smelled like a mixture of a McDonald's and a firing range.[14]

Attempts to gain financial support for the project were initially met with ridicule, while scientists who read the *Nature* article describing the invention hooted with laughter.[15] But the amusement soon faded. The gene gun was commercialised and this way of generating transgenic plants became widespread, leading to three major breakthroughs: the creation of transgenic maize and the genetic transformation of two key organelles in the plant cell – mitochondria and chloroplasts (the site of photosynthesis).[16]

Although all of these techniques solved the problem of delivering recombinant DNA into the cell, for that DNA to be adequately expressed it needed to be preceded by a sequence called a promoter – one of the most widely used promoters, 35S, is from a cauliflower virus. This process also needed a terminator sequence, often taken from *Agrobacterium*, to tell the cell when to stop processing the message.[17] Furthermore, following a development by Chilton's group in 1983, recombinant plants were often identified by including a gene that made them resistant to an antibiotic – this was a convenient technique for working out which plants contained the GM construct, but it meant that as well as the desired transgene, the plants were at least initially resistant to certain antibiotics.[18] All these additional genetic components of the GM process were later to prove significant in shaping public attitudes to transgenic crops.

*

Monsanto's role in the development and application of plant transgenics was fundamental, but the chemical company's initial objective, as seen by Monsanto researchers and high-ups at the time, might surprise you. In the 1960s, with its historic trio of symbolic products – the insecticide DDT, plastic grass (Astroturf) and Agent Orange, the powerful defoliant used in the Vietnam war – Monsanto seemed to represent a horrifying world of artificiality and the death of nature. Genetic engineering was supposed to change all that and enable the company to get out of pesticides completely. As Dick Mahoney, the CEO of Monsanto, put it in the late 1980s: 'I don't ever want to be in chemicals again ... that's why we're in biotechnology.'[19] A few years later, this was fleshed out in the vision of Robert Shapiro, who became CEO of Monsanto in 1995. In the stuffy conformist world of US big business, Shapiro passed for a radical – he worked in an open-plan office, often went tieless, was known to wear lime-green shirts and was said to have the air of everyone's favourite university professor.[20] 'The system is unsustainable', he proclaimed to Monsanto staff, before going on to show how the company would profit from the growing crisis: 'This world will not go quietly toward extinction. The world is going to be prepared to pay people who can help it survive.'[21]

One of Monsanto's projects was to enable farmers to massively reduce their blanket pesticide use by creating transgenic plants that expressed substances that killed or deterred caterpillars, but not other insects. These substances were derived from a naturally occurring soil bacterium, *Bacillus thuringiensis* (the insecticide is therefore generally known as Bt), which had been used for some time by organic farmers as a natural pesticide. Monsanto was not the only group chasing this prize – the first proof that a Bt gene provided protection against an insect came from Marc Van Montagu's Plant Genetic Systems company in Belgium.[22] *

Virtually as soon as Monsanto's Bt cotton plants were approved

* Van Montagu is a lifelong socialist; in 2021 I asked him why he had set up a company to turn his discoveries into products. He told me that he had no option, as the Belgian government refused to be involved, arguing that commercial exploitation should be left to the private sector.

for use in the United States, the sales campaign received an unexpected boost. A few months earlier, in the summer of 1995, US cotton farmers in Alabama and Mississippi had been hit by a terrible outbreak of tobacco budworm caterpillars. Faced with massive losses, farmers bought all of the newly approved plant Bt cotton seeds. One farmer recalled:

> I don't know how much cotton would have been left in Mississippi if we had another year like that ... It was devastating. When you have a loss like that, it takes 8 to 10 years to get over it. Bt really made a difference. ... I'd say it was the critical part of us staying in production.[23]

The advantages were not simply economic. Bt crops enabled farmers to reduce their pesticide use by around 80 per cent and the deployment of transgenic crops involved a sophisticated sustainable approach – farmers were discouraged from planting too much Bt seed, in order to allow some pest insects to survive.[24] This counter-intuitive advice was based on the realisation that random mutations would inevitably produce a small number of insects that could resist the plant's pesticide; if those animals mated, a resistant strain would emerge. But if crops in part of the field did not express the Bt gene, the resistant insects would be more likely to mate with ordinary insects that had grown on the normal plants, and the resistance genes would be diluted in the population and would slowly disappear.

There were large sums of money at stake – the depredations of the corn borer caterpillar in the United States alone cost around $1 billion per year, accompanied by the financial and ecological cost of spraying vast quantities of broad-spectrum insecticide from the ground and the air.[25] But the undoubted effectiveness of the product did not mean that the extra premium farmers had to pay was always worth it – Bt maize, for example, only produced clear benefits for farmers in years when the corn borer moth boomed, about one year in four.[26] Even the highly effective Bt cotton could not produce benefits if there were no moths to deter. That depended on the whims of nature, but luckily for Monsanto in the case of cotton that was not something farmers were willing to bet against. As a result, US coverage with Bt cotton went from 2,000 hectares in 1997 to 700,000 in 2000, comprising 20 per cent of total US cotton acreage. By 2015, 96 per cent of US cotton was Bt.[27]

While the approach of agribusiness to Bt crops showed hints of ecological awareness, the same could not be said of what became GM's biggest success story – Roundup Ready crops. Roundup is the trade name of Monsanto's herbicide, glyphosate. First marketed in the mid-1970s, glyphosate kills any plant that it touches by blocking the activity of an enzyme called EPSPS that is involved in amino acid synthesis. It is now the number-one herbicide used in US agriculture and is a best-seller around the globe. This success is not simply down to the efficiency with which glyphosate kills plants; it also flows from Monsanto's bright idea of creating crops that would be tolerant to glyphosate, enabling farmers to spray a field with the herbicide, killing all the weeds but leaving the crop intact. This would reduce time and energy spent on weeding, enable crops to be grown at a greater density thereby increasing yields, and would also cut the need for tilling between crops, a practice that destroys soil structure.

The scheme was initially met with horror by many in the industry. In Switzerland, CIBA-Geigy had discussed a similar proposal for their own patent herbicide but stopped the project in its tracks because of predicted public opposition.[28] In the early stages of the development of Roundup Ready plants there was similar hostility within Monsanto itself – Robb Fraley is reported as saying: 'If all we can do is sell more damned herbicide, we shouldn't be in this business.'* But despite Fraley's apparent scruples, Monsanto soon came around to the idea and began refining the perfect way of selling more damned herbicide.

After a number of false starts, in 1989 Monsanto discovered that it had inadvertently created a bacterial strain expressing precisely the tolerance gene it was after – it was living in the waste pools of one of the company's manufacturing plants.[29] This new gene, which produced a slightly different form of EPSPS that was not affected by glyphosate, was eventually integrated into a range of plants, including maize and soy. And of course, it was patented.

---

* In 2021 Robb Fraley told me he could not recall making this comment.

"And while we're at it, let's sue the bees for illegal distribution of our intellectual property."

The problems began for Monsanto – and for other companies – when it came to commercialising its products. Uncomfortable with the traditional world of US seed companies, Monsanto came up with a cunning plan. It would sell its GM seeds at the same price as normal seeds, but there would be an additional 'technology fee' which would allow the farmer to use the product.[30] In other words, the farmer would be licensing the gene. And, in a move that was to prove decisive, Monsanto made the farmers sign an agreement that they could not keep any seed produced by their plants. This was particularly significant in the case of Bt plants because the Bt gene was dominant – this meant that the seed produced from Bt plants would also express the gene, even if they were produced by a cross with a wild-type plant. But if you lease something, you do not own it. The farmers did not own the magic genes in those seeds, Monsanto did.

*

According to Robb Fraley, Monsanto wanted to become 'the Microsoft of agriculture'.[31] By this he meant not only that it would have a dominant, near-monopoly position, but also that it would sell a whole industrial ecosystem, tying users into complex deals that required them to use other Monsanto products. At the time Fraley said this, Microsoft was being investigated under US anti-trust laws precisely because of its dominant position and its unfair practices. No matter how attractive the parallel with Microsoft might have sounded to investors, running the risk of ending up in the same sticky legal position was perhaps not a good strategy.

As with many other companies at the time, a key element of Monsanto's business plan was its systematic patenting of all its discoveries. To protect its intellectual property – and its market position – the company became obsessed with pursuing breaches of patent. It hired private investigators to spy on farmers, encouraged farmers to spy on each other and to report miscreants on a free phone number, and it aggressively prosecuted those suspected of saving seed to sow in subsequent seasons. In the final years of the twentieth century, Monsanto investigated 500 cases of alleged seed piracy and threatened legal proceedings in sixty-five cases. Virtually all the farmers paid up.[32]

The most notorious case occurred in Canada in 1998, when Percy Schmeiser was sued because his canola seed (canola is a type of oilseed) contained the Monsanto Roundup Ready gene, but he had not paid the company to use it. Schmeiser claimed the gene had got into his seed by wind pollination from neighbouring crops; Monsanto argued he had been illegally storing seed. After much legal rigmarole, culminating in a Canadian Supreme Court hearing in 2004, Monsanto won the case. But to the general public the company appeared vindictive, a multinational trying to crush a small farmer, who in 1998 had made a mere $20,000 from his illegal crop.[33] The decision to sue Schmeiser cemented Monsanto's reputation as the bully boy of global agribusiness.

By 1998, Shapiro had guided the company through a massive series of acquisitions of seed companies and rivals worth a staggering $8 billion. Although this saddled the company with massive debt, it achieved the desired aim. Monsanto had become a planet-bestriding colossus that influenced every aspect of food production – 'dirt to dinner' was the term that was used. Less attractively, it was widely considered to crush small farmers, eradicate rivals and spread its patented engineered genes across the planet. The capitalists' dream was the anti-capitalists' nightmare.

*

The application of genetic engineering to agriculture has proved controversial, for complex scientific, political and cultural reasons. This occurred from the very beginning. In the 1970s, scientists had found that naturally occurring bacteria could worsen frost damage by inducing the formation of ice crystals through the expression of a particular protein. By spraying plants with genetically modified 'ice-minus' bacteria that did not express the protein it was expected that strawberry plants would be able to resist temperatures as low as –7°C. University of California scientist Stephen Lindow, who had been studying the problem of bacterially induced frost damage for a decade, worked with DNA start-up Advanced Genetic Sciences (AGS), which marketed the modified bacteria as Frostban.

In 1983 an initial field trial of Frostban, with no plants involved, was blocked by the California courts following a campaign led by Jeremy Rifkin. All sorts of lurid fears were conjured up, including the suggestion that the bacteria might migrate into the upper atmosphere and alter the weather. In a sign of things to come, after the court eventually approved the field trial in April 1985, Environmental Protection Agency workers who took air samples at the site were dramatically clad in hazmat gear – white Tyvek suits and half-face respirators – even though there was no threat to anyone. Shortly afterwards the field was trashed by protestors. Lindow had previously carried out field experiments with strains of naturally occurring bacterial mutants that also did not express the protein, but because they were not regulated nobody paid any attention.[34]

When Rifkin ran out of legal road in 1987 and his case was dismissed by the courts, AGS finally began tests of Frostban on strawberries and potatoes at two locations in California; both sites were immediately trashed and the project was eventually abandoned. The product has never been commercialised, partly because of a lack of appetite among growers.[35] The main consequence of the discovery has been the use of bacteria containing (rather than lacking) the frost-inducing protein to produce artificial snow in ski resorts and to maintain uniform freezing in food processing.[36] Lindow later deplored the public's response to Frostban and wearily concluded that trying to apply his discovery was 'the stupidest thing I've ever done. I wouldn't recommend it to anyone else – not until people are more educated.'[37]

As various transgenic crops were developed in the second half of the 1980s and early 1990s, there were repeated political arguments over how they should be regulated. As well as the same kind of debates about environmental security and the health of researchers that had characterised the use of recombinant DNA in the 1970s, there was now an additional layer of concern – these products were to be eaten, either by humans or animals, and therefore had to meet food safety standards. In the United States, the Reagan administration eventually produced a coordinated framework for the regulation of GM products which did not involve new legislation and which provided a relatively light touch, much of which was voluntary. There was still one problem, however. Whatever the legal situation, companies wishing to introduce GM products into the food chain had to convince the public that their product was safe.

The first transgenic food to be sold, the ostentatiously named Flavr Savr tomato, had to go through a great many regulatory hoops before it was finally allowed onto the supermarket shelves in the United States, only for the company selling it to collapse soon thereafter. Developed by a small biotech start-up called Calgene that had been created in the golden year of 1980 and whose CEO did not even like tomatoes, the Flavr Savr was designed to increase shelf life by blocking the production of a particular enzyme.[38] In the 1980s, cloning and manipulating the gene that produced the enzyme was the focus of an intense international race, with publications in the leading science journals.[39] Once the basic biology was done, Calgene eventually began commercialising its invention. Calgene's tomato contained a complementary version of the enzyme-producing gene (it is described as 'antisense') inserted in the genome – the mRNA it produced would bind with the mRNA produced by the gene that encoded the enzyme, thereby blocking gene expression as soon as it began (this process is called antisense RNA inhibition). The idea was that in the absence of the enzyme tomatoes would not only have an increased shelf life, they would also taste more tomatoey.

From the first field trials in 1988 the project hit problems, not least because the company voluntarily subjected itself to the intense regulatory scrutiny of the US Food and Drug Administration (FDA) in order to alleviate any potential public concern. This process of investigation took four years, during which time the company accrued debt and struggled not only with bureaucracy, but above all with the complex, low-margin world of tomato farming. Rather than licensing its product to people who knew what they were doing, the bosses at Calgene decided to disrupt the whole farming business. In the end, the only thing they disrupted was their own company.

In her 2001 book *First Fruit*, Calgene molecular biologist Belinda Martineau vividly conveyed the complex, chaotic world inside Calgene as it sought to develop the product (cynically branded as 'MacGregor's' to sound rural and authentic).[40] Martineau described in detail the rocky road from basic biological discovery to the creation of a product. In particular, she emphasised the vagaries of the transgenic transformation of plants – a reality that is far more frustrating and haphazard than the breezy descriptions I have been giving here for the sake of simplicity. For example, to create the plants that were used to get FDA approval, 21,250 individual transformation experiments were completed, only 960 of which were effective.[41] Of those plants, in only 167 cases (less than 20 per cent) did the antisense RNA block the enzyme to the 95 per cent level that was required by the Calgene protocol. Of *those* plants, only 130 produced enough seed to be able to carry out the next test, which involved checking for the number of copies of the antisense gene – they wanted only one copy of the inserted gene in each plant. Only forty-four plants met that criterion. All that had taken a year. Later on, it turned out that many of the plants that seemed to be successful in fact had multiple insertions of other genes and they too were eliminated. In the end, only eight plants met all the criteria, out of over 20,000 initial attempts. It was later discovered that some of these plants had bits of *Agrobacterium* DNA floating about in them, left over from the initial transgenic experiment, and a new round of purification had to begin.[42]

The bathetic end to the Flavr Savr came when, despite a glowing FDA approval in 1994, healthy sales, and positive responses from both public and food journalists, the company went under in 1995 for reasons that had nothing to do with the safety or the palatability of the product. As well as a series of naive decisions, such as not planting crops at key moments, the company was hampered by the fact that the added value contained in the tomato – longer shelf life – made little difference to customers. Blocking the enzyme did not really improve the taste of the tomato; it was OK, but no more. There was no real prospect of making millions because, well, it was just a tomato. As Wall Street started to short Calgene shares, Monsanto swooped, buying up the company. Calgene disappeared along with its poorly spelled product.

While there was little public opposition to the Flavr Savr – in fact, quite the opposite as customers were intrigued by the new product, which was fêted by the media – the first transgenic product put on the market by Monsanto soon encountered problems with public acceptance. The company used recombinant bacteria to produce bovine growth hormone (BGH) – related to the human somatostatin that had been Genentech's first success in 1977. This hormone was purified and injected into cows, leading to increased milk production. There was substantial opposition from dairy producers and campaigners who feared that the milk from the cows might be contaminated with the hormone, but in 1993 the FDA declared that BGH had no serious effects on human health. Above all, the Bush administration prevented producers who did not use the hormone from labelling their product 'BGH-free' – no milk of any kind contained BGH, and it was not possible to detect if a given sample of milk came from a cow that had been administered BGH. This labelling issue partly explains the relatively high levels of GM products in the US food chain. People simply do not know it is there.

Monsanto CEO Richard Mahoney later admitted that the company had blundered into the issue of public acceptance of transgenic technology without thinking about what it was doing, beyond technical and financial questions: 'There wasn't even one discussion of the social implications. I never thought of it', he said.[43] But the lesson was not learned, and through the 1990s and beyond, Monsanto's reputation for unthinking ambition grew and grew.

Inevitably, as the new products came closer to market, the public began to find new ways of describing them. In December 1989, the *Sunday Times* ran an article by food critic Egon Ronay under the headline 'Frankenstein Food', brandishing one of the oldest and most potent symbols of fears about artificial life forms. Ronay was not writing about GM crops but about an anodyne proposal to allow the mild irradiation of food in order to reduce microbial contamination; nevertheless, the alliterative phrase was soon adopted as a way of describing GM food.[44] Three years later, one Paul Lewis wrote a brief letter to the *New York Times*, in which he apparently coined a term that would stick to GM products like mud: 'If they want to sell us Frankenfood, perhaps it's time to gather the villagers, light some torches and head to the castle.'[45]

To an extent, scientists had brought this on themselves, by a mixture of overexcitement and ill-considered hyperbole. In 1981, at the beginning of the commercial application of genetic engineering, the president of the International Plant Research Institute, the fortuitously named Martin Apple, promised the readers of the *New York Times*, 'We are going to make pork chops grow on trees.'[46] Quite how much Apple was either joking or misquoted does not really matter – with repeated promises of an amazing future, it was hardly surprising that the public began to take those promises seriously and even to fear what they might lead to.

*

Before the advent of GM crops, genetic engineering seemed to have become routine. The fears of what might escape from laboratories slowly evaporated as nothing went wrong, and the existence of recombinant organisms became accepted – in the laboratory at least – and a promising future seemed to beckon. This transition could be seen in popular music. In their 1978 song 'Genetic Engineering', British punk band X-Ray Spex had warned of a potentially fascist future:

Ein, zwei, drei, vier!
Genetic engineering could create the perfect race
Create an unknown life-force that could us exterminate
Introducing worker clone as our subordinated slave
His expertise, proficiency will surely dig our grave
It's so very tempting, will biologists resist?
When he becomes the creator, will he let us exist?

Five years later the British new wave band Orchestral Manoeuvres in the Dark also released a single called 'Genetic Engineering'. But whereas Poly Styrene had screamed her warning against a background of raw guitars and a honking saxophone, OMD's futuristic pop song included synthetic speech from a Speak and Spell machine and cheery voices proclaiming a manifesto that was apparently intended seriously:

Efficient, logical, effective, and practical.
Using all resources to the best of our ability.
Changing, designing, adapting our mentalities.
Improving our abilities for a better way of life.[47]

This period also saw the appearance of a new cultural metaphor for the power and the potential perils of genetic engineering that soon rivalled *Frankenstein* – *Jurassic Park*. Michael Crichton's technothriller, published in 1991 and turned into a blockbuster Steven Spielberg film two years later, features dinosaurs that were created by genetic engineering, their DNA having been extracted from blood sucked up by mosquitoes that were trapped in amber millions of years ago. The dinosaurs are kept in a huge amusement park by a benevolent but foolish billionaire and – spoiler alert – it all ends badly. Leaving aside the fundamental flaw in the plot (my colleagues have shown that DNA does not survive in insects trapped in amber even from the 1960s[48]), the key moral message was contained in a pithy speech made by a sceptical 'chaos theoretician', Dr Ian Malcolm, played in the film by Jeff Goldblum:

Your scientists were so preoccupied with whether or not they could, they didn't stop to think if they should.[49]

The problem, Crichton was saying, was not the technology, but its misuse by overconfident scientists (so do not recreate nasty bitey dinosaurs, but other uses of genetic engineering are probably fine).*

New, significant metaphors related to genetic engineering were adopted during this period. In 1981, Arthur Riggs and Keiichi Itakura wrote a review of their synthetic DNA method in which they made the earliest use I have found of a metaphor that has recently come back into fashion – they described their technique as facilitating 'gene editing'.[50] And in 1988, in his popular science book *Invisible Frontiers*, Stephen Hall took the widespread idea that DNA is a text and drew a metaphor from contemporary technology, suggesting that genetic engineering was an 'editorial enterprise' that could best be understood by thinking about photocopying, and literally cutting and pasting bits of paper.[51] Although cutting and pasting metaphorically on personal computers had been popularised by Apple from 1983, it had obviously not reached a wide audience by the time Hall was writing. The original, physical meaning of 'cut and paste' has now been forgotten and its current use in DNA editing refers to the way we use computers, adding a layer of deceptive simplicity to the power of the metaphor. 'Recombinant DNA' had initially seemed like a neutral choice to describe the result of genetic engineering, but it was opaque to ordinary people. Furthermore, once it became the focus of public concern, there was a slow shift to another, apparently neutral phrase – genetically modified organism (GMO).[52] The brave new world had its new words.

---

*Crichton tried to repeat the trick in 2006 with what was to be the final book published in his lifetime, *Next*. This rambling and complicated novel, which features a hyperintelligent parrot and a transgenic talking chimp, took aim at the genetic engineering industry, the corruption of academia and in particular at the patenting of genes. It was not a great success.

– NINE –

SUSPICION

All around the world, public concerns about the main visible products of genetic engineering – GM crops – reached a paroxysm in the second half of the 1990s, through a cascade of events that mixed science, politics, culture and deep fears, interacting with other political developments around the planet, in particular the politics of globalisation.[1] Global views have largely remained locked in this state over the last quarter century, although the debate has become much less heated with the subsequent spread of GM crops and the widespread presence of GM products in many countries.

One of the key events that precipitated the crisis in confidence had nothing to do with GMOs. In 1990, the UK government revealed the existence of a lethal neurodegenerative condition affecting British cattle – bovine spongiform encephalitis (BSE), or as the media soon dubbed it, 'mad cow disease'. Despite growing public concern at the potential danger of eating beef from affected cows, the British Conservative government was consistently reassuring.[2] In 1996, the bombshell dropped: a terrifying new neurodegenerative disease had appeared in humans, variant Creutzfeldt–Jakob disease (vCJD). Largely affecting young people and invariably lethal, like BSE this disease was caused by a prion – an infectious protein that alters the shape of proteins with an identical amino acid sequence, transforming their shape from a normal to a pathology-inducing form.[3] It was widely suspected, but has still not been formally proved, that vCJD was caused by eating BSE-contaminated beef.[4] * British beef sales slumped, and exports of British beef to Europe were banned for ten years. More significantly, confidence in government, in experts and in food safety plummeted everywhere.

In February 1997, fears that biotechnology might be taking the world in an unwelcome direction were reinforced when one of the long-standing promises – or threats – of genetic engineering was finally realised, in the form of Dolly the sheep, a clone, or genetic copy of another sheep.[5] Cloning of frogs went back over forty years, and the first mammals – three sheep – had been cloned in 1986 by introducing DNA from an embryonic cell into an egg cell from which all nuclear components had been removed.[6] Dolly was different because her DNA came from an ordinary skin cell. (The name 'Dolly' was coined by a stockman, after the singer Dolly Parton, because the cell was taken from the mammary gland of her ... donor? predecessor? self? mother? – we still do not have the words to describe their relationship.[7])

Although the 1986 cloning report had gone largely unremarked, for some reason Dolly caused a global media storm. *Science* magazine proclaimed her as 'breakthrough of the year' and all the fears about the cloning of humans resurfaced as President Clinton demanded the US National Bioethics Advisory Commission discuss the matter.[8] US political debates about stem cells, IVF and abortion all got rolled up into overheated concerns about cloning.[9] More prosaically, in autumn 1998 Greenpeace activists tried to kidnap Dolly. The plot failed, allegedly because the raiders were unable to identify their target – all the sheep looked the same.[10] Suffering from arthritis, Dolly was put down in 2003 and her stuffed body can be seen in the National Museum of Scotland in Edinburgh. In a way, she lived on far beyond that date – after being frozen as embryos, four clones

* By 2015, 226 cases of vCJD had been identified worldwide; 81 per cent of the patients had lived for at least six months in the United Kingdom. At a global level it looks like we dodged a bullet, but on a personal scale this was a terrible tragedy – all of these people died horribly.

from the same cell line that gave rise to her were born in 2007 and survived for nearly a decade.[11]

In the end, Dolly had rather less impact than expected. The whole point of the experiment had been to find a way of easily reproducing genetically modified mammals – as *Science* magazine promised, the technique was intended 'to quickly create herds of identical animals that churn out medically useful proteins'. None of this happened.[12] Although cloned babies have been born to nearly two dozen mammalian species, the technique continues to resist widespread application because the yield is so low (over 250 attempts were made before Dolly succeeded) and it does not work in birds because of the structure of the avian egg. The rich can get their mammalian pets cloned (Barbara Streisand has two dogs that are genetically identical to her defunct pet, Samantha) but the use of cloning in agriculture has been relatively limited, as has the use of cloned race horses.[13]

The situation in humans is more complex. After some fraudulent claims by South Korean researchers at the beginning of the century, human cells were finally cloned by Shoukhrat Mitalipov's non-federally funded team in Oregon in 2013 and 2014, but no attempt was made to implant the resultant embryos.[14] No cloned human babies have been born. As the US Panel on Scientific and Medical Aspects of Human Cloning reported in 2002, 'Human reproductive cloning should not now be practiced. It is dangerous and likely to fail.'[15]

*

With the vCJD scandal stoking fears about food safety and Dolly raising the prospect of copies of animals and even humans, 1998 became the *annus horribilis* for advocates of GM food, as a series of unfortunate events decisively shifted public opinion. The first click of the ratchet came when a US patent was granted to the United States Department of Agriculture and the Delta Pine and Land Company for the crossing of two types of GM plant, such that a gene that was not expressed in either parent was nevertheless expressed in offspring. Buried deep in the document was one particular application: 'the present invention involves a transgenic plant or seed which, upon treatment with an external stimulus produces plants that produce seed that cannot germinate'.[16] Biotech companies would no longer need to hire private detectives to rummage through farmers' waste bins to uncover cases of seed piracy. The seeds would commit suicide.

Cunningly labelled 'Terminator technology' by opponents, the patent rapidly became linked with Monsanto, which had launched a takeover of Delta Pine and Land and inadvertently acquired the patent. Monsanto had no particular interest in the technology, and although the public – and many farmers – found the idea of the Terminator deeply shocking as it appeared to be an ingenious way of tying farmers into Monsanto's business plan, it had for decades been routine for farmers to discard seeds produced by highly productive hybrid crops, created by crossing between ordinary plants, because the hybrids did not breed true. As part of the 'green revolution' that powered the expansion of global agriculture in the second half of the twentieth century, farmers all over the world had long accepted that to get the benefits of a hybrid crop, the seeds would have to be bought anew each year.[17] But as 1998 progressed, the Terminator – which only ever existed on paper – came to embody the supposedly inhumane nature of genetic engineering in general and of Monsanto in particular.

In June 1998, Prince Charles, ever attentive to the needs of plants, complained in unoriginal terms about the way the world was going: 'this kind of genetic modification takes mankind into realms that belong to God, and to God alone', the future Defender of the Faith proclaimed.[18] As Terminator technology became a lightning rod for opposition to GM crops, in 2000 the UN Convention on Biological Diversity adopted a de facto moratorium on the idea. In 2015, Monsanto's unwanted patent lapsed; not one Terminator seed has ever been planted, not even in a laboratory greenhouse.

The next step in the growing global fears about GM food came in August 1998, when Arpad Pusztai of the Rowett Research Institute in Aberdeen claimed on British television that rats fed GM potatoes showed pathological changes in their gut lining and had suppressed immune systems. The same thing could happen to humans, Pusztai warned: 'If you gave me the choice now, I wouldn't eat it', he said.[19] The programme caused huge consternation, leading the Royal Society to investigate the claims (unsurprisingly, it dismissed them). The *Lancet*, which eventually published Pusztai's weak research,

This image was used by anti-GM campaigners in Canada with the slogans 'Defend Food Sovereignty' and 'BAN Terminator Seeds'.

boasted that 'a debate about the science, rather than unsupported claims about that science, has now begun'.[20] But there was no debate, just conflict.

Pusztai's study involved only six rats in each experimental group and failed to properly control for the nutritional content of the GM potatoes – the rats may simply have been undernourished. The response of the scientific establishment to this piece of shoddy science was heavy-handed – the outcome of the Royal Society's investigation looked like it had been decided in advance, while Pusztai was suspended by his employer and his work was 'audited'. Eventually his contract was not renewed and he returned to his native Hungary. For those on the lookout for an establishment cover-up, this seemed like victimisation.

Following Pusztai's claims, a cynical UK tabloid press campaign hyped up fears of 'Frankenstein food'. Faced with the furore, in March 1999 Marks & Spencer removed products containing GM soy and maize from its shelves, prompting all UK supermarkets and Europe-wide food processors Unilever and Nestlé to follow suit. Millions of cans of clearly labelled GM tomato paste had been sold by Sainsbury's and Safeway, but despite the product's popularity (the price point was deliberately low) and its safety, it too was withdrawn. The tide was turning against GM food, and by 1999 opposition to GM food increased in all European countries, becoming the majority view in most.[21]

This was true even in the United States, where public opinion had long been supportive or indifferent about genetic engineering but now began to become more suspicious. By 2003, 55 per cent of American consumers considered that scientifically modified fruit and vegetables were 'bad'.[22] The 1990s had seen a major expansion in groups around the United States that were mobilising against GM crops, including well-funded campaigning by advocates of transcendental meditation, who, for example, in 1999 distributed half a million copies of a well-received brochure called *Safe Food News* which claimed that GM foods would damage 'natural genes', reduce the effectiveness of antibiotics and damage the nutritional content of food.[23]

In France, the opposition to GMOs increasingly had a name and a face – José Bové, a grinning farmer with an Astérix-style moustache and a pipe. At the beginning of 1998, together with three comrades from the Confédération Paysanne trade union, Bové broke into a warehouse containing GM maize seed and set fire to the grain.[24] During his subsequent trial, Bové outlined his unfounded fear that humans could become resistant to certain antibiotics by eating GM crops containing antibiotic resistance marker genes. His arguments counted for little and together with two comrades he was given an eight-month suspended jail sentence. Eighteen months later, in an extension of the domain of struggle, Bové and around 300 campaigners invaded a construction site where a new McDonald's restaurant was being built and attacked the half-finished building in the name of the fight against *la malbouffe* (junk food). *Le Monde* summed up the heartfelt, if somewhat incoherent, clash between two worlds:

> On the one hand, the World Trade Organization, the agro-chemical industry, GMOs, indoor intensive farming, growth hormones and dodgy animal feed; on the other, resistance by a 'peasant' agriculture, rural heritage, quality food and grass-rearing.[25]

After spending some time on the run in the hills, Bové was arrested, convicted and imprisoned, prompting demonstrations all over the country and even greater support from rural communities. As *Le Monde* noted in September 1999:

> There is an amazing movement of solidarity around José Bové. Yesterday he was a complete unknown, today he is a national hero of the resistance against the importation of growth hormone beef and against unpredictable GMOs.[26]

Even the socialist Prime Minister, Lionel Jospin, expressed sympathy with Bové's 'just cause'.[27] The new technology was awakening deep psychological fears and resonating with historical images of peasant struggles down the ages and rural resistance during the Second World War, as well as deep-rooted hostility to the growing strength of US culture. Against the steamroller of globalisation, the small farmers claimed the right to *l'exception française* as a defence of French culture, a view that was immediately embraced by much of the population.

Events were taking a similarly activist turn on the other side of the Channel. In the first of a series of 'decontamination' stunts, Greenpeace activists assembled at a field in Oxfordshire where a field trial of GM oil seed rape had begun. Under the glare of press and television cameras, protestors clad in white Tyvek suits and face masks flew biohazard flags and ripped up the plants as a police helicopter whirred overhead. Clack-clack-clack went the chopper, click-click-click went the cameras.[28] The media lapped it up and churned it out to an increasingly alarmed public – 'Mutant Crops Could Kill You' and 'Gene Crops Could Spell Extinction for Birds' were headlines in the *Daily Express* and *The Guardian*, respectively.[29] In the kind of tabloid jape that passes for cutting-edge journalism in the United Kingdom, the *Daily Mirror* printed a picture of Tony Blair photoshopped as Frankenstein (he was the 'Prime Monster') with the headline 'I EAT FRANKENSTEIN FOOD and It's Safe'.[30]

All over the world, opposition to GM crops blurred into hostility to globalisation as the newly created World Trade Organization (WTO) attempted to remove tariff barriers, opening all markets to the free movement of capital and goods, including removing restrictions on GM crops. In response, anti-GM activists broadened their angle of attack. A typical example was a November 1998 campaign against a field trial of Monsanto Bt cotton in India, which used the slogans: 'Stop genetic engineering. No patents on life. Cremate Monsanto. Bury the World Trade Organization.'[31]

Monsanto eventually realised which way the wind was blowing, and in October 1999 its CEO Robert Shapiro accepted an invitation to address a Greenpeace conference. Speaking over a videolink, Shapiro cut a sorry figure. Monsanto, he admitted, had 'forgotten to listen'. From the outset, the company had mistakenly regarded opponents as 'wrong or at best misguided.'[32] In a BBC *Horizon* documentary 'Is GM Safe?', Shapiro pursued his self-criticism:

> Our confidence in this technology and our enthusiasm for it has, I think, widely been seen and understandably so, as condescension or indeed arrogance.[33]

Shapiro's contrition came too late, both for his company and for himself. A few months later, faced with growing financial problems, Monsanto merged with US pharmaceutical giant Pharmacia. Within a year, Shapiro had left his job. Monsanto, which Pharmacia turned back into a separate group in 2002, retained the sulphurous reputation it acquired in the 1990s. This public hostility played a part in its sudden decision in May 2004 to drop plans to market GM wheat in the United States and Canada, even though it had invested a large amount of time and effort in developing the product and then defending it against criticism from activists, farmers and politicians.[34] The company name finally disappeared in 2018, following a $66 billion takeover by Bayer.

*

Faced with these public concerns over health and biosecurity, the European Union instituted a temporary moratorium on the approval of GM crops. In response, the major GM producers – the United States, Canada and Argentina – successfully used the WTO to impose the export of GM products to Europe.[35] Hundreds of GM crops, including cotton, maize, oilseed rape, soybean and sugar beet are now registered for use in the European Union but not for human consumption, only as animal feed.[36] European farmers' groups have responded to GM technology in different ways: the French Confédération Paysanne opposes any cultivation of GM crops, whereas in Austria the focus has been on protecting organic crops.[37] In the European Union, all GM feed is labelled as such – an important proportion of maize and soya in the diet of European farm animals is GM. There is no evidence that this has any ill effect on the animals, nor on the humans who eat the animals or their products.[38]

The framework adopted by the European Union to govern its regulation of GMOs is the Cartagena Protocol on Biosafety, part of the UN Declaration on Biodiversity. Signed by over 170 countries – but not the United States – the key aspect of this protocol is the precautionary principle: 'Where there is a threat of significant reduction or loss of biological diversity, lack of full scientific certainty should not be used as a reason for postponing measures to avoid or minimise such a threat.' Unfortunately, this avoids the question of how to estimate whether a particular organism is indeed a threat, or even what 'threat' means.[39]

Global suspicion of GM crops was soon strengthened by the discovery that stuff happens – GM products may turn up where they are not supposed to be. In 2000 a newly formed European company, Aventis, found itself at the centre of a US scandal involving its version of Bt maize, known as StarLink, which was approved only for use in animal feed. In September 2000 anti-GM campaigners found traces of StarLink maize in the massively popular Taco Bell taco shells. The manufacturer Kraft quickly moved to protect the integrity of its brand and withdrew all of the shells nationwide.[40] The incident proved hugely damaging for Aventis and its bosses, as millions of dollars were paid out in compensation to companies and individuals; three leading figures in the company were fired a few months later.[41]

In 2006 genes from Bayer's experimental LibertyLink herbicide-tolerant rice contaminated some US long-grain rice. Japan and Russia promptly banned imports of the grain and Bayer had to pay out $750m in compensation to US rice farmers.[42] Similar worries have swirled around Bt maize, which for a while was planted in Mexico, the historical home of maize, where the plant has great cultural significance, particularly for indigenous people. In 2001, researchers reported in *Nature* that transgenic components from GM crops had found their way into Mexican landraces of cultivated maize ('landrace' is the term for a domesticated animal or plant that has become adapted to local conditions).[43] Rival researchers criticised the methods and interpretation, and less than a year later *Nature* published an Editorial Note in which it concluded that 'the evidence available is not sufficient to justify the publication of the original paper', but confusingly did not issue a retraction.[44] Eighteen further surveys have been carried out since the original publication; with only one exception, they have all found evidence of transgene contamination in Mexican traditional maize.[45] Whatever the source – probably illegally planted transgenic maize – it is clear that native landraces of Mexican maize have been contaminated by transgenes acquired through cross-pollination. *Nature* has not seen fit to revise its surprising judgement of the 2001 paper.

For French historian Christophe Bonneuil and his colleagues, the row over Mexican maize constituted 'a political and cultural struggle, a new episode of a contentious cultural encounter between worldviews, values, forms of knowledge, and forms of life'.[46] No doubt, but as so often in debates about GM agriculture, this shifts the argument to the more general question of the capitalist drivers of globalised agriculture (and academic publishing). It skips over the biological implications of the presence of transgenes in maize landraces and what should be done about it. In reality, the transgenic components will have no lasting effect on the small proportion of Mexican maize plants in which they are currently found. With no advantage over the indigenous landraces, they may linger at low frequency in the population if they have no effect on plant survival and reproduction; if there is even a slightly deleterious effect they will soon disappear. Despite this reassuring reality, the worries about GM maize are a justified cultural fear relating to a globally significant resource. As a result, Mexico has completely banned the planting of transgenic maize since 2013.[47]

Bt maize has also been implicated in the decline of the monarch butterfly, one of North America's conservation icons which each year makes an astonishing multigenerational migration from its wintering sites in Mexico up into the United States and back again. In 1999 it was reported – again in *Nature* – that monarch caterpillars fed on milkweed leaves dusted with Bt maize pollen showed a 40 per cent reduction in survival.[48] Although based on a small sample, this result was alarming because the maize pollen season coincides with the period when monarch caterpillars are feeding. Two years later, large-scale field tests suggested that the use of insecticides on non-Bt crops was a far more significant threat to the monarch.[49] Furthermore, the version of GM maize used in the initial experiment had unusually high Bt levels and was subsequently withdrawn. Exactly how researchers can determine the potential effects of Bt crops on the monarch, given its continent-spanning mobility and its complex year-long cycle, remains a matter of debate.[50] Meanwhile, monarch butterflies continue to decline and the causes are still unclear.[51] The direct or indirect involvement of Bt pollen cannot be excluded. Around 90 per cent of US maize, soy and cotton crops are now GM, and many of these crops encourage farmers to increase their herbicide use, which in turn may affect milkweed, the monarch caterpillar's sole food plant. The decline of the monarch is probably due to many factors; one contributory component may be not so much GM crops themselves, but the farming practices they encourage.

*

From the very outset, one of the arguments used by advocates of genetic engineering has been that GM crops can help overcome the apparently lower levels of productivity found in African agriculture. In reality, things are not quite so clear.[52]

In Burkina Faso, enthusiasm for Bt cotton soon declined when it became apparent that the GM plants did not produce the quality of cotton that buyers desired. Having initially been hailed as a role model for its adoption of GM crops, by 2018 Burkina Faso was growing no Bt cotton at all.[53] In many countries Bt cotton is not resistant to local pests and seeds require additional insecticidal treatment, increasing costs and discouraging farmers from adopting the crop. In South Africa, the production of GM cotton and the number of farmers growing the crop both dropped over the first fifteen years of the century as disillusionment set in.[54]

Similar problems have affected GM maize.[55] In South Africa, between 80 and 90 per cent of all maize is now GM, but this take-up has mainly been by large-scale commercial farmers. Very few of the country's two million subsistence maize farmers have adopted GM crops. The reasons are complex, but they include higher fees for GM maize and the substantial annual variability in the benefits the crops provide, which depend on the vagaries of the pests they are designed to combat. Another major factor is the need to plant refuges to ensure that Bt resistance does not develop in local insect populations – this is easier to do on a large farm than on a smallholding, and as a result insect Bt resistance has appeared in areas where smallholders predominate.

On a vast continent with multiple pasts and presents, the response of African countries to the promise of GM crops has varied enormously. At the end of the 1990s, South Africa and Egypt legalised the planting of Bt cotton, while in contrast between 2002 and 2006 Zambia refused US aid deliveries of GM maize despite a massive agricultural crisis and famine.[56] Sometimes, anti-GM campaigners have gone way overboard in their description of the supposed dangers of GM crops – ActionAid Uganda claimed that GM crops cause cancer, while in Ghana a leading group of activists campaigned around the slogan 'GMO/Ebola out of Africa'.[57] In South Africa, anti-GM campaigners were able to achieve a minor victory in terms of labelling of GM products and above all shifted the debate on the continent towards a much more critical stance. This could be seen in 2007, when at its foundation, the Alliance for a Green Revolution in Africa, an organisation set up to support smallholder farming throughout the continent, with support from the Bill and Melinda

Gates Foundation and the British government, both of which have a record of advocating GM crops, nonetheless emphasised the importance of traditional crop breeding instead.[58]

A good summary of the situation is provided by Canadian environmental geographer Matthew Schnurr in his perceptive book *Africa's Gene Revolution*:

> First-generation GM crops were developed to succeed within the confines of industrial agriculture – large-scale, heavily capitalised, mechanised monoculture. Transplanting this technology to small-holder African farmers who operate in a completely distinct mode of production has failed. Expect this trend to continue into the future.[59]

Recently, second-generation GM crops have been developed by philanthropic organisations and distributed by public–private partnerships. These crops are those grown by smallholders (for example, sorghum, sweet potato, maize, cowpea) and the characteristics that have been introduced into them (insect-, drought- and disease-resistance) are sometimes desired by farmers, although in general it is the donors that decide priorities, not local people.[60] Unfortunately, many of these programmes repeat the errors of previous well-meaning but unsophisticated attempts to provide a one-size-fits-all solution to a complex problem. For example, Nigeria has recently approved the cultivation of a Bt version of cowpea – the black-eyed pea – the most important legume crop in West Africa. This plant will resist attack by the cowpea pod-borer, but the need to plant refuge areas to avoid the evolution of resistance will either limit uptake by smallholders, who do not have the land to set aside, or will lead to the appearance of resistant strains where farmers do not use the crop in the right way.[61] In 2017, sociologist Rachel Schurman summarised the kinds of factors that will determine the future of GM crops in Africa:

> Farmers' decision-making will prove crucial to determining the fate of the technology. We can expect this decision-making to reflect local agrarian histories as well as contemporary realities, household and farm characteristics (from land and

> labour access to gender relations), the availability of crops that farmers find useful and appealing, and the cost of GM seeds.[62]

Faced with the complex realities of agriculture spread across a whole continent, the enthusiasm of GM advocates such as the late Calestous Juma, a Kenyan-born Harvard academic, can seem to lack nuance. Juma argued that 'GM foods in emerging countries have the potential to revolutionize the lives of suppliers and consumers', but things are not so simple.[63] As Schnurr rightly remarks: the underlying problems of African agriculture are multiple and are not amenable to a single technofix. They require more radical solutions – 'changing a plant's genome is an easier task than changing an entire system', Schnurr states.[64]

A similar complex view of GM crops and their regulation emerges from Aniket Aga's detailed account of the situation in India, *Genetically Modified Democracy*.[65] Ninety per cent of cotton farms in India grow Bt cotton, but over two-thirds of landholdings are less than 1 hectare in size – far too small to make it possible to use Bt or herbicide-tolerant crops without resistance emerging. India is not a passive recipient of this technology – GM crop innovation in India has been carried out both by state research organisations as well as by Western companies, in particular Monsanto. Furthermore, attempts by opponents to restrict the deployment of GM crops have primarily invoked national and local legislation, rather than appeals to international agreements such as the United Nations Cartagena Protocol. It was a detailed use of national regulations that led to the 2010 moratorium on the growth of Bt brinjal (aubergine or eggplant), which still holds.

Towards the end of his book, Aga emphasises the 'multiple, deep crises' affecting Indian agriculture and wisely concludes:

> I am struck by the frequency with which policy makers harp on GM crops as a solution to the deep distress. The fascination with GM crops among policy makers seems to be disproportionate to the actual capacity of the rDNA technology to help matters. ... I cannot shake away the feeling that the GM debate, irrespective of which side 'wins', is much too narrow

> to address the pressing problems of sustainable agriculture, nutritious, culturally appropriate food, and dignified lives and livelihoods with some semblance of equal opportunity.[66]

*

Doubts and suspicions about GM crops are found all over the world. Since the beginning of the century a debate over GM crops has been taking place in China. In 1988, the first GM crop in the world was planted – illegally – in China; once administrative hurdles were overcome and its growth was approved, this virus-resistant tobacco crop was commercialised in 1992 and exported to the United States for use in Marlboro cigarettes. The United States became concerned that sales might be affected by the presence of GM tobacco in their products (the carcinogenic properties of tobacco were not a problem, it appears), and put diplomatic pressure on China to stop production of the GM crop.[67] In a sign of how things have changed in subsequent decades, the Chinese government acceded to this request – it is hard to imagine the same thing happening today.

In 1997, China was growing 1.6 million hectares of GM plants and nearly fifty new transgenic crops were being developed. By 2012 the plant biotechnology sector consisted of 7,500 companies, employing 250,000 people, including 2,500 laboratory scientists.[68] Such is China's financial power and its interest in the application of genetic engineering that in 2016 the massive ChemChina company moved to acquire the Swiss-based Syngenta, a major developer of GM crops, for a staggering $43 billion. The Chinese Communist Party Central Committee has regularly discussed GM technology, but despite initial enthusiasm from both farmers and the Party, many approval certificates for locally developed GM crops have been allowed to lapse as Chinese agribusiness met similar problems to those encountered elsewhere – many of the crops were not demonstrably better than traditional varieties. Only Bt cotton and papaya are currently commercially available in China.

The GM debate in China has been focused on the possibility of creating transgenic versions of the country's main staple and powerful cultural symbol: rice.[69] Around one-third of the public funds for the development of GM crops were initially devoted to developing Bt rice and a local Bt rice was approved in 2009. This soon led to public concerns, with the Party allowing Greenpeace China a surprising amount of freedom to campaign against the proposed commercialisation of the crop. Although China has the most stringent legal framework for labelling GM products – any GM contribution must be indicated, whereas in Europe only levels of GM >0.5 per cent have to be labelled – in such a vast country with a long tradition of farmers ignoring central instructions, GM crops are illegally produced and sold.[70] In 2014 a Chinese television investigation revealed that unlabelled Bt rice was on sale in many Chinese supermarkets, leading to concerns from countries that imported Chinese rice, including those in the European Union.[71] (The United States has never commercialised GM rice, for fear of losing its export markets.)

Chinese suspicions of GM rice involve a complex mixture of health concerns and nationalist fears of US influence, as well as attitudes to food as a cultural good (there is little opposition to Bt cotton, which soon saw smallholder yields increase by 6 per cent and insecticide use decline by 80 per cent).[72] Furthermore, public confidence in Chinese government regulation of food safety has been shaken by repeated catastrophic failures of food standards that had nothing to do with GM crops, such as the 2008 scandal of baby formula milk that had been adulterated with ground-up plastic in order to increase its nitrogen content; hundreds of thousands of babies were affected, tens of thousands were hospitalised and some died.[73] Other scandals have included artificial eggs made of chalk and chemicals, tainted noodles composed of wax and ink, and ersatz rice fabricated out of potatoes and plastic resin.[74]

In 2010 over a hundred scholars petitioned the National People's Congress against the commercialisation of Bt rice as it was 'likely to endanger the Chinese nation and China's national security'. The debate has not always been so decorous. At a lecture given by a leading GM scientist in Beijing, a middle-aged woman interrupted, shouting: 'You are a traitor! You are using 1.3 billion Chinese as guinea pigs to serve your American master!' On other occasions, protestors shouting 'Traitors! Liars!' have thrown mugs and placards at GM advocates.[75]

Baseless conspiracy theories have been spread by quite senior figures, focusing on the foreign links of GM scientists, many of whom trained in the United States. Major General Peng Guangqian of the People's Liberation Army, who is also deputy secretary-general of the National Security Forum, has been a leading opponent of GM technology, claiming that GMOs are part of a Western conspiracy, cause cancer and are a source of infertility. With the usual conspiracy theorist's love of innuendo and rhetorical questions, he stated:

> Interest groups such as Monsanto and DuPont have been unusually generous to dump GM crops to China. What is exactly behind this extremely abnormal activity on the American side? Is it a pie or a trap?[76]

These kinds of outbursts are not only surprising to those of us in the West. For Chinese scholar Cong Cao, author of the eye-opening *GMO China*, it is 'beyond comprehension' that the Chinese government tolerates promotion of 'anti-government-policy views' on the internet and permits the defamation of scientists and government officials with regard to GMOs. Cao suggests that this unusually permissive attitude reveals deep divisions in the Chinese ruling bureaucracy over GM technology and fears of major civil opposition, driven by our powerful, often irrational, attitudes towards food.

*

Decades after the widespread introduction of GM crops around the world, no country's population has enthusiastically embraced the new food.[77] The vast majority of GM crops, which target pests, not food quality, have no direct benefit for consumers, beyond an implied promise of lower prices through improved agricultural efficiency. To the understandable frustration of scientists and agribusiness, the fears that still characterise many peoples' attitudes to GM crops are apparently rooted deeper than facts can reach.

In 1996, in the immediate aftermath of the vCJD scandal, researchers from Lancaster University carried out a series of focus groups about GM food, uncovering attitudes that are still widespread around the globe. For example, one working mother said:

> It sounds dangerous and unnatural. … I get the impression that all the food's been meddled with in a laboratory before it reaches the supermarket. It doesn't interest me at all. It's like all these, you know these fruits that they inject with stuff to keep the apples redder for longer and things. I want food to be fresh, I don't want it to have all this stuff in it.[78]

Underlying many of these views were doubts about the motivations of those who wanted to introduce GM crops. A housewife expressed deep suspicion of both business and politicians:

> Because you don't know. You don't know anything. The Government can tell you anything, the manufacturers can tell you anything, the supermarkets can tell you anything. And you have no way of knowing what they are saying, or it's like their hidden agenda isn't it? We don't know why they're doing it or what they're doing.

A quarter of a century later, nothing much has changed, even if the intensity of those feelings has apparently diminished and some of those irrational suspicions may now be focused on, say, COVID vaccines. In 2015, GM crop pioneer Marc Van Montagu and his colleagues wearily explored what they called 'the intuitive appeal' of opposition to GM crops, showing how 'folk biology, religious intuitions, and emotions such as disgust leave the mind readily seduced by representations of GMOs as abnormal or toxic'.[79] Even in the United States, where GM consumption is the highest, companies have successfully lobbied against open labelling of GM products, fearing it will lead to a decline in sales.*

Not all GM crops have been developed to improve agriculture.

---

*Since 2016, US consumers have been able to find out if a product contains GM material, but only by using an unhelpful QR code or ringing a telephone number on the product label.

Golden Rice, a GM crop that has yet to be grown in fields despite decades of development, is intended to provide a way of treating the vitamin A deficiency that affects many people living in the tropics, in particular in southeast Asia. Perhaps one million people a year die and another 500,000 become permanently blind because of this easily treatable deficiency. The solution is relatively simple: supplement the diet of those affected with additional vitamin A. Billions of capsules of vitamin A have been distributed to this end, but such programmes inevitably fail to reach everyone in need; a more direct solution would be to fortify some consistent part of the diet, such as rice, with the missing ingredient, much like folic acid is added to flour to prevent babies being born with spina bifida.

The idea goes back to the very beginning of GM crops. In April 1984, the International Rice Research Institute in Manila hosted an informal discussion in which researchers mused about what gene they would introduce into rice if anything was possible. Veteran rice breeder Peter Jennings chose a gene for β-carotene, a pigment found in many plants (it makes carrots orange) and which is converted into vitamin A in the human body.[80] At the time, this was a pipe dream – it was still not known if genetic engineering would work in rice – but researchers gradually identified the genes responsible for producing carotenoids in plants and discovered ways of reliably producing transgenic rice. Central to this work was Ingo Potrykus, who had long been thinking about activating the β-carotene gene in rice endosperm – the white grain and virtually the only part of the rice plant that does not contain β-carotene.[81] Joined by Peter Beyer of Freiburg, Potrykus looked for ways to produce a rice plant with a golden endosperm. After a number of false starts, the first Golden Rice grains were produced in February 1999 (the catchy name was coined a few months later).[82]

Success soon gave way to controversy. Greenpeace has consistently opposed Golden Rice, highlighting the low level of β-carotene in the earliest version, maintaining its hostility as various more productive forms of the transgenic rice were developed over subsequent years.[83] Initial claims by opponents that the project was simply a ploy by big agribusiness were undermined in October 2004 when Syngenta, which had ended up owning the patent, abandoned its commercial interest in the plant and handed over all samples, technical information and know-how to a charitable foundation. Undeterred, Greenpeace continues to claim that the project is a waste of money, that conventional rice crops will become contaminated, that the transgenic process itself might have produced unexpected changes in the rice genome (this did happen with one version of the crop) and that the effects on human health are unknown.[84] Golden Rice, they say, is 'environmentally irresponsible' and 'a disservice to humanity'. Potrykus has riposted in kind, accusing Greenpeace of 'a crime against humanity' because of the hundreds of thousands of deaths that have not been prevented, while in 2016 over 100 Nobel laureates signed an open letter to Greenpeace urging the organisation to withdraw its objections to GM crops in general and to Golden Rice in particular.[85]

In the last few years Australia, New Zealand, Canada and the United States have all approved Golden Rice for human consumption, not to allow commercialisation – there would be no benefit as their populations do not suffer from vitamin A deficiency – but rather to permit rice imports that might accidentally contain the GM grains.[86] In July 2021, the government in the Philippines issued the first commercial cultivation permit for Golden Rice, but this does not mean that farmers can now grow the crop. Further necessary steps include getting the variety registered and the crucial issue of having enough seed to be able to plant. Lurking in the background is the pressing need to use 'heritage' landraces of rice that are adapted to specific local ecologies, but which have largely been overlooked following the massive expansion in rice culture since the 1960s.[87] In the meantime, non-genetic approaches to the underlying problem have proved successful. Basic public health education, capsule supplements and nutritional education have driven down the rate of vitamin A deficiency in the Philippines from over 40 per cent in 2003 to less than 16 per cent by 2019.[88] That is still far too many people being affected by a devastating and easily curable condition, but it shows that Golden Rice is not the only possible solution.

For the moment, none of the countries that could potentially benefit from Golden Rice have fully approved it, and no commercial crops have been produced. It is possible that none of the current versions will ever reach the bowl, in part because of those fears that the way that the transgenic plant was created might have led to unknown mutations. That does not mean that the idea of biofortifying rice with β-carotene is dead – scientists at the University of California have already used CRISPR gene editing to produce carotenoid-enriched rice.[89] It remains to be seen if this experimental organism can become a viable crop, and if so, whether GM opponents will drop their criticism. The former is possible, the latter seems unlikely.

*

One of the lessons of the last two decades, and of Golden Rice in particular, is that researchers need to recognise that their inventions may encounter opposition and, if possible, accommodate those criticisms. Simply complaining that opponents are not being scientific and urging everyone to get on with things ('Stop worrying; start growing' as a group of Swedish agricultural scientists put it testily in 2012) will not succeed.[90] A better approach was shown by UK scientists at Rothamsted Research who engineered a strain of wheat to release an aphid alarm pheromone; in the laboratory the plants repelled disease-transmitting aphids and attracted their parasitoid wasp predators.[91] In 2011, the British government gave permission for a field trial of the GM plants.[92] The researchers sought to defuse opposition by engaging with protestors and appearing on debates in the media. It worked. Although anti-GM campaigners threatened to 'decontaminate' the site, only a handful of protestors turned up when the trial began; the experiment continued undisturbed until autumn 2013.

Despite this sophisticated public engagement strategy, when the results of the trial were published in 2015, they were disappointing – under field conditions, the GM wheat produced no change in the behaviour of aphids or parasitoid wasps.[93] The explanation seemed to be that the pheromone is normally released by aphids in pulses, while in the GM plant it was released continuously, changing the meaning of the chemical to the insects or altering their ability to detect it. (This issue had been highlighted by German researchers in 2010 and was one of the criticisms made by anti-GM campaigners before the project was approved.[94])

It is not only apparently cunning GM technology that has not turned out as hoped for. Even the simple increase in productivity that was predicted through the use of Bt or herbicide-tolerant GM crops has not appeared. According to the US Department of Agriculture, the introduction of GM characters into existing varieties of plants did not increase yields in the fifteen years following their commercialisation.[95] Virtually all GM crops focus on just two characteristics – they are either Bt or are herbicide-tolerant or both. Only a handful of crops have been engineered to resist disease (among the few exceptions are GM papaya, which resists the ringspot virus, and a GM squash that resists plant mosaic viruses). Whether this is because of difficulties in creating such crops, or lack of financial support from agribusiness companies seeking to get a good return on their investment, is not clear.[96] The result is that, globally, 12 per cent of the acreage of GM crops is Bt, 47 per cent is herbicide-tolerant and 41 per cent consists of 'stacked' crops, with both characteristics.[97]

GM animals have been even less successful.* For example, thirty years ago, a transgenic Atlantic salmon was created by US company AquaBounty, involving a growth-hormone regulatory gene from the Pacific Chinook salmon and a promoter gene from a distantly related fish called the ocean pout. The result is a salmon that grows twice as fast as normal, allowing fish farmers to get it to selling weight in half the time. Although in 2015 the salmon became the first GM animal to be approved for human consumption in the United States and commercial production has begun, the courts are still arguing over the possibility of the salmon escaping, even if they are grown in tanks on land.[98] No one has yet eaten one of these fish in a restaurant, and no one will be making a fortune from the salmon any time soon. It has been calculated that AquaBounty's current facilities would enable

---

*In 1956 French researchers claimed to have altered ducks by injecting DNA from one type into another. The results did not withstand criticism and the experiment turned into folklore, with Sydney Brenner joking that the experiment had in fact involved crossing a duck and an orange to facilitate French cuisine. Benoit, J. et al. (1960), *Transactions of the New York Academy of Sciences* 22:494–503.

the company to make around $1 million profit each year. Those facilities cost $60 million to build.[99]

Exotic GM animal hybrids have been created, but with little consequence outside of the laboratory. In 2002 researchers at Nexia Biotechnologies in Canada announced they had been able to express genes for immensely strong spider silk in mammalian cells, thereby producing silk as a first step towards industrial production (it is not possible to farm spiders; each arachnid produces tiny amounts of silk and they tend to eat each other).[100] Eventually, researchers were able to create two goats that secreted spider silk in their milk.[101] The spider-goats excited a great deal of attention but there were too many technical obstacles and Nexia soon went bust (the goats were briefly displayed in a museum before disappearing).[102] Although Randy Lewis of Utah State University has continued the spider-goat project with financial backing from the US Navy, it seems likely that CRISPR-based recombinant DNA production of spider silk in silkworms, which have been used for millennia in silk production, will prove more successful, if less bizarre.[103]

*

Despite these sobering examples, the adoption of GM crops – in particular of Bt maize and cotton – has been remarkable, partially realising the bold vision of Monsanto's Dick Mahoney from the 1980s, who saw GM technology as a way of reducing pesticide use. In 2020 it was estimated that the adoption of GM plants had led to an 8 per cent reduction in the total amount of pesticide being sprayed onto fields – 775 million kg less.[104] In the decade following the introduction of Bt maize in the United States there was a clear decline in the use of insecticides, and a consequent increase in the number of non-target insects.[105] Given that widespread pesticide use is almost certainly a contributory factor in the global decline of insects, it is hard to see a downside to this. Furthermore, spraying involves the extensive use of fossil-fuel-powered vehicles, so there will also have been a concomitant reduction in CO_2 emissions from farm machinery. By combining use of Bt cotton and the release of sterile insects, it has been possible to eradicate the pink bollworm from the cotton-growing areas of the United States and Mexico, removing an invasive species and contributing to an 82 per cent reduction in the amount of insecticide used.[106] However, none of these gains will be permanent as pest plants and insects are inevitably developing resistance to herbicides and to the Bt insecticide.[107] As the molecular biologist Leslie Orgel put it in the 1960s: 'Evolution is smarter than you are.'

It has been claimed that the introduction of Bt cotton in India was accompanied by a massive wave of suicides among farmers – there have been hundreds of thousands of deaths over the last decades – as seen in Micha X. Peled's 2011 documentary *Bitter Seeds*. But the very real financial difficulties faced by Indian farmers preceded the introduction of Bt cotton and were exacerbated by government policies rather than by GM crops – suicide rates varied enormously in different Indian states between 2001 and 2011, with no apparent link to the adoption of GM crops.[108] The key factor in explaining the pattern of suicides in rural areas appears to be poverty, in particular where debt-laden people grow market-sensitive crops on marginal smallholdings.[109] The exact financial benefit – if any – provided by Bt crops in India over the long term as resistant insect strains have emerged remains a matter of debate.[110]

Things are equally complicated with herbicide-tolerant GM crops. Although US herbicide use was constant from 1995 to 2010, there was a shift to the use of glyphosate to reap the benefit of GM crops, which included not having to till, thereby preserving soil structure. But not all the environmental effects have been good – run-off from treated fields has serious effects on freshwater aquatic systems and amphibians.[111] Furthermore, by 2013 glyphosate-tolerant weeds had been found in eighteen countries, leading Monsanto to advise its customers to mix glyphosate use with other herbicides and with extra tilling, thereby undoing some of the good it had done to soil structure and returning herbicide use to a spectrum, with broader impacts on the environment.[112]

GM crops have neither saved the planet nor have they destroyed it; they have not transformed our diets for good or ill (the vast majority of crops are non-GM), nor have they led to disease or to antibiotic resistance in humans or animals. Claims in 2012 by a

French researcher that consumption of glyphosate-tolerant maize treated with Roundup led to cancer in two dozen rats were widely debunked at the time and in 2018 were disproved by another group of French researchers who fed hundreds of rats eight different GM diets over a six-month period with no ill effects.[113] Because of stringent food safety legislation – partly introduced in response to GM fears – we can safely eat GM food. That does not mean that spraying Roundup is necessarily safe. US courts have repeatedly found that farm workers suffering from non-Hodgkin lymphoma can claim damages from Bayer, which now owns both the product and the liabilities. Although Bayer insists that glyphosate is not a carcinogen, it is currently considering withdrawing glyphosate from public sale in the United States to avoid further massive claims.[114]

Genetically engineered crops may have been less successful than their advocates promised, but they have certainly been far less catastrophic than their opponents feared. Neither utopia nor catastrophe has materialised. In 2013, an editorial in *Nature* perceptively concluded: 'With GM crops, a good gauge of a statement's fallacy is the conviction with which it is delivered.'[115] Although many people around the world remain suspicious of the idea of eating GM food, anti-GM protests have waned as far more serious threats to global biodiversity have appeared in the shape of climate change, rising CO_2 levels and the widespread use of pesticides. The involvement of GM crops in global food production has become almost routine, partly because of the success of the regulatory frameworks that GM advocates are often frustrated by, and perhaps because, like the rest of agricultural activity, the reality is hidden from most of us.

1. Bob Pollack in his office at Cold Spring Harbor Laboratory in 1971. The telephone he used to call Paul Berg is on the bench.

2. Herb Boyer (l.) and Paul Berg (r.) at Asilomar. The young man, top right, was dentified on Twitter as UK plant geneticist Ray Dixon, who is still an active researcher.

3. The Asilomar conference, February 1975. From top left, clockwise: Maxine Singer; David Baltimore (l.) and Norton Zinder (r.); Soviet scientists Vladimir Engelhardt (l.) and Alexander Bayev (r.); Sydney Brenner addressing the conference.

4. David Baltimore (l.) and Wally Gilbert (r.) discussing with the public during an outreach event in 1976.

5. Protestors at an American Association for the Advancement of Science forum on recombinant DNA, March 1977. Maxine Singer is on the panel, second from the left; NIH Director Donald Fredrickson is on the far right.

6. City of Hope Hospital and Genentech researchers who produced insulin in a bacterium using recombinant DNA. Left to right: Keiichi Itakura, Arthur Riggs, David Goeddel and Roberto Crea. Dig Riggs' facial hair.

7. Ananda Chakrabarty in front of the Supreme Court in 1980. He is holding his recently awarded patent and two flasks of oil, before (left) and after (right) treatment with his patented genetically engineered microbes.

8. Label for Flavr Savr GM tomatoes, marketed as 'MacGregor's', 1994. The contrast between the folksy name and image and the technology behind the plant is striking.

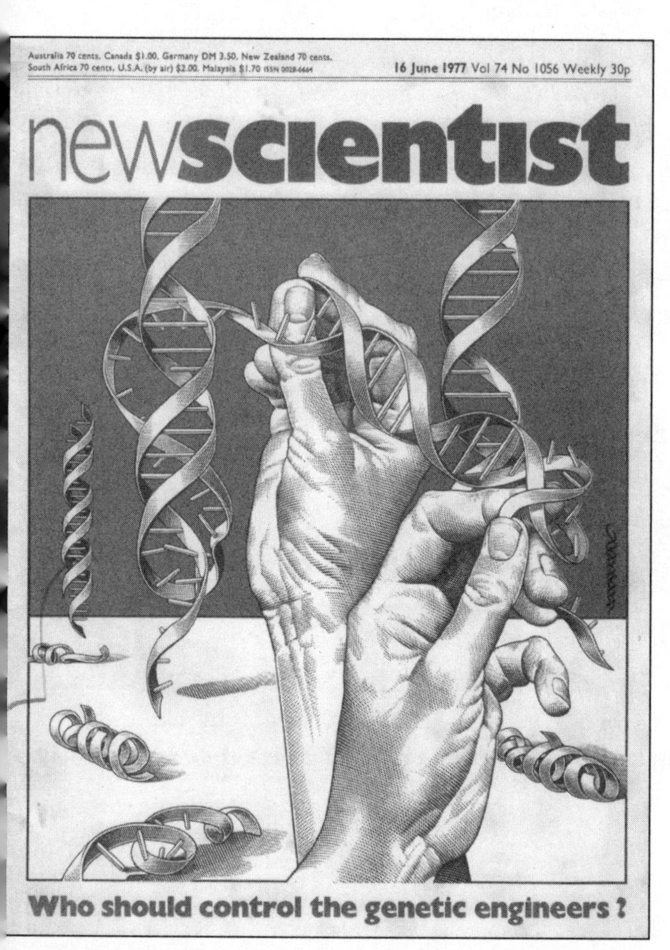

9. (Left) *New Scientist* covered the recombinant DNA controversy extensively. This memorable cover by an unnamed artist was from June 1977.

10. (Bottom left) Cover of the *Mirror*, February 1999.

11. (Bottom right) August 2001 cover of the satirical French weekly, *Charlie Hebdo*: a new GM maize 'resists the Confédération Paysanne' in the person of José Bové. The drawing is by Charb, who was one of those murdered in the 2015 terrorist attack on the magazine's offices.

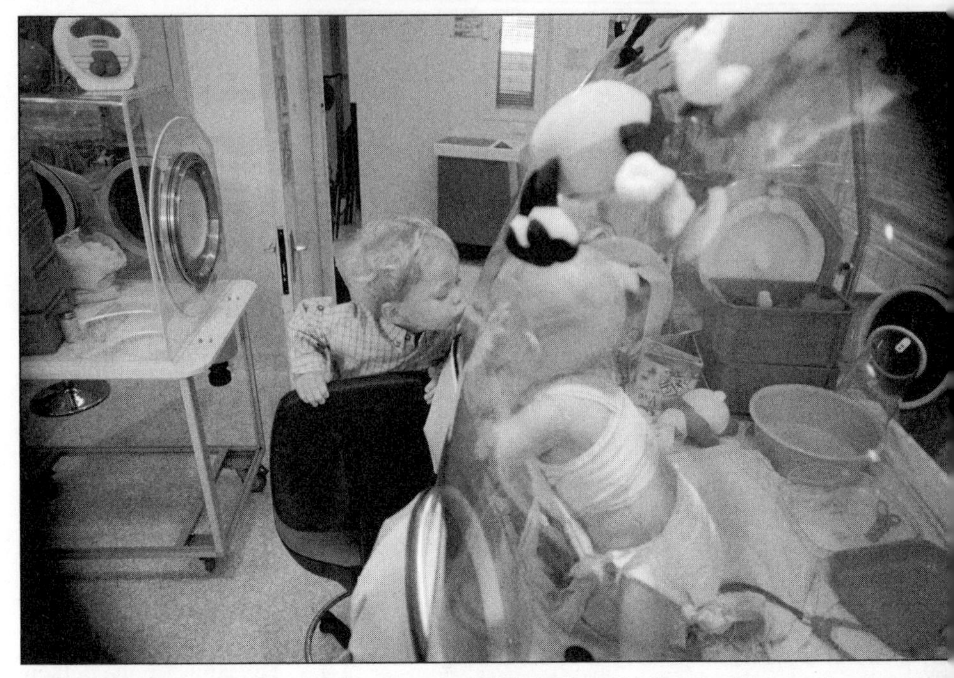

12. Wilco, a 'bubble boy' who was cured of his SCID-X1 disease by gene therapy in Alain Fischer's clinic in Paris, in 2002, gives a kiss to Yacine, a child who is still ill.

13. A mosquito-catcher employed by Target Malaria in Burkina Faso during the experimental release of marked wild mosquitoes in 2019.

14. Anti-GM demonstration in Oxfordshire, 1998.

15. The announcement of the 2020 Nobel Prize in Chemistry to Emmanuelle Charpentier and Jennifer Doudna for their invention of CRISPR gene editing.

16. Chinese geneticist He Jiankui of the Southern University of Science and Technology in Shenzhen, China, speaking during the Second International Summit on Human Genome Editing at the University of Hong Kong.

17. Anti-GM demonstrators in China protesting against ChemChina's plan to acquire Swiss company Syngenta, 2016.

- TEN -

THERAPY

On a cold day in November 1999, a couple of dozen mourners trekked up Mount Wrightson, around eighty kilometres south of Tucson, Arizona. When they got to the top, Dr Steve Raper read some words from Thomas Gray's 1750 poem 'Elegy Written in a Country Churchyard':

> Here rests his head upon the lap of Earth
> A youth to Fortune and to Fame unknown.
> Fair Science frown'd not on his humble birth
> And Melancholy mark'd him for her own.

Family and friends then scattered the ashes of 18-year-old Jesse Gelsinger, the first casualty of gene therapy. The medical promise embodied in genetic engineering, the prospect that had enthralled researchers and the public for decades, had inadvertently claimed the life of this youth, no longer to Fame unknown.[1]

Two months earlier, Raper had injected Jesse – a rebellious young teenager who loved professional wrestling – with genetically modified viruses.[2] Jesse suffered from OTC, a genetic liver disease, and the viral vectors carried a correct copy of the gene that was defective in Jesse. Jesse himself was not particularly ill – an intensive pharmaceutical regime of thirty-two pills a day and a strict diet enabled him to keep his disease under control. But he was keen to help the research project, and enthusiastically signed up for the procedure. The aim of the experiment was to see if the treatment was safe, as a first step to developing gene therapy for babies suffering from the disease. Tragically, within hours of the injection, Jesse's body responded to the vector with a massive immune reaction. Within twenty-four hours he was in a coma and after four days he was dead.

After the tragedy, Paul Gelsinger, Jesse's father, initially supported the research team – 'These guys didn't do anything wrong', he said.[3] But an inquiry revealed that one of the researchers, James Wilson, had a financial stake in a company that stood to gain if the trial were successful, while adverse reactions in both animals and humans had not been revealed to Jesse and his family before he signed up. Following a five-year legal battle between the US Department of Justice and the institutions involved, over $1 million in fines were paid – to the government, not to Jesse's family.[4]

Jesse's death not only made headline news around the globe, it also tarnished the reputation of gene therapy for years. Regulatory authorities all over the world immediately strengthened their guidelines, and a chill descended over the widespread optimism that had existed about the technique. Twenty years later, Guangping Gao, who worked with James Wilson on the viral vector that killed Jesse, recalled:

> For the entire field, it was a drop from a peak to a deep valley. We experienced 10 years of dark ages for gene therapy.[5]

*

For decades, deliberately altering human genes in order to cure – or augment – had been something between a fantasy and an objective for some geneticists. Within four years of Ed Tatum's 1958 Nobel Prize speech in which he predicted 'the improvement of all living organisms by processes which we might call biological engineering', wife and husband team Elizabeth Szybalska and Wacław Szybalski tried exactly that. They were studying human cell lines in which the

HPRT gene was completely inactive. To correct the mutation, in 1962 Szybalska and Szybalski added DNA from human cells with normal *HPRT* genes to the mutant cell culture, in a parallel with Avery's transformation of bacteria in 1944. Astonishingly, the activity of the mutant cells was transformed and they produced the HPRT enzyme normally.[6] This experiment was purely scientific, with no thought of therapy, but in 1964 a neurological disorder was discovered in children that was caused by the absence of the HPRT protein. Half a century later Szybalski claimed that their 1962 experiment represented the first model of human gene therapy.[7]

Without reliable and effective tools to manipulate DNA, medical applications of the new genetics remained in the realm of dreams. There were plenty of bold dreamers. In 1968, a young paediatrician with an interest in genetics, W. French Anderson, gave a lecture in Chicago in which he predicted that the first attempt at altering a defective human gene would occur within the next few years. A year later, molecular biologist Vasken Aposhian of the University of Maryland claimed that 'within the next five to ten years, certain inborn errors of metabolism will be treated or cured by the administration of the particular gene that is lacking'. Aposhian also coined the term that came to define the whole field: gene therapy.

Not everyone was so sanguine – in December 1969, Bernard Davis of Harvard Medical School gave a lecture in which he complained bitterly about the irresponsible hyperbole that surrounded the non-existent field.[8] That tension between an audacious and exciting medical vision and the chill reality of difficult science would characterise the whole history of gene therapy, down to the present day.

No sooner had Aposhian put a name to this hypothetical approach than Stanfield Rogers, a physician at Oak Ridge National Laboratory, decided to turn it into reality. For some years, Rogers had been considering the possibility of using a virus to introduce new genetic information into a human cell.[9] He was particularly interested in the Shope papilloma virus, which induces horn-like warts in rabbits – he had found that laboratory workers accidentally infected with the virus had low levels of the amino acid arginine and concluded that the virus produced arginase, an enzyme that breaks down arginine. Injecting someone with the virus would be a way of getting their body to produce arginase, should such a thing be required, Rogers reasoned.

At the end of the 1960s Rogers' hypothetical cure found its disease, when three young sisters in Germany were discovered to be suffering from a unique genetic condition: they were unable to make arginase. Arginine built up in their bodies, producing mental retardation and convulsions. Rogers believed that injecting the little girls with Shope virus might cure them, so in collaboration with a group of German physicians he went ahead with the experimental treatment in 1970. It made headlines all over the planet.[10] Surprisingly, there was no scientific publication to explain what had been done – it would be five years before Rogers and his colleagues published a brief, three-page article on the experiment, laconically admitting that the procedure 'did not influence the underlying metabolic disease'.[11]

In 1992, a review of the Rogers affair concluded that 'the biochemical and virological grounds for the experiment were incorrect, the experiment failed and little or no useful information was obtained'.[12] The Shope virus genome turns out not to contain a gene for arginase; instead infection with the virus prompts the body to produce the enzyme, but there is no evidence that this happened in the three children.[13] The fate of the girls remains unclear; although they were not apparently harmed by the procedure, their genetic defect will have had severe consequences. Two years after the story broke, Rogers justified his actions:

> When one has a patient with a progressively deteriorating disease that is known not to respond to dietary or other known measures, and one has a possible means of stopping the progression of the disease with an agent that has been extensively investigated for 40 years, there appears to be little alternative other than to try it.[14]

Not everyone was convinced. *Science* published an editorial outlining its reservations about the experiment, using arguments that would be repeated over and again in the subsequent half-century:

> The promises offered by the proponents of gene therapy largely

> ignore its limitations and hazards. To mislead the public in this regard risks another period of disappointment and reaction. We are still primarily in a descriptive phrase in our understanding of human genetics, with little, if any, idea of how to intervene safely at any level. Let us now seek public support for research towards a better understanding of normal and abnormal human biology, rather than promise quick glamorous cures.[15]

A year later, in 1972, a major article in *Science* highlighted the challenges faced by the non-existent field of gene therapy. For treatment to be effective, DNA would have to be introduced into the right cell, be integrated into the genome without disrupting any other gene and be expressed in the right way at the right time. These problems largely remain today. One of the potential developments the authors proposed was to engineer a virus so that it would infect a cell with the 'correct' version of a gene. This was at the same time as Berg was developing his modified version of SV40 to introduce genes into *E. coli*, but before anything had been published. The concept was clearly in the air, but in the absence of any reliable way of delivering a gene to a mammalian cell, all the hopes, and the doubts, remained hypothetical. That was one of the reasons given for excluding all discussion of potential human gene therapy or genetic manipulation from the agenda of the February 1975 meeting at Asilomar. In fact, this was precisely the time when such issues should have been discussed, because as soon as the techniques of recombinant DNA spread from bacteria to animal cells, physicians became impatient to apply them to humans.

In July 1980, Martin Cline, a Professor of Haematology at the University of California at Los Angeles (UCLA), used plasmids containing the human gene for β-haemoglobin to transfect bone marrow cells. These cells had been taken from two women patients – one in Italy, aged sixteen, the other in Israel, aged twenty-one – who suffered from β-thalassaemia, a genetic disease in which insufficient haemoglobin is produced.[16] The altered cells were then reintroduced into the patients and … nothing happened. As with Rogers' experiment a decade previously, the announcement was made in the press, not in a scientific journal. As far as anyone could tell, the genetically modified cells had no apparent effect – neither good nor, thankfully, bad. There was no indication that the new gene functioned at all.

Cline's experiment provoked an outcry – Richard Axel of Columbia said 'There is simply no scientific basis for expecting this experiment to work in people', while Tom Maniatis, who had provided the globin DNA without the slightest idea how it would be used, said that none of the previous animal experiments suggested that the procedure would work.[17] A subsequent enquiry revealed that although Cline had obtained approval from the relevant ethics authorities in both Israel and Italy, the UCLA Human Subjects Use Committee that oversaw his work had rejected his proposal until further animal experiments had been conducted.[18] He went ahead anyway. His later description of the approval process was revealing:

> I felt the members of the committee were ill qualified to judge human experimentation because many of them had never done it. The people who constituted that committee were not stupid, but they were not intellectual giants, and they certainly had no experience in the field of human clinical research.[19]

Not for the last time, an arrogant medic seemed to consider that his brilliance and his audacious vision overrode established ethical safeguards. A decade earlier Rogers had escaped without sanction; Cline was not so lucky. He had to resign from his UCLA leadership positions and lost two of his NIH grants; in a humiliating first, the NIH also required him to obtain special permission before carrying out any further recombinant DNA research.[20]

*

Even before the Cline affair, President Carter had convened an eleven-person Presidential Commission to explore bioethical issues associated with gene therapy, in response to the growing power of recombinant DNA technology and the assumption that it would eventually be applied to humans. In 1982 the Commission published its conclusions – *Splicing Life: A Report on the Social and Ethical Issues*

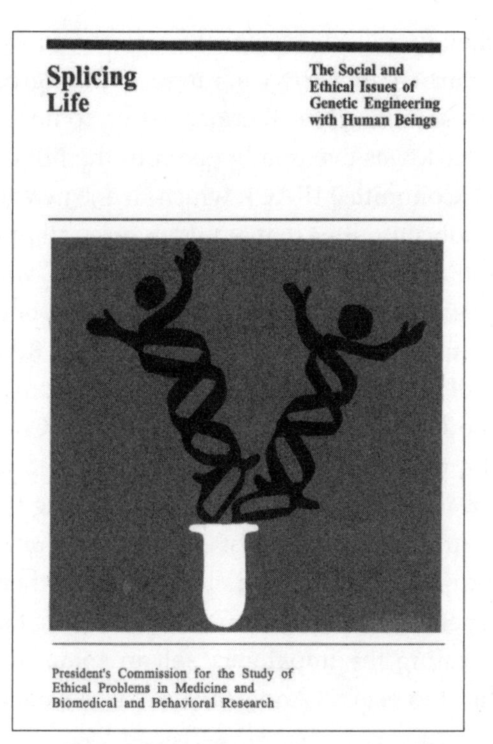
Splicing Life

The Social and Ethical Issues of Genetic Engineering with Human Beings

President's Commission for the Study of Ethical Problems in Medicine and Biomedical and Behavioral Research

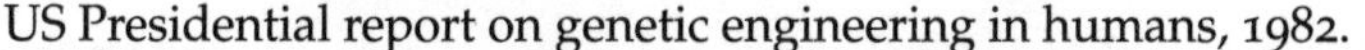

US Presidential report on genetic engineering in humans, 1982.

of Genetic Engineering with Human Beings.[21] From the opening pages of its report, the Commission dismissed fears that scientists 'might remake human beings, like Dr Frankenstein's monster' as 'exaggerated' (despite this, the cover showed joyous double-helix humans emerging from a test tube).

However, the Commission also warned that 'if beneficial rather than catastrophic consequences are to flow from the use of "God-like" powers, an unusual degree of care will be needed with novel applications'. Even more perceptively, the report contained a warning to twenty-first-century advocates of using CRISPR on humans:

> If genetically engineered changes ever become relatively easy to make, there may be a tendency to identify what are in fact social problems as genetic deficiencies of individuals or to assume that the appropriate solution to a given problem, whether social or individual, is genetic manipulation.[22]

The solution they proposed was hardly radical – an oversight body composed of participants from a range of backgrounds should scrutinise proposals for gene therapy. After some organisational argument, this task was eventually given to the NIH Recombinant DNA Advisory Committee (RAC), which in 1983 created a Human Gene Therapy Subcommittee that would oversee all proposals in the area.[23] From the outset, it was clear that the RAC would not consider any proposals for heritable germline changes, only for somatic gene therapy. This position was maintained by the RAC, essentially unchanged, until the committee was disbanded in 2019.[24]

Neither the Cline affair nor the Presidential Commission did anything to diminish the appetite of physicians to develop gene therapy, nor did it quench the burning desire among the most ambitious to be the first to do so. In November 1980, French Anderson, who twelve years earlier had predicted that gene therapy was just around the corner, co-authored an article in the *New England Journal of Medicine* revealing the impatience felt in some medical circles. 'When is it ethical to begin?' Anderson and his colleague asked:

> We argue that gene therapy for some genetic disorders will be possible in the foreseeable future. … We recognise that the urgent desire to treat patients having lethal or extremely serious genetic diseases may lead the attending physician to attempt promising, though not fully tested, new regimens.[25]

Animal studies soon provided the kind of tools that would be required. In 1983 modified retroviruses – RNA viruses that use reverse transcriptase to copy their genome into DNA, which is then integrated into the host's genome – were employed to introduce corrected versions of DNA into human cells. The gene involved was *HPRT*, which had been studied by Szybalska and Szybalski two decades earlier.[26]

The following year saw one of the first animal demonstrations of the power of gene therapy, when US researchers reversed the effects of a mouse gene mutation called *little*, which led to reduced levels of growth hormone and particularly small mice. By injecting the rat growth hormone gene into a *little* mouse single-cell embryo they were able to produce mice that were much heavier. Too heavy. The resultant recombinant rodents were over 50 per cent larger than normal, highlighting how tricky it is to ensure that an introduced gene is properly regulated.[27]

Despite these problems, Anderson and his colleague Michael Blaese decided to develop gene therapy for a form of severe combined immunodeficiency (SCID). They chose this disease partly because it was a simple genetic condition involving a known faulty enzyme, and partly because of public awareness. In its most extreme form SCID makes the newborn child extremely susceptible to disease – at the time, the only solution was for the infant to live in a totally sterile environment. The most famous patient with the disease was David Vetter, called 'the bubble boy' by the US press. Vetter was born in 1971 and lived inside a sterile chamber; he eventually died in 1984 after a bone marrow transplant from his sister tragically led to him developing lymphoma. Vetter's condition – shared by only a few dozen children around the world – captured the popular imagination and shortly after his death was turned into a television drama starring John Travolta. In 1986 the singer Paul Simon wrote 'The Boy in the Bubble', a semi-ironic song about 'the days of miracle and wonder' produced by modern technology, while a 'bubble boy' is the unseen protagonist of an unusually dark and complex 1992 episode of the sitcom *Seinfeld*. This level of public awareness of the disease would heighten the impact of successful therapy.

The type of SCID that Anderson and Blaese decided to treat was caused by a defective enzyme called adenosine deaminase (ADA). Their idea was to introduce a functioning *ADA* gene into the T cells of an ADA-deficient patient in a test-tube procedure and then inject the patient with the altered cells. With luck, the patient's body would start producing the correct form of ADA.

In 1987 Anderson and his colleagues felt sufficiently confident in their proposed experiment to submit a 500-page preclinical data document to the Human Gene Therapy Subcommittee. Derisively called the phone book* because of its size and tedium, the protocol

*Note to younger readers: the phone book was a very thick, large-format paperback containing the names, addresses and landline telephone numbers of

was never actually used. Not only was there pushback from the Subcommittee about the procedure, even the members of Anderson's own research group had their doubts. When Anderson asked his colleagues if they would go ahead with the experiment if it involved their own child, only a third of them said 'yes'. Anderson himself later admitted that had he been on the Subcommittee, 'I probably would have voted against us too.'[28]

Faced with these criticisms, the ADA project was put to one side and in July 1988, Anderson and Blaese, together with oncologist Steven Rosenberg, put in a new proposal for the first application of gene therapy. This proof-of-principle trial merely involved using a retrovirus to introduce a marker gene into the DNA of tumour-infiltrating lymphocytes (TILs) – white blood cells that were thought to attack tumours, but whose exact behaviour was unclear. The experiment was intended to show that gene therapy was safe, with the added bonus of perhaps revealing more about the behaviour of the TIL cells in cancer.

Although the proposal was relatively innocuous, it was initially rejected by the RAC Subcommittee, which was suspicious of Anderson's sudden change in strategy. In particular, the molecular biologists on the panel considered that it was premature to apply these techniques to humans. They feared that any catastrophic outcome, such as the TILs turning cancerous, perhaps killing the patients, 'would produce such a backlash that all our research would be affected'. One researcher said prophetically 'we would be dead for ten years'.[29] Faced with detailed experimental criticisms in October 1988, Anderson repeated his argument from eight years earlier in response to those who called for caution:

> I was asked at the break 'What's the rush in trying to get your protocol approved?' Perhaps the RAC members would like to visit Dr Rosenberg's cancer service and ask a patient who only has a few weeks to live: 'What's the rush?' A patient dies of

---

everybody in a particular region, printed in tiny print – in a large city it could consist of many volumes. These were given free to every telephone subscriber; they began to go extinct in the early years of the twenty-first century.

> cancer every minute in this country. Since we began this discussion 146 minutes ago, 146 patients have died of cancer.

Despite the fact that the protocol was designed to have no direct effect on the patients concerned, the committee was apparently swayed by this crude rhetoric and approved the experiment.[30] One of the more serious but grim arguments in favour of the proposal was that the patients all suffered from severe terminal cancer and had at most three months to live. With such a limited lifespan, it would not particularly matter if the retrovirus used in the protocol produced cancerous TIL cells.[31]

Anderson and his group were still not out of the woods – the NIH director made the proposal go through a few more hoops, meaning that it eventually gained unanimous approval from seven different regulatory committees, involving fifteen separate appearances by the would-be gene therapists.[32] Approval was not finally made public until the end of January 1989. After a further delay caused by a lawsuit from Jeremy Rifkin, the experiment eventually went ahead on 22 May 1989.

Maurice Kuntz, a 52-year-old lorry driver, was the first volunteer to undergo the procedure.[33] As expected, the vector was well tolerated – shortly after the injection, Kuntz joked 'I haven't grown a tail yet' – and the marker gene functioned as hoped. But while the protocol proved that introducing a foreign gene into the human body did not necessarily have any immediate adverse effects (this happens every time we have a viral infection), it was not clear what it showed about the behaviour of the targeted cells.[34] One RAC member said that the results revealed nothing beyond the fact that the cells could be marked and then found in the bloodstream for a certain period of time.[35] Over the following two months, four more patients were treated; three of them, including Mr Kuntz, died in less than a year. Two survived for more than two years, but not thanks to the gene therapy, which had no effect on the course of their disease.[36] Anderson achieved his ambition; he was the first to carry out gene therapy in a human, even if there was actually no therapeutic aspect to the procedure.

The following year, Anderson and Blaese returned to their original idea of using retroviruses to alter T cells in order to treat the ADA form of SCID. But the children who would be treated were in reality already relatively stable – in 1987 a drug treatment called PEG had been developed for the disease and thirteen ADA-deficient youngsters in the United States and Europe had been successfully treated.[37] However, the cost of this treatment was crippling. For example, in 1995 the bill for PEG treatment incurred by one family with two boys who suffered from the disease, Rhett (aged four) and Zach (aged two), was $40,000 a month. At the time, the family's insurance policy had a limit of $1 million, half of which had already been eaten up.[38] There was immense pressure on parents to embrace the potential cure promised by gene therapy.

After several rounds of argument with the RAC and the FDA it was agreed that the children would continue taking PEG throughout the gene therapy experiment. After all, they were sick children, not laboratory rats. This meant that the procedure would probably be scientifically meaningless, as it would be impossible to conclude if the new gene had any effect – any improvement might be due to PEG. Nevertheless, on the day the project was approved, the RAC Chair hyperbolically announced that 'Doctors have been waiting for this day for a thousand years'.[39] Doctors, not patients.

*

On 14 September 1990, Ashanti DeSilva, who suffered from SCID-ADA and had celebrated her fourth birthday less than two weeks earlier, received an injection of her modified T cells. The procedure went well, and the little girl was soon able to return home. In subsequent years, Ashanti became a symbol of the apparent success of gene therapy. In September 1994 she appeared before the Science House Committee of the US Senate, where Democrat George Brown Jr declared that she was 'living proof that a miracle has occurred', conveying the overheated atmosphere that surrounded gene therapy at the time.[40] Ashanti's progress was also covered by the *New York Times* and she was regularly followed by television cameras. As an adult with a normal life, she eventually became a genetic counsellor. In 2015 she recalled the significance of her diagnosis, and her gene therapy treatment:

> My parents finally received an answer as to why I couldn't walk more than a few feet without collapsing, and why I was constantly catching infections without full recovery. ... The gene therapy was not the cure we had all hoped for, but my cell counts continued to increase. If you ask my parents or myself, it worked, because my fate was written with my diagnosis, yet here I stand today.[41]

Although both DeSilva and another patient tolerated the treatment well, and their ADA levels did increase and were sustained for several years, Anderson and Blaese admitted that the effects on the patients' well-being were difficult to measure.[42] The children were not cured; DeSilva is well, but only because she still takes PEG to replace the version that is deficient in her body. However, in an astonishing demonstration of the potential efficacy of gene therapy, twenty-six years after the procedure, up to 15 per cent of her T cells still had the correct version of the gene.[43]

At the same time as the ADA protocol was approved by the RAC, Rosenberg got approval for a rather different use of gene therapy, which came to dominate the field. Instead of targeting a genetic disease, Rosenberg decided to use gene therapy to treat cancer. The idea was to repeat the pioneering TIL marker protocol but to introduce into the TILs a gene that would kill the tumours targeted by these cells by producing a protein called tumour necrosis factor (TNF). The procedure was first carried out at the beginning of 1991 on a 29-year-old woman known as Suzanne Marotto and a 42-year-old man described as Robert Antrim (those were not their real names). Both patients had advanced melanoma that had resisted all other treatments; like Maurice Kuntz their life expectancy was less than three months. They had nothing to lose.

Although only 5 per cent of the injected TIL cells actually expressed TNF, three weeks later no viable tumour cells were recovered from the injection sites, suggesting that the experiment had worked.[44] But neither patient made a full recovery, and Mr Antrim died within months. Rosenberg carried out the procedure on a total of nine patients, with unconvincing results. One ex-colleague said that Rosenberg had crossed an ethical line by suggesting the protocol was therapy rather than research and in October 1992 an NIH oversight panel froze his funding for two years.[45]

Despite such controversies, these experiments were hailed by researchers and the media. 'Human gene therapy comes of age' proclaimed Dusty Miller, one of the key developers of retroviruses as ways of delivering DNA to human cells.[46] In reality, gene therapy was still very much a laboratory technique rather than an established medical procedure. Nevertheless, the excitement in the press and among the public was palpable. On learning of Anderson's proposed gene therapy trial, *Le Monde* announced that it was 'Year one of gene therapy',* while Anderson trumpeted in his journal, *Human Gene Therapy*:

> *The time has come.* Gene therapy is here; the press covers it extensively; it does offer 'hope'. ... Two little girls who had a life-threatening disease are now leading essentially normal lives. This is wonderful.[47]

In 1995, journalists Jeff Lyon and Peter Gorner published *Altered Fates*, a best-seller that accurately caught the frenzied atmosphere of ambition and vainglory that surrounded gene therapy.[48] The centre of their attention was Anderson, who they considered to be 'a visionary to the bone', a man on a crusade. In a particularly purple passage, Anderson and his colleagues were described as 'modern-day Magellans who are charting the vast molecular unknown of the human body. They are the polar and African explorers of our time, and their field is wide open ... and ripe for magnificent findings.'

Widely seen as the father of gene therapy, Anderson was involved in twenty-six of the thirty-seven gene therapy protocols that were approved by mid-1994.[49] That same year, he was runner-up for *Time* magazine's coveted Man of the Year award, and it seemed as though his future would be paved with gold. After receiving the prestigious Lasker Prize, Anderson denied dreaming of even greater

---

* This was a reference to the French revolutionary calendar, introduced in 1793, in which (after a lot of squabbling) the foundation of the Republic in 1792 became Year 1.

rewards: 'I don't really care about the Nobel Prize. If I get it, I'll go to Stockholm, but I don't really care one way or the other.'[50] You can believe that if you want.

Researchers began investigating potential gene therapy treatments for other diseases, including HIV, familial hypercholesterolaemia – a genetic disease that produces high cholesterol – and glioblastoma, which causes over half of brain tumours.[51] Some took the growing enthusiasm for gene therapy a step further and argued that researchers and physicians should consider using the technique to alter the human germline in order to prevent genetic disease.[52] In a prefiguration of later transhumanist fantasies, people started to discuss the idea of enhancing human abilities by genetic engineering, an idea that was enthusiastically advocated by *The Economist* in 1992, although the author showed little grasp of either ethics or genetics:

> Should people be able to retrofit themselves with extra neurotransmitters to enhance various mental powers? Or to change the colour of their skin? Or to help them run faster, or lift heavier weights? Yes, they should. Within some limits, people have a right to make what they want of their lives. ... People may make unwise choices. Though that could cause them grief, it will be remediable. That which can be done, can be undone; people need no more be slaves to genes they have chosen than to genes they were born with.[53]

Leaving aside the dodgy biology – relatively few characteristics are susceptible to gene therapy, and what is done cannot necessarily be undone – this libertarian attitude was very much a minority position. At around the same time, the UK government set up a committee on the ethics of gene therapy chaired by Cecil 'Spike' Clothier, which recommended that gene therapy be limited to life-threatening disorders and should not be used to change or enhance normal human traits, or to alter the germline – the DNA in egg and sperm that is passed to the next generation.[54]

Despite such occasional doses of cold water, the atmosphere was heady and the physicians had a strong view of their place in history. Gary Nabel of the University of Michigan boasted:

> This is the golden age of biology. In the same way that Einstein and Planck and Heisenberg were around for the golden age of physics earlier in this century, we are present at a very special point in time.[55]

*

Reality began to bite in 1995, when the new director of the NIH, Harold Varmus, set up a commission to explore the reality of gene therapy and to determine if the $200 million a year the NIH devoted to the field was money well spent. Varmus suspected that gene therapy researchers had created what he described as 'the mistaken and widespread perception of success'.[56] That was exactly the conclusion of the commission chairs Stuart Orkin and Arno Motulsky in December 1995.[57] Even before the report appeared, some gene therapists surveyed their field with brutal honesty. Oncologist Drew Pardoll admitted:

> There's been an emphasis on glitz. It has produced a culture in which getting into clinical trials – getting into the club – has been more important than getting a meaningful result.[58]

In the five years since poor Maurice Kuntz had received the first experimental treatment, 106 clinical protocols for gene therapy had been approved in the United States, and 597 patients had received gene transfers. Nevertheless, real progress had been minimal. Orkin and Motulsky pulled no punches:

> Clinical efficacy has not been definitively demonstrated at this time in any gene therapy protocol. ... Significant problems remain in all aspects of gene therapy. Major difficulties at the basic level include shortcomings in all current gene transfer vectors and an inadequate understanding of the biological interaction of these vectors with the host. In the enthusiasm to proceed to clinical trials, basic studies of disease pathophysiology ... have not been given adequate attention.'

The report explained that physicians saw clinical experiments as prototype therapies; as a result they seemed to have forgotten the point of what they were doing:

> Only a minority of clinical studies, illustrated by some gene marking experiments, have been designed to yield useful basic information ... very little research effort is focused on understanding the mechanisms that govern maintenance or shutoff of gene expression following gene delivery in gene therapy experiments. Available data are largely anecdotal.

In a brutal summary, the report argued that the field was characterised by insufficient attention to research design, poorly defined molecular and clinical endpoints, and lack of rigour. Ouch.

Amid the subsequent media excitement, Orkin highlighted the problems associated with even the most famous treatments such as the 'miracle' of Ashanti DeSilva and the other ADA-SCID patients who were taking PEG when they underwent gene therapy. 'The kids were not so sick', he said. 'They were well then and they are still well today.'[59] Furthermore, Orkin argued, the focus on genes could lead medicine and the pharmaceutical industry to ignore significant alternative therapies:

> If the gene for hypercholesterolaemia had been discovered ten years ago we might have focused exclusively on the gene and missed out on a highly effective class of drugs.[60]

Those drugs were statins, a cheap treatment for high cholesterol that has saved millions of people around the planet from the effects of heart attack or stroke. Not all world-changing treatments are as complex or as sexy as gene therapy.

Following publication of the report some physicians were abashed, while the NIH decided to focus its funding on the basic research that had been cruelly lacking.[61] In 1997, an editorial in *Nature Biotechnology* warned that gene therapy revealed 'what happens when an entire field begins to believe its own press'.[62] The editorialist did not mention it, but that press included the *Nature* stable of journals, which had published some of the most uncritically enthusiastic articles about gene therapy.

Despite the drubbing from the Orkin–Motulsky report, the field continued in more or less the same way, although there was growing interest in the use of adenoviruses (these cause colds, diarrhoea and various other infections and are also used as a component in several types of vaccination) which would deliver the new genetic material into the cell and then use the cell's machinery to reproduce, rather than inserting themselves into the genome, with potentially mutagenic consequences. One of the researchers who began using adenoviruses was Jim Wilson. In 1998 a group of leading gene therapists and bioethicists enthusiastically discussed the possibility of carrying out human germline transformation – introducing genetic changes that would be passed on to offspring – both for therapeutic reasons, and for what they called enhancement.[63] The gene therapists had learned nothing and forgotten nothing.

And then, in September 1999, Jesse Gelsinger died.

*

The tragedy led to another round of breast-beating. The FDA increased protection for participants in clinical trials, alarmed by the realisation that, as RAC Human Gene Therapy Subcommittee member Ruth Macklin put it, 'gene therapy is not yet therapy'.[64] An article in the FDA's magazine for the public was scornful: 'the hyperbole has exceeded the results… little has worked'.[65] Even the academic journals devoted to the new medicine were critical of the world they had helped create – in 2001, an editorial in *Molecular Therapy* stated bluntly:

> Several of the most highly visible early studies were described and carried out in overly expectant and optimistic terms, promising far more than they could deliver and thereby carrying with them the seeds of their own failure. The miscalculation can be seen among exaggeratedly optimistic investigators, research institutions, interest groups, and scientific and lay press.[66]

One of the reasons for this exaggeration was that lots of money was involved. At the time Jesse Gelsinger received his fateful injection, 331 gene therapy protocols involving over 4,000 patients had been approved. Just forty-one procedures were aimed at diseases caused by single genes – the rest were for cancer or AIDS. There would be little money to be made from treatment of diseases with only a handful of patients, whereas cancer and AIDS were both massive markets. In reality, however, almost all of these apparently more lucrative approaches turned out to be dead ends.[67]

In 2009, Jim Wilson, the scientist who had co-led the adenoviral study that killed Jesse Gelsinger, reflected on what had led researchers astray in the 1990s:

> (i) a straightforward, if ultimately simplistic, theoretical model indicating that the approach 'ought to work'; (ii) a large population of patients with disabling or lethal diseases and their affiliated foundations harbouring fervent hopes that this novel therapy could help them; (iii) unbridled enthusiasm of some scientists, fuelled by uncritical media coverage; and (iv) commercial development by the biotechnology industry during an era in which value and liquidity could be achieved almost entirely on promise, irrespective of actual results.[68]

The drive to make potentially lucrative progress was so strong that virtually as soon as Jesse's ashes were scattered on Mount Wrightson, the field began to look for reasons to be cheerful. In 2000, researchers in Paris led by Marina Cavazzana-Calvo and Alain Fischer reported that they had used retroviral gene therapy to treat two young children (eight and eleven months old) who suffered from SCID-X1 – the 'bubble boy' disease: 'Both patients left protective isolation at days 90 and 95 and are now at home 11 and 10 months, respectively, after gene transfer without any treatment. Both enjoy normal growth and psychomotor development', the researchers said.[69] Four months later, similar results were reported with a third patient and the treatment programme was soon extended. The result was dramatic – as the historian Michel Morange has put it:

> This cure was symbolised not by any single change in the patients' physiology but instead by their spectacular release from their isolation bubbles.[70]

Despite this success Fischer emphasised that 'We are still in an experimental phase. We are checking the security and appropriateness of the technique, analysing the risk–benefit ratio.'[71]

Fischer was right to be circumspect. Within six months, just days after *Le Monde* reported that the first two children to be treated were in perfect health with a normal immune system, one of the dozen infants who had been treated with gene therapy developed leukaemia.[72] A second case was discovered at the beginning of 2003, and two more appeared subsequently – one of the children eventually died. The French SCID-X1 trial was put on hold, while in the United States the FDA suspended thirty or so retroviral clinical trials as a precaution.[73] The problem was the vector: the mouse retrovirus used to deliver the correct version of the gene turned out to preferentially insert itself into a gene that, when disrupted, caused leukaemia. The researchers had inadvertently created a disease-inducing mutation. As two leading gene therapists wrote in 2009:

> What was once a promising new start for gene therapy became an enormous set-back. Vector safety had always been a pressing issue – now it had become a yawning chasm.[74]

Although the French team insisted that mortality caused by gene therapy was substantially less than the risk of a bone marrow transplant – the only other available treatment – the death of Jesse Gelsinger cast a long shadow.[75]

*

The first gene therapy protocol – rather than an experimental trial – to be approved anywhere in the world was Gendicine, a Chinese therapy developed for the treatment of head and neck squamous cell cancer by expressing a tumour-suppressing gene in transduced cells, which was first used routinely in 2004. It has since been given to over

30,000 patients with substantial improvements over standard treatments, and with high rates of five-year survival, but it has not been approved in Europe or the United States.[76] The first gene therapy to be approved in Europe was Glybera, a cure for an ultra-rare lipoprotein lipase deficiency, which became available in 2012. But with few potential patients, and a cost of €1 million, the treatment was eventually withdrawn in 2017. It was given to only thirty-one patients worldwide and had been prescribed to only one person outside of a clinical trial.[77] In the same year, the FDA approved the use of Luxturna to treat inherited retinal disease due to mutations in the *RPE65* gene; Europe followed suit in 2018.[78] Two further US gene therapy treatments, both for rare cancers, were also approved in 2017.[79]

Up until this point the US biotech industry had invested over $10 billion in developing gene therapy but had generated a revenue of zero.[80] With the sequencing of the human genome, which was finally announced in June 2000, the potential for gene therapy had become broader, in that it opened the road to large-scale surveys of genetic contributions to a wide range of diseases, with the potential for developing therapies. But it also became narrower, in that the significance of rare genetic diseases (conditions that affect fewer than one person in 2,000) began to grow. Charities devoted to supporting a handful of patients have increased their influence by grouping together to raise funds and develop new approaches or acting as a larger lobby to influence biomedical policy.[81]

One tiny but significant example of the power of this new genomic technology can be seen in the case of a boy from Wisconsin called Nic Volker. When he was a toddler Nic developed appalling physical symptoms as his immune system riddled his gut with painful holes. His physicians were bewildered and the poor child spent much of his life in hospital, enduring scores of operations. Then, in 2009, when Nic was aged five, his treatment team gained access to the very latest sequencing equipment to study his exome – those stretches of the genome that produce protein or RNA. That did not immediately help: Nic's exome had a bewildering 16,124 variants compared to the 'standard' genome (this level of variability turns out to be not unusual), any one of which was potentially responsible for his condition. His physicians finally found the needle in this genomic haystack – a single-base-pair change in *XIAP*, a gene involved in mobilising the immune system. None of the 2,200 human genomes that had been sequenced at the time shared Nic's mutation, which remains incredibly rare, but armed with the knowledge as to what was wrong, Nic's doctors went ahead with a straightforward bone marrow transplant which effectively cured him.[82] In the future, if a child is found to have this genetic change, it might be possible to use gene therapy to correct the inborn error. The lesson of Nic's case is that genomic sequencing can enable rapid, life-saving intervention; in 2021 physicians in the United States used sequencing to diagnose a five-week-old baby's disease and prescribe appropriate treatment within an astonishing thirteen hours following admission to hospital.[83] Neither of these examples involved gene therapy, but they led to therapy based on genes.

Persistent difficulties with gene therapy have arisen from the precise technology involved, in particular the variety of vectors that have been used to deliver the correct form of the gene to the relevant cells. Most gene therapy protocols use plasmids or viruses to insert DNA into the cell or into the genome; as shown by the cases of Jesse Gelsinger and the French SCID-X1 patients, these vectors may prove dangerous. There are now a variety of vectors available, each with their advantages and disadvantages, presenting clinicians and researchers with a difficult set of choices. In 2013 a survey of the ten types of vectors being used in gene therapy in clinical settings concluded:

> It is obvious that human gene therapy as a treatment modality has been more complex than expected.[84]

By 2020 the molecular bases of over 3,600 genetic diseases had been identified. Over 4,000 gene therapy clinical trials had been carried out on five continents, the vast majority in the United States and over two-thirds of them relating to cancer. These trials included 300 phase 3 studies – the decisive, final step when researchers discover if a treatment is truly effective, as well as being safe. And yet despite all this work, there are only a handful of gene therapy treatments available. Only seven protocols have been approved in Europe

(Glybera was withdrawn); of the remaining six treatments, half are for various kinds of cancer (melanoma, leukaemia), while the other three are for the kind of genetic disease that was originally seen as being the target of gene therapy – retinal dystrophy, β-thalassaemia and ADA-SCID.[85]

These therapies are undoubtedly successful, but until recently relatively few patients have had their lives transformed. Between 2000 and 2011, eighteen children were treated with Strimvelis for ADA-SCID; they all survived – without treatment, babies habitually die by the age of two.[86] In 2021, a joint US–UK study reported that a more sophisticated version of the original approach – using a virus to deliver the *ADA* gene to stem cells outside of the patient, and then injecting the transformed cells – had effectively cured forty-eight out of fifty young patients, with no side effects.[87]

Greater impact has been achieved through the use of chimeric antigen receptor T cells (CAR-T), which have been genetically engineered to target canceous cells. In the case of leukaemia, over 1,000 patients have been treated with CAR-T in Europe alone and the long-term efficacy of the treatment is extraordinary. A study of two of the first people to be treated with this approach showed that more than a decade after infusion with the genetically engineered cells both patients are not only cancer-free, but that the engineered CAR-T cells are still present and active.[88]

*

Throughout the early years of the century there were repeated attempts by the gene therapy community to get its mojo back, to demonstrate that the field was truly about to become firmly established. Researchers needed this boost because they were dealing not only with the unforeseen safety issues that emerged both in the Gelsinger case and in the SCID-X1 trials, but also with a shocking turn of events that had nothing to do with gene therapy. In 2004, Anderson, the 'father of gene therapy', the man who felt he was 'a molecular Captain Kirk', was arrested for child abuse.[89] He was subsequently convicted and sentenced to fourteen years in prison.

As the handful of safe and effective gene therapy protocols were gradually authorised, researchers and journalists began to repeat the boastful tropes that had characterised previous decades. In 2011 *Nature Biotechnology* published an article entitled 'Gene Therapy Finds its Niche' in which the author claimed to detect 'a growing sense that the promise of gene therapy is – after several false starts – finally being realised'.[90] The same claims were made seven years later, when *Science* magazine ran an article under the title 'Gene Therapy Comes of Age', explaining that 'gene therapy is now bringing new treatment options to multiple fields of medicine'.[91] With so few therapies actually available, that was an exaggeration.

Strikingly, there were virtually no films, television series or major novels looking at what could go wrong – or right – with gene therapy. Hollywood's biggest attempt to address the issues associated with human genetics and genetic engineering is Andrew Niccol's 1997 film *Gattaca*, set in a near-future dystopia in which society is divided into 'valids' and 'in-valids', depending on the genes that each group possesses – those wishing to have a 'valid' child carry out an unspecified form of genetic engineering in order to have the child they desire. The 'volunteer scientific consultant', brought in to 'make sure the science wasn't absurd' was, inevitably, Anderson. The underlying themes of the film – parallels with race and a critique of genetic determinism (the plot revolves around one character's ability to 'pass' as valid despite having in-valid genes) – mean that it remains a favourite among those teaching bioethics courses.[92] The film was backed by a brilliant spoof newspaper advert for designer babies – 'Children made to order. How far will you go? How far will your child go?'

However, very few of the genetic bases of the traits on the fake advert are properly understood, even now, a quarter of a century later. For example, eye colour, which we learn in school is a simple trait, turns out to be horribly complex – a 2021 study of 192,986 people revealed that sixty-one genes are involved in determining eye colour, including fifty that were previously unidentified.[93] Although wild claims are often made in popular articles and books about the possibility of creating superhumans, the reality is rather different.[94] Because many traits, in particular those that people are interested in, such as height or intelligence, have very complex

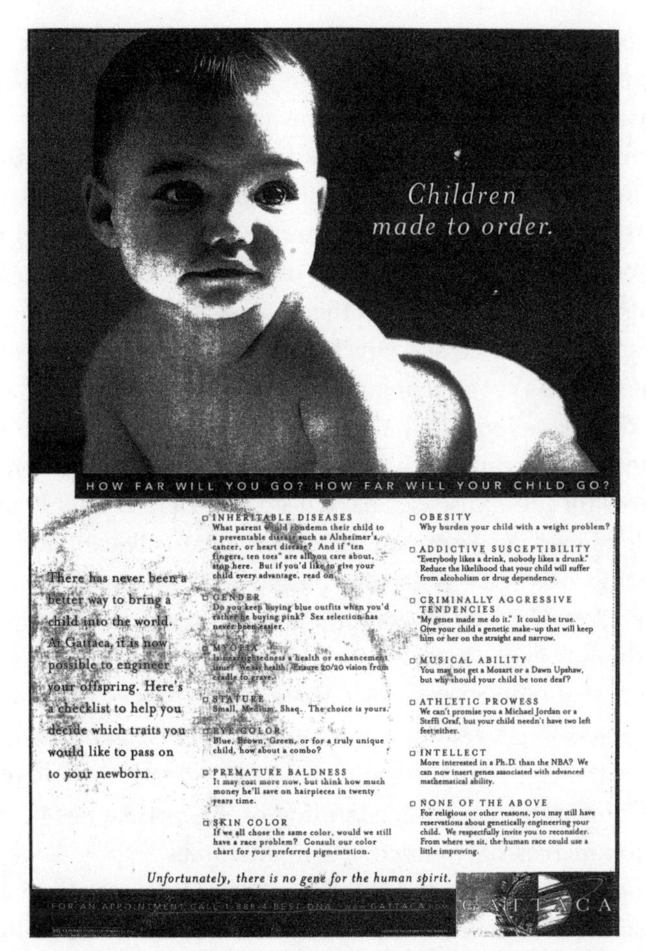

Spoof advert for the film *Gattaca*, which appeared in *The New York Times*, *Washington Post* and elsewhere in September 1997.

genetic contributions and a massive environmental component, selecting among embryos for 'better' offspring would barely make any difference and would probably increase the danger of less desirable characteristics being selected.[95]

In fact, gene therapy has still neither found its niche, nor come of age. The hopes it generated have not yet been realised. There is progress – as well as the recent success with ADA-SCID, a recent four-year clinical trial for gene therapy against OTC deficiency, the disease that Jesse Gelsinger suffered from, proved positive.

Eleven patients were treated, with no ill effects, and three people have apparently been effectively cured.[96] The United Kingdom has recently approved Zolgensma, a gene therapy that treats spinal muscular atrophy in children by using an adenovirus to deliver the gene for the missing protein to the infant's motor neurons. Zolgensma is immensely expensive – £1.79 million per patient – but a single dose can save a child's life. About sixty-five babies born in the United Kingdom each year suffer from this disease.

Despite such successes, which are transforming the lives of some, gene therapy has yet to change the lives of most of those affected by genetic diseases – indeed, the way that the biotechnological research is funded and organised has meant that until recently proposed treatments for these diseases have been less well funded than those targeting cancer. Even a genetic disease as relatively well understood as cystic fibrosis – its molecular basis was first revealed in 1989 – has proved recalcitrant to gene therapy, with no prospect of a true cure, although a simple description of how the disease works would suggest that should be possible. The complexities of safely delivering a replacement version of the damaged *CFTR* gene to the right tissues in the right quantities have so far proved insurmountable; instead, our molecular understanding of the disease has been used to develop increasingly effective drug therapies.[97]

This relative failure to fulfil the promise is true even in an area that has caused immense controversy around the world: mitochondrial transfer, which produces what is misleadingly known as a 'three-parent baby'. Strictly speaking this procedure is not genetic engineering, but it is a genetic therapy that alters the inheritance of an individual. When an egg is fertilised, the father's sperm contributes only its genetic material; the fertilised egg's energy-producing mitochondria and the rest of the cellular contents come solely from the mother and are passed down in an exclusively maternal lineage. Mitochondria, which are found in all eukaryotes, have their own genome – they are remnants of one component of a chance fusion between two wildly different kinds of single-celled organisms that occurred about two billion years ago, leading to the appearance of eukaryotes. In some cases, women with genetically defective mitochondria can find pregnancy impossible or the child may have a genetic disease of varying severity. In mitochondrial transfer all of the mother's mitochondria in one of her egg cells are replaced with those from a healthy donor; the egg is then fertilised by the father's sperm, using IVF. The first baby created through mitochondrial transfer was born in Mexico in 2016 and there are now several dozen apparently healthy children around the world who have been born as a result of this procedure. As an indication of how rare the condition is, and how complex the therapy is, despite all the debates in the United Kingdom in the run up to the 2015 change in the law that permits this technique, there is still no news that a child has been born in the United Kingdom through this procedure.

After over three decades the state of gene therapy highlights a very real problem in how we view novel and incredibly complex medical advances. As the French gene therapist Patrick Aubourg put it in 2011:

> Scientists in particular have to remember we are only doing medicine, which is an art, not a science – and not a perfect art.[98]

– ELEVEN –

EDITING

In many of its key applications, from GM plants to gene therapy, late twentieth-century genetic engineering was hampered by a major technological hitch. Novel DNA sequences could be introduced into cells with relative ease, but the DNA was often associated with genetic material from the vector, and insertion into the genome was uncontrolled, sometimes causing problems when a significant gene was inadvertently disrupted. The best solution would be not to add a genetic instruction from the outside, as in recombinant DNA, but instead to simply rewrite the existing genes so that they make the cell do something else. This is the kind of stuff that is easy in fiction, but hard in reality. And yet that is basically where we are today.

At the heart of this radical change has been the discovery of what happens in cells when the DNA double helix is broken. Damage to a single strand of the double helix is easy for the cell to repair – the sequence on the intact strand is used as a way of synthesising the required complementary bases. In the 1980s it was discovered that mammalian stem cells can respond to a double strand break by using the corresponding sequence on the other chromosome as a template to repair the missing bases. In this repair mechainism, called homologous recombination, the cell can be tricked into using an artificial DNA template to synthesise a new DNA sequence, which now forms part of the cell's genome. This method of gene targeting eventually enabled the creation of recombinant mammals and raised the possibility of carrying out gene therapy in stem cells from the blood, skin or liver.[1]

But a number of problems remained. This procedure worked only for mammalian stem cells and had a very low level of efficacy. The ideal way of changing genes precisely and effectively would be to cut the double helix at a desired point in the genome, in a cell that was in an appropriate physiological state to use homologous recombination to alter the damaged sequence in a precise fashion, using some introduced DNA as a template.* To take the first step – cutting the DNA only at one particular point out of three billion base pairs in a human – required a nuclease that would recognise perhaps dozens of base pairs, to ensure only one point in the genome was targeted. Ordinary restriction enzymes recognise sequences of only a handful of bases, which are present at many points in the genome, so they cut the DNA into dozens or hundreds of pieces. Although the natural world turned out to contain many highly specific nucleases – known as meganucleases – that would target dozens of bases and would therefore alter only one point in the sequence, they needed to be made to work in mammals, not just in the bacteria and fungi where they were originally found.[2]

In 1994, Maria Jasin and her colleagues from the Sloan-Kettering Institute in New York took a meganuclease from yeast that made a double strand break at one particular sequence of eighteen base pairs and persuaded the enzyme to function inside mammalian cells. The cells then used the sequence of a piece of template DNA that had been introduced into the cell to synthesise a new stretch of DNA at that point through homologous recombination.[3] This is the essence of gene editing, and why it is different from the various techniques of recombinant DNA, where a foreign piece of nucleic acid, along with various marker genes, is introduced more or less at random into

* For complex reasons that remain obscure, different cells in different organisms at different points in their cycle tend to preferentially use either homologous recombination or non-homologous end-joining, in which the cell simply tries to stick the two ends together. If this does not go perfectly it can lead to the deletion or addition of bases.

the genome. In gene editing, the aim is to change only the desired sequence – the template and the nucleases will soon be metabolised, leaving no trace of the intervention.

For genome editing to become practical in mammals, meganucleases needed to be precisely targeted at desired locations – they needed to be programmable. From the mid-1980s onwards cell biologists had become interested in the function of transcription factors – proteins that bind to specific DNA sequences and thereby affect how (or if) a gene is transcribed. Many of these proteins have parts that require the presence of zinc ions to form claw-like structures, poetically described as zinc fingers.[4] In 1996, in an astonishing feat of biochemistry, Srinivasan Chandrasegaran of Johns Hopkins University in Baltimore combined these molecular properties with an existing nuclease called Fok1 to create a hybrid enzyme that would home in on a precise DNA sequence encoded in the zinc finger structures (each zinc finger recognises three base pairs, so by stringing together a number of these molecules a high degree of sequence specificity can be achieved), cutting the double helix at that point.[5] These artificial zinc finger nucleases were described as chimeric, after the chimera, a hybrid animal in Greek mythology.

In 1999, Chandrasegaran foresaw the future:

> The availability of chimeric nucleases, a new type of molecular scissors that target a specific site within the human genome, will likely contribute and greatly aid the feasibility of genome engineering and, in particular, ex vivo gene therapy using stem cells.[6]

Programming chimeric zinc finger nucleases was a complex affair – Chandrasegaran later described it as tedious and time-consuming, because the relation between the shape of the zinc finger and the DNA sequence it targeted was not easy to work out.[7] Nevertheless, by 2003 these artificial enzymes had been used to get *Drosophila* cells to turn one version of a gene into another and had successfully targeted genes in human cells.[8]

The advent of programmable meganucleases held out the promise of a more precise and safer version of gene therapy as only the gene of interest would be altered. Chandrasegaran soon showed it was possible to target *CCR5*, a human gene associated with susceptibility to HIV, while researchers at Sangamo Biosciences in California successfully altered a gene involved in human severe combined immunodeficiency (SCID).[9] In 2008 Dana Carroll, who had worked with Chandrasegaran, predicted that the system would be used in gene therapy trials within five years.[10] This turned out to be pretty accurate: in 2014 the first clinical trial of zinc finger nucleases was completed by Sangamo Biosciences researchers, targeting the *CCR5* gene. The aim was to alter the gene so that it corresponded to a naturally occurring version that provides some people with a degree of immunity to HIV by preventing the virus from entering their cells.[11] Twelve patients with HIV were treated with T cells in which the *CCR5* gene had been modified by zinc finger nucleases; some of them showed reduced levels of HIV and in one case the virus became undetectable, an apparent cure for HIV through gene therapy.

But at the end of 2009 the zinc finger bandwagon was derailed just as it was starting to roll. Two groups, from Martin Luther University Halle-Wittenberg in Germany and from Iowa State University, simultaneously described a far simpler system of gene targeting.[12] The starting point was a plant pathogenic microbe that possessed 'transcriptor-like effector' (TALE) sequences that produced a protein that targeted a precise part of the plant genome. The two groups of researchers cracked the code that enabled the TALE protein to bind specifically (the titles of their papers used the buzzwords 'code' and 'cipher') and then created artificial TALE molecules that targeted new sites in the genome. One of the articles highlighted the advantages of this approach over zinc finger nucleases and predicted a great future for the system:

> TAL effectors use a DNA binding code that can be exploited to generate DNA binding domains for any DNA target.[13]

By linking TALE sequences to the Fok1 nuclease, creating a system known as TALEN (the N stands for nuclease), any gene could be targeted with relative ease, heralding a new era in genetic engineering.

So significant were these developments that a new phrase gradually emerged – the process involved was now called gene or genome *editing*. Although 'editing' already referred to various natural processes that occur inside the cell as the genetic message is processed, the term gene editing had first been used to describe the experimental introduction of deliberate changes into genes as long ago as 1980.* It had never really caught on, probably because the changes that could be introduced at the time were coarse and uncontrolled, which did not fit with the precision of the editing metaphor. In the era of programmable meganucleases that could induce very precise changes, 'editing' began to look quite apt.[14] The semantic shift was signalled by *Nature* in 2005, when Fyodor Urnov and his colleagues described the application of zinc finger nucleases to correct a SCID mutation. The *Nature* editors noticed Urnov's novel use of the phrase 'genome editing' in the article and put it on the cover of the journal, pushing the term into the mainstream. Five years later Urnov and his colleagues published a major review of zinc finger nucleases, using the words in the title. The shift in terminology was complete.[15]

Despite all the excitement around zinc finger nucleases, and their effectiveness in clinical trials, TALENs were now the way to go. They were far easier to programme, showed lower levels of toxicity and produced fewer off-target effects where the wrong piece of DNA was edited. As an example of what the new approach offered, in August 2012 a group of Harvard researchers submitted an article describing the use of TALENs to change DNA such that cells would express genes involved in hepatitis C infection, motor neuron disease and diabetes, thereby potentially enabling these diseases to be studied in cell lines, rather than in humans. The scientists confidently concluded: 'we anticipate TALEN-mediated genome editing of human cells becoming a mainstay for the investigation of human biology and disease'.[16]

It never happened.

Just one week earlier, Emmanuelle Charpentier, Jennifer Doudna and their colleagues had published an article in *Science* showing that

---

*By Arthur Riggs and Keiichi Itakura, as described back on page 167 in case you weren't taking notes.

a bacterial immune system could be programmed to easily and efficiently cleave DNA at specific locations. They accurately predicted that their discovery 'could offer considerable potential for gene-targeting and genome-editing applications'.[17] Although no one had been arguing that the world needed a better gene editing system, that was exactly what had been invented. The CRISPR revolution had begun.

*

The term CRISPR is odd. Most acronyms reveal something when their component words are spelled out, but CRISPR stands for 'clustered regularly interspaced palindromic repeats', which gives no hint of a role in gene editing and only really means anything to those with an interest in the structure of genomes. Which is not surprising, as those are the people who came up with the term.

In 2001, Francisco Mojica of the University of Alicante in Spain was exchanging emails with Dutch microbiologist Ruud Jansen, one of the few people on the planet who shared his interest in some intriguing repetitive DNA sequences that were found in the two branches of the microbial world, bacteria and archaea (these two superficially similar groups split shortly after life appeared on Earth). Japanese researcher Yoshizumi Ishino had first noticed these sequences in the bacterium *E. coli* in 1987 but could only conclude that 'the biological significance of these sequences is not known' and did not pursue the matter.[18] From 1992 Mojica, a young PhD student who studied archaea found in salty lagoons near Alicante, had become obsessed with these strange stretches of DNA. At the turn of the century, Jansen's group also noticed these odd repeats, and the two groups agreed to come up with a common terminology. They finally settled on the snappy acronym CRISPR to describe the repeated sequences, avoiding Jansen's alternative of SPIDR (SPacers Interspersed Direct Repeats).[19] CRISPR was first used in the scientific literature in 2002, in the same year that it was suggested that nearby CRISPR-associated (or Cas) genes, which appeared to encode enzymes capable of unwinding DNA and cutting it, might be involved in DNA repair in these microbes.[20]

Asunto: Re: Acronym
Fecha: Wed, 21 Nov 2001 16:39:06 +0100

```
Dear Francis

What a great acronym is CRISPR.
 I feel that every letter that was removed in the alternatives made it
less crispy so I prefer the snappy CRISPR over SRSR and SPIDR.
Also not unimportant is the fact that in MedLine CRISPR is a
unique entry, which is not true for some of the other shorter
acronyms.
```

Email from Ruud Jansen to Francisco Mojica, November 2001, approving the CRISPR acronym.

As the CRISPR acronym indicates, these repeated sequences tended to be clustered in the genome, were vaguely palindromic and were regularly interspaced with other stuff. Initially, the sole aim of Mojica and Jansen was to understand this bit of microbial genomic weirdness. That began to change in 2003, when Mojica searched the early genome databases and discovered that the DNA between the CRISPR sequences came from bacteriophage viruses, suggesting that the bacteria had acquired these sequences after viral infection – CRISPR sequences were a natural form of recombinant DNA. Mojica thought about why such sequences might be so widespread and decided that CRISPR and their accompanying interspaced bits of DNA and the Cas genes somehow represented a bacterial immune system – a way of recognising and resisting viral infection. Having gathered experimental evidence in support of this hypothesis, in autumn 2003 Mojica submitted an article to *Nature* describing his discovery. The editor was unimpressed and rejected the paper forthwith, claiming that all this was already known (it was not). It took two years and rejection by three more journals before the article finally appeared in the *Journal of Molecular Evolution*.[21] Mojica was rightly frustrated by the delays – within months of this article appearing, two other groups, working on different microbes, reported similar findings.[22]

None of these researchers were able to show exactly what these genes did, nor how the viral DNA got into the regions between the

CRISPR sequences. That was partly resolved in 2007 by a group working for Danisco, the Danish agribusiness company, led by Philippe Horvath and Rodolphe Barrangou (Danisco was interested in the phenomenon because the bacteria involved in producing yoghurt were repeatedly attacked by bacteriophages, and CRISPR might be a way of protecting them). They showed that after infection with a bacteriophage, bacterial colonies acquired new 'spacer' DNA from the virus; removing spacer DNA affected the ability of the bacteria to resist the virus.[23] As Mojica had suspected four years earlier, CRISPR was involved in acquired resistance to bacteriophages.[24]

The following year, researchers showed that bacteria turned the viral DNA trapped in their genomes into RNA and used that sequence to guide Cas enzymes to attack invading viral nucleic acids.[25] It was still unclear whether the Cas enzymes cut DNA or RNA; although most researchers suspected that the target was DNA, some favoured the idea that they blocked the activity of RNA molecules. (Craig Mello and Andrew Fire had recently won the Nobel Prize in Physiology or Medicine for their discovery of the influential RNA interference technique, which may have skewed assumptions.)

CRISPR was beginning to attract attention – there were publications in leading journals, the sequences were found throughout the microbial tree of life, which suggested that they had a major evolutionary significance, and yet the way the system worked was still unresolved. In 2008, the first CRISPR conference took place in Berkeley, attended by around thirty-five people. This might not seem very many, but Mojica, who was brought over from Spain as a keynote speaker, later recalled his excitement at seeing so many people interested in the topic he had pioneered fifteen years earlier. It was huge, he said.[26]

A few months later a hint of CRISPR's future was given when Luciano Marraffini and Erik Sontheimer of Northwestern University showed that CRISPR and its associated Cas sequences targeted viral DNA, not RNA.[27]* Marraffini and Sontheimer drew out the immense practical implications: 'the ability to direct the specific addressable destruction of DNA that contains any given 24- to 48-nucleotide

---

*Some CRISPR systems have now been discovered that do target RNA.

target sequence could have considerable functional utility, especially if the system can function outside of its native bacterial or archaeal context'.[28]

That last point was a big 'if' – it would require persuading the ensemble of CRISPR genes to target a sequence in the nucleus of a eukaryotic cell (all multicellular organisms are eukaryotes), something that had never happened in billions of years of evolution. Marraffini and Sontheimer were so sure this was possible that they submitted a patent in which they claimed CRISPR could be applied to DNA in a eukaryotic cell, including 'in a subject' – gene therapy.[29] Although the patent was later abandoned because it lacked the detail to support its claims, Marraffini and Sontheimer's article alerted scientists to the possibility that the enzymatic activity of CRISPR might be used to target not invading viruses, but any DNA sequence, in any organism. At the same time as TALENs were replacing zinc finger nucleases, CRISPR was looking like it might be a lot more significant than the mere microbial oddity some considered it to be.*

*

The final transformation of CRISPR into an unexpected revolutionary tool for genetic engineering took place in two years of furious work and intense excitement as growing numbers of researchers became involved. In 2011 the essential components of the process were described, in the shape of four key developments.[30] First, a major review was published by most of the leading CRISPR researchers – no one from Jennifer Doudna's group at the University of California Berkeley was involved, even though her interest in CRISPR went back to 2006. This article classified the different kinds of CRISPR systems found in various microbes and came up with an agreed nomenclature for the enzymes involved, clarifying concepts and focusing everyone's minds on what remained to be discovered. Second, Emmanuelle Charpentier's group – split between Vienna and Umeå in Sweden – identified an essential RNA component of the system, called tracrRNA, although its exact function remained

*Guilty.

uncertain. Then, Doudna's team, in collaboration with her colleague Eva Nogales, described the crystalline structure of the molecules involved, making possible a precise understanding of the process. Finally, a group led by Virginius Šikšnys of Vilnius University in Lithuania, in collaboration with Horvath and Barrangou, transferred the CRISPR system from one species of bacteria to another and suggested that one Cas enzyme, Cas9, was the nuclease that cut the DNA molecule. It was also in 2011 that Doudna began to collaborate with Charpentier. Doudna says that when they met for the first time, at a conference held in Puerto Rico that year, she felt a shiver of excitement as the two agreed to work together; Charpentier has not reported a similar feeling, giving an indication of their very different personalities.[31]

The global interest in this field highlights how much science had changed in the four decades since Paul Berg made his breakthrough. Unlike the discovery of recombinant DNA and gene cloning, or even the development of zinc finger nucleases, all of which were purely American affairs and initially involved a very small number of people, CRISPR was truly global. As Doudna later recalled of her work with Charpentier and their teams: 'we would be quite the international group: a French professor in Sweden, a Polish student in Austria, and a German student, a Czech postdoc and an American professor in Berkeley'.[32] Scattered across the world, the researchers discussed and communicated their results with each other via Skype and FedEx.[33]

2012 was the decisive year, as two interlinked races took place. First, there was a competition among the microbiologists, until recently the only people interested in CRISPR, to show exactly how the system worked and to demonstrate that it could be manipulated in microbes, targeting the CRISPR enzymes at any DNA sequence. That was the race won by Doudna, Charpentier and their colleagues with their paper in *Science* published online at the end of June 2012.[34]

The losers in this race were Šikšnys, Horvath and Barrangou, who saw their article published in the *Proceedings of the National Academy of Sciences* ten weeks later, even though it had been submitted before Doudna and Charpentier's paper – indeed, in April it had first been submitted to *Cell* but rejected without review (the editor has since admitted that they got that call wrong).[35] But even before that race was over, there was another competition, in which researchers who had only just caught a whiff of the potential of CRISPR decided to make the system work in what for many was the holy grail of genetic engineering – mammalian cells, and thence in humans.

The articles from Doudna–Charpentier and from Šikšnys contained very similar elements – they both showed that the ensemble of CRISPR genes directed the Cas9 enzyme to cut both strands of a DNA sequence at a point defined by an RNA sequence, which in natural conditions was that of a virus. Above all both groups showed that this sequence could be simply replaced by a sequence of interest, thereby programming the CRISPR–Cas9 system. Because both strands of DNA were cut, this could lead to homologous recombination if an additional template DNA was supplied to the cell. As with zinc finger nucleases and TALENs, the system exploited the cell's DNA repair mechanisms to persuade the cell to synthesise a novel DNA sequence at a precise location. The major difference was that programming the CRISPR system was far easier than in the two previous approaches.

The reason why Doudna and Charpentier have been fêted, while Šikšnys has tended not to be, is not the weeks that separated publication (although sadly this kind of thing does impress some scientists). Rather, the Doudna and Charpentier paper contained a significant discovery. They revealed that CRISPR activity normally involved two RNA molecules fused together – the CRISPR RNA and the mysterious tracrRNA Charpentier had described a year earlier. Following an idea of Martin Jinek, Doudna's postdoctoral researcher, they then showed it was possible to use a single RNA molecule – known as guide RNA – thereby making programming the system even more straightforward.

This reveals a second difference between the two articles. Although both papers highlighted the potential of CRISPR, Doudna and Charpentier were much more clear-sighted because their additional step using a single guide RNA molecule allowed them to fully grasp the implications of their discovery. Where Šikšnys and his colleagues concluded 'these findings pave the way for the development of unique molecular tools for RNA-directed DNA surgery', Doudna and Charpentier emphasised the programmable nature of the system even in the title of their article, and pointed to the importance of their use of a single RNA molecule, which had 'potential utility for programmed DNA cleavage and genome editing'.[36] In their final sentence they contrasted their method with those of the existing genome editing techniques:

> We propose an alternative methodology based on RNA-programmed Cas9 that could offer considerable potential for gene-targeting and genome-editing applications.[37]

One weakness shared by both papers was that they did not demonstrate that CRISPR would work in a eukaryotic cell. This was the focus of the second CRISPR race of 2012 and was briskly resolved in a matter of months as two groups from Harvard published back-to-back articles on the *Science* website at the beginning of January 2013.[38]

Feng Zhang, a young researcher from the Broad (rhymes with 'road') Institute had become interested in CRISPR in 2011 (somewhat alarmingly, Zhang was inspired to work in genetic engineering after watching *Jurassic Park* as a teenager[39]). With his colleagues, Zhang showed that CRISPR could work in human and mouse cells. He did this by slightly altering the structure of the guide RNA molecule, substantially improving its activity. The other paper was published by George Church's group, which had a long-standing interest in genetic engineering (Church had been Wally Gilbert's PhD student in the 1970s).[40] By the end of January 2013 Doudna's group had published its own, slightly less effective version of CRISPR editing in mammalian cells, as had the Korean group of Jin-Soo Kim.[41] Doudna later recalled the ease with which the final step was taken as she worked with Martin Jinek:

> In just a few simple and routine steps, Martin and I had selected an arbitrary DNA sequence within the 3.2-billion-letter human genome, designed a version of CRISPR to edit it, and watched as the tiny molecular machinery followed through with its new programming – all inside living human cells.[42]

Within a decade CRISPR had been transformed from a strange piece of bacterial genomics to an astonishingly powerful and flexible system for editing genes in any organism. A piece of pure research, which initially interested only a handful of unknown researchers, was set to transform science and medicine.

*

In the year following the breakthrough papers of 2012 there were over 250 scientific articles published on CRISPR as researchers rushed to show how it could have a massive impact on fundamental, applied and medical science – it was used to create mutant zebrafish and flies, to manipulate genes in pigs, cows and goats, to change human stem cells and to repair disease-causing genetic errors in laboratory cultures of human cells.[43] This explosion of activity had an effect on the industries that had grown up around the nascent world of gene editing. Cellectis, a French biotech company that had been selling bespoke TALENs to researchers, noticed that in early 2013 sales growth began first to flatten and then to sharply decline as researchers turned to the far simpler and cheaper CRISPR system. According to the founder of the company, the arrival of CRISPR soon hit their business like a tsunami.[44]

To get some idea of the scale and pace of this change, consider that at the end of 2013, Zhang's group reported they had used CRISPR to target 18,080 human genes (about 80 per cent of the total) one by one in a cell line, using 64,751 unique guide sequences, in order to explore the effects of the loss of each and every one of those genes on resistance to a leading cancer drug. This is just staggering. This feat would have taken years and years using either zinc finger nucleases or TALENs, and almost certainly would never have been attempted.[45]

Probably the most bittersweet of these early studies was carried out by those Harvard researchers who, a few months previously, had published the results of their laborious application of TALENs to create cellular models of key diseases. Now they compared CRISPR, which they had mastered in a matter of weeks, with the hard-won results of their previous study. As they reported in a brief article, the effectiveness of TALENs varied from 0 to 34 per cent, while CRISPR

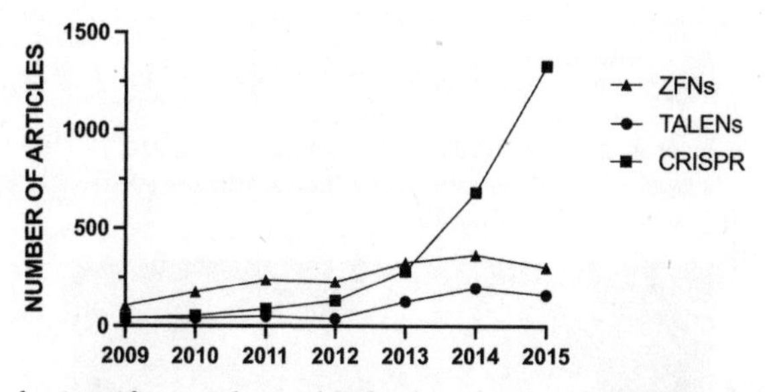

Number of scientific articles published each year on ZFNs, TALENs and CRISPR, 2009–2015. By 2021 the figure for CRISPR was over 7,000.

had an effectiveness of 51–79 per cent.[46] The data did not lie. In 2019, the lead researcher on this project recalled:

> Effectively I was conceding that all of the work we had done with TALENs was now obsolete and should be ignored, and that everyone should start using CRISPR–Cas9 instead … my laboratory never used TALENs again.[47]

In the middle of 2013 *Science* magazine announced that scientists were gripped by the CRISPR craze.[48] They still are.

From 2014 the number of publications on CRISPR far outstripped those on those gene editing has-beens, TALENs and zinc finger nucleases, increasing exponentially and passing 7,000 papers in 2021. Exactly as hoped, CRISPR proved a game-changer, allowing genetic manipulation of any organism. The precise details of the system continue to evolve as researchers explore the huge range of CRISPR systems to be found in nature, searching for new enzymes or new hints of how even greater ease and precision might be obtained. Hoping to imitate the influence of the CRISPR acronym, researchers have come up with dozens of acronyms to describe techniques involving CRISPR. My colleague Antony Adamson of the University of Manchester Genome Editing Unit has been tweeting a list of these sometimes frivolous acronyms, which range from SPAMALOT to CASANOVA to FUDGE. At the time of writing the thread contains over 100 entries.[49]

The explosion in the number of research groups using CRISPR

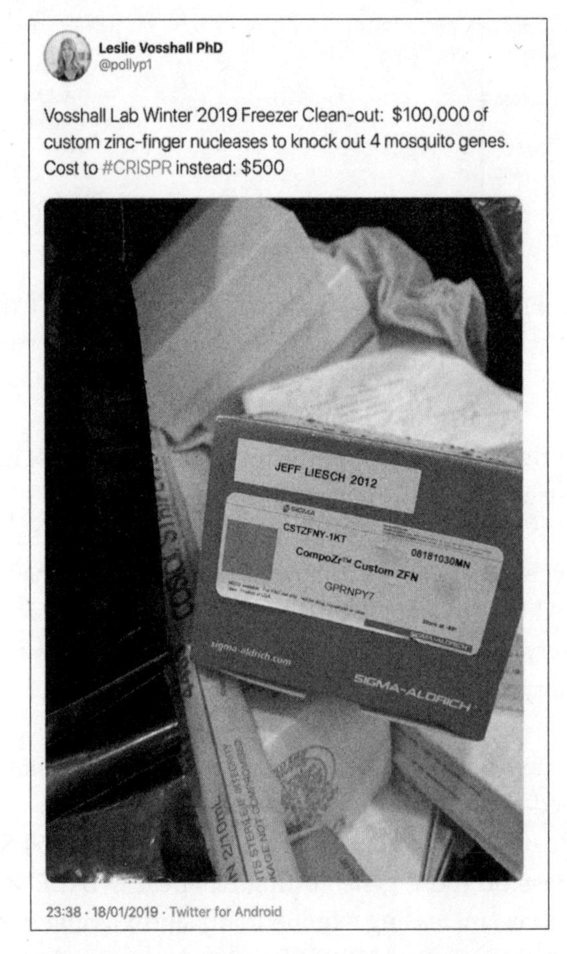

Tweet by Professor Leslie Vosshall of Rockefeller University, January 2019.

is not only due to its ease and widespread application, but also to its cost. At the beginning of 2019, my good friend the neurobiologist Leslie Vosshall organised a clear-out of a laboratory freezer and tweeted a photo of a pack of custom-made ZFN enzymes she had purchased in 2012 to alter mosquito genes at a cost of $100,000. The equivalent cost for CRISPR was $500.[50]

As many people have pointed out, CRISPR has democratised gene editing – it is now within the reach of every laboratory. There is now no financial or biological obstacle to any research group manipulating the genes of its study organism. If you know the genetic sequence you want to alter, and you can rear the organism in the laboratory and deliver the CRISPR components to the relevant tissue, you can genetically alter any organism.

My personal favourite piece of research from this period was the appearance in *Cell* – a hugely influential journal devoted to the hottest discoveries in cell biology – of the use of CRISPR on two species of ants (this was the only research article *Cell* has ever published on ants).[51] Using CRISPR to knock out the ants' olfactory receptors, the researchers revealed the unsuspected role of olfactory neurons in shaping the structure of the ant brain. Previously, studies in *Drosophila* had persuaded scientists that the olfactory parts of the insect brain were largely hard-wired because knocking out the fly's olfactory receptors had no effect on brain structure; the scrambled brains of the CRISPRed ants revealed that what is true in *Drosophila* is not necessarily true of all insects. There will be many similar instances in the future – by allowing scientists to investigate all sorts of organisms with interesting ecologies, rather than the usual laboratory suspects of bacteria, worms, flies, zebrafish and mice, CRISPR will lead to surprising new discoveries.

All this work was made possible because pharmaceutical companies big and small leapt onto the bandwagon, selling CRISPR kits. The commercialisation of CRISPR also saw the creation of a series of spin-off companies, sometimes involving the key researchers. First out of the blocks was Jennifer Doudna, who in 2012 set up a company called Caribou (a mash-up of 'Cas' and 'ribonucleotides') with her former student Rachel Haurwitz. The aim of the start-up was to use the Cas6 enzyme to diagnose viral infections such as HIV and hepatitis C. This was before the CRISPR craze and they found little enthusiasm among the Bay Area venture capitalists, so they set up the company with their own money.[52] It is still going strong and in 2021 went public, raising $300 million.

In 2013 an attempt was made by the Broad Institute and the University of California Berkeley to create a common CRISPR commercial structure, in the shape of Gengine Inc., which soon changed its name to Editas Medicine. Backed by a number of US venture capital companies, the CRISPR supergroup soon split due to irreconcilable financial differences when Doudna discovered that Zhang and the Broad had paid a few hundred dollars extra to fast-track their patent for the use of CRISPR in mammals, stealing a march on a more ponderous and cheaper claim by herself and Berkeley. Soon the key inventors of CRISPR gene editing were involved in three different companies – CRISPR Therapeutics (Charpentier and Novak), Editas Medicine (Zhang and Church), and Intellia Therapeutics (Doudna, Barrangou, and the two men who had first sniffed the commercial potential, Sontheimer and Marraffini).

By the beginning of 2017 it was estimated that over $1 billion had been invested in various CRISPR biomedical start-ups.[53] So far, the activity of all those companies revolves around potential. No CRISPR health products have yet been brought to the market, not surprising given the long path from discovery to application and the strictures of regulation. A new drug takes on average fourteen years to develop, with a success rate as low as 1.3 per cent.[54] That zinc finger nuclease treatment for HIV, shown to be effective in a clinical trial in 2014, has still not become an approved therapy.

The issue of who owns the patent rights to CRISPR is becoming particularly acute. Although attention generally focuses on the US patent dispute between Zhang and the Broad Institute/Harvard/MIT on the one hand, and Doudna and Berkeley on the other, this is merely one part of the vast range of complex legal squabbles in the United States and around the world.[55] In 2016 a total of 591 different CRISPR components were claimed as inventions, including 212 related to guide RNA (there are tens of thousands of guide RNA libraries in laboratories around the world; only some of these are considered either original or worthwhile enough to patent). The US Patent Office is currently dealing with about 6,000 CRISPR patent applications, with 200 being added to the list every month. Only one-third of those applications are from commercial companies; the remainder are from universities and research establishments.[56] Nevertheless, as in previous rounds of biotech excitement, these patents can pass from company to company as takeovers take place – DuPont recently became a big player in the world of CRISPR patents, simply by buying Danisco.[57]

At the time of writing, the central US dispute between Berkeley and Broad/Harvard/MIT seems to have been won by the east coast consortium while a joint claim by Doudna and Charpentier has been accepted in the United Kingdom, Europe, China, Japan, Australia and a number of other countries. However, as long as the researchers' institutions have money to throw at their lawyers, it seems unlikely that the issue will be resolved any time soon. In 2021, *Nature* called for universities around the world to find ways of allowing patented CRISPR technology to be accessed free of charge for research purposes.[58] That may eventually happen, but it is hard to imagine any organisation, anywhere, giving up the potential riches that may flow from commercial applications.

As with all previous discoveries relating to genetic engineering, the final resolution of the patent disputes will probably not stifle scientific discovery – researchers will still be able to use the method freely, with fees required for commercial exploitation. In 2021, Wageningen University & Research in the Netherlands announced that they were waiving their CRISPR patents for non-profit organisations to develop non-commercial food and agricultural products. It seems that CRISPR has simply gone too far for the gates to suddenly come crashing down.[59] Furthermore, biologists have a long tradition of sharing resources and reagents, in particular once they have been published. Perhaps surprisingly, that has applied to the potentially lucrative area of CRISPR research too. Addgene is a non-profit repository that, for a small fee, distributes plasmids containing various genetic constructs, in particular CRISPR. Set up in 2004 by Melina Fan, her husband and her brother, it really took off in 2012, when the first CRISPR plasmids were deposited. Describing itself as Amazon for plasmids, in the four years following the breakthrough publications of 2012 Addgene distributed 60,000 CRISPR plasmids to 20,000 laboratories in eighty-five different countries.[60]

The financial whirlwind that has swept through the world of CRISPR is interacting with the rapid pace of scientific discovery and its communication in a way that was not present during earlier bouts of speculative excitement regarding recombinant DNA or gene therapy. For example, on 5 January 2018 a preprint (an unreviewed manuscript) appeared showing that 65 per cent of people have antibodies to Cas9 and warning that if CRISPR were to be used in humans it might result in significant toxicity.[61] On this news, shares in the companies set up by the CRISPR inventors – CRISPR Therapeutics,

Editas Medicine and Intellia Therapeutics – fell by around 10 per cent. Five days later, on 10 January, another preprint appeared from a different research group, describing a cunning way to circumvent this problem.[62] The shares bounced back up.

Whatever the technical obstacles that lie in the road to CRISPR being used for practical purposes, there is enough money floating about and there are enough very smart people in the field to ensure that, eventually, those problems will be resolved.

*

Despite the excitement in the scientific community from 2012 onwards, global media outlets were less savvy – it was not until November 2013 that the general public got any inkling that something was happening. The first newspaper coverage came in the United Kingdom, from Steve Connor in *The Independent*; other UK newspapers from *The Times* to *The Sun* soon picked up the story, followed by the *Boston Globe*, which inevitably had a Harvard-based angle.[63] A few weeks later, *Le Monde* published an article by two French geneticists explaining the significance of the technique, opening with an apt quote from Sydney Brenner: 'Progress in science depends on new techniques, new discoveries and new ideas, probably in that order.'[64] * In the United States, the mainstream media were largely silent about the excitement that was agitating the world of science, even though *Science* declared CRISPR to be a runner-up in a list of the breakthroughs of 2013 (the winner was cancer immunotherapy).[65] †

---

* In fact, Brenner did not quite say this, and he spent some time tracking down what exactly he did say where and when. As with 'Play it again, Sam', 'Alas, poor Yorick, I knew him well' and so on, the inaccurate popular version is better. Brenner, S. (2002), *Genome Biology* 3:comment1013.1–1013.2.

† *Science* magazine's 2012 Breakthrough was the detection of the Higgs boson at CERN; CRISPR was not even a runner-up, overshadowed by the application of TALENs – luridly described as 'genomic cruise missiles' – to pigs. CRISPR finally became *Science*'s Breakthrough of the Year in 2015, three years after … the breakthrough. In a simultaneous public vote, CRISPR was beaten by the New Horizons mission to Pluto. Anonymous (2012), *Science* 338:1525–32; Travis, J. (2015), *Science* 350:1456–7.

In March 2014 – nearly two years after the key experiments were performed – the *New York Times* finally noticed the scientific frenzy unleashed by Doudna and Charpentier's invention, revelling in the 'editing' metaphor: 'a genome can be edited, much as a writer might change words or fix spelling errors'.[66] The *Washington Post* took even longer to prick up its ears – the first mention of CRISPR was in November 2014, when it made a brief appearance as one of '3 breakthrough science ideas you'll be talking about in 2015'.[67] Even the *San Francisco Chronicle* – Doudna's local paper, which had closely covered the recombinant DNA revolution in the 1970s – did not produce a detailed article until September 2014.[68]

The same was true for other global media – the BBC website began to have in-depth coverage on CRISPR only in 2015, while the first BBC programme devoted to CRISPR, *Editing Life*, was broadcast on the World Service and Radio 4 at the beginning of 2016 (it was written and presented by me and produced by Andrew Luck-Baker).[69] Television took even longer to catch up – in 2019 Netflix released a three-episode mini-series on CRISPR, *Unnatural Selection*, while the BBC broadcast *The Gene Revolution: Changing Human Nature* a few months later, seven years after the experiments by Zhang, Church, Doudna and Kim showed that human genetic engineering with CRISPR was possible.

From the earliest days, media and scientific attention focused on Doudna and Charpentier as the key figures in the CRISPR revolution, even though they worked together for less than two years. When Steve Connor broke the story in *The Independent* – over fifteen months after the 2012 *Science* paper was published – his article was accompanied by a large photo of Doudna, while in 2016 *Le Monde* described the pair as the Thelma and Louise of biology.[70] In the four years following their 2012 publication they received over thirty awards and honorary degrees, to the extent that they were beginning to lose count.[71] The prizes have kept on coming, but some have been more memorable than others. In 2015 they won the Breakthrough Prize worth $3 million; in 2018, together with Šikšnys, they received the Kavli Prize worth $1 million; and in October 2020 they achieved what most would consider to be the pinnacle of any scientific career – the Nobel Prize.

Although it is generally invidious to give scientific prizes to a small number of people, and particularly so in the case of CRISPR, given how many scientists made significant contributions, the scientific community appeared to be pleased at the award, partly because women have so rarely been recognised by the Nobel committee. Accounts of the history of CRISPR have appeared in books, both personal (Doudna's 2017 memoir *A Crack in Creation* and Walter Isaacson's 2021 biography of Doudna, *The Code Breaker*) and in many popular science books (if you can only read one it should be Kevin Davies's 2020 *Editing Humanity*).[72]

At the beginning of 2016 an apparently arcane academic squabble over CRISPR history exploded onto the pages of the world's newspapers and their websites. The flashpoint was an article in *Cell*, 'Heroes of CRISPR', by Eric Lander, the head of the Broad Institute.[73] Although Lander was meticulously even-handed in his coverage of the early years of CRISPR (he mentioned many more people than I have here), when it came to the decisive steps – indeed, when it came to defining what those steps were – Lander caused a scandal. Many readers considered that he systematically downgraded the contributions of Doudna and Charpentier, while emphasising those of Broad Institute researcher Feng Zhang. In particular, Lander focused on Zhang and Church's use of CRISPR in mammalian cells as being the key breakthrough, rather than the demonstration that CRISPR could be turned into a programmable gene-editing system.

A Twitter storm erupted; blogs were written – some vituperative ('The villain of CRISPR'), others considered ('A Whig history of CRISPR') – emphasising flaws in Lander's description, starting with his failure to declare his institution's interest, both scientific and financial, in Zhang's work, in particular at a time when there was a furious US patent dispute between the Broad Institute and Berkeley.[74] A more subtle bias was revealed in a map Lander included in his article, showing the location of the various research groups. In Lander-world the Atlantic Ocean had mysteriously shrunk, leaving Boston at the centre of the planet, the focus of what Lander saw as the ultimate victory, with all the other locations subliminally presented as peripheral participants. The Korean laboratory of Jin-Soo Kim was not even included.

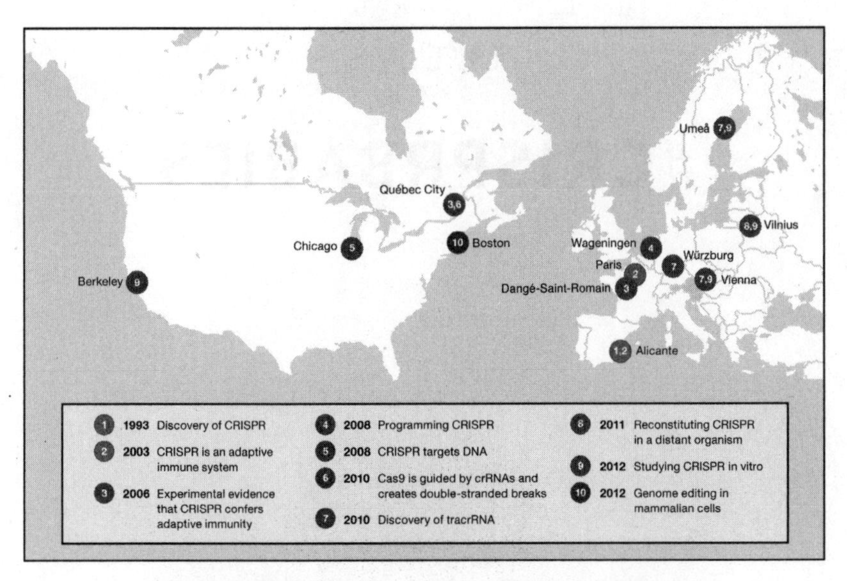

Map from Eric Lander's 2016 article in *Cell*, 'Heroes of CRISPR', showing how the history of CRISPR 'unfolded across twelve cities in nine countries'.

Twitter had a field day before rapidly moving on to something else, while the mainstream media easily mined the fracas to produce cheap articles, but there was a serious point at stake. Lander's one-sided account (his view was sharply, if succinctly, criticised by Doudna, Charpentier and Church[75]) was clearly aimed at shaping views of how CRISPR came to be. His motivations were less obvious – it is not clear that the Nobel Prize committee are influenced by such things (if that was the intention, it clearly failed) and patent offices are equally not renowned for listening to academic histories. Nevertheless, as George Orwell put it in *Nineteen Eighty-Four*, 'Who controls the past controls the future', and it looked like Lander had been attempting to do exactly that.

But when it came to dystopias, CRISPR soon conjured up not the bleak dictatorship of *1984* but the genetic apartheid of *Brave New World*. In 2015 it became apparent to scientists and the general public that CRISPR was so simple that it would almost inevitably be used in humans. The genie of heritable human gene editing was pushing its way out of the bottle.

– TWELVE –

#CRISPRBABIES

In spring 2014, Jennifer Doudna had a nightmare. In her dream she was asked to explain to someone how gene editing worked; the someone turned out to be Adolf Hitler with a pig's face. He was sitting there, pen poised, ready to take notes:

> Fixing his eyes on me with keen interest, he said, 'I want to understand the uses and implications of this amazing technology you've developed.'

She awoke with a jolt, her heart pounding.[1]

Although the immediate cause of the nightmare might have been too much cheese before bed, the underlying reason was straightforward – scientists had recently taken a decisive step towards using CRISPR to edit human embryos. A few weeks earlier, in February 2014, Chinese researchers led by Jihao Shah had explained in *Cell* how they had used CRISPR to edit three genes in single-cell monkey embryos.[2] Although the researchers claimed that the procedure was efficient and reliable, with no unintentional edits elsewhere in the genome, the details of the experiment revealed the difficulties involved. In total 186 embryos were injected; of these, eighty-three were implanted into twenty-nine surrogate monkeys, leading to only ten pregnancies and only one successful birth of two monkeys, cutely called Lingling and Mingming.[3] Furthermore, only two of the three desired mutations were actually induced and both the monkeys were mosaics – not all their cells had the same versions of their genes. Some cells were edited, others were not. CRISPR is described as 'editing', but this looked like someone had been using a word processor while wearing boxing gloves.

Mosaicism can be benign – many people who have different coloured eyes are mosaic (not David Bowie) – but it can also have extremely serious consequences, such as in congenital heart disease.[4] Without adding some genetic marker, reliably identifying mosaic embryos is impossible; you would have to sequence the genome of every cell in the embryo, which would involve destroying it. Checking the efficacy of CRISPR in somatic cells, manipulated outside the body, is relatively straightforward – those cells are not unique and precious, and some can be sacrificed to measure the frequency of mosaicism. The same cannot be said of an embryo. Given our ignorance of the consequences of mosaicism, for clinical purposes a mosaic frequency of zero would be necessary for germline editing to be truly safe in humans.

The reasons why those two monkeys were mosaics reveals the biochemical complexity of CRISPR that lurks behind the simplicity of the 'editing' metaphor. Although the editing occurred when the embryo was a single cell, a zygote with only one paired set of chromosomes, the zygote is a dynamic, structured place. Different areas are already destined to do different things, and above all the cell is in motion and, compared to the tiny molecules involved in CRISPR, vast. The CRISPR components have to rapidly find the chromosomes, get access to the bases which are on the inside of the double helix and then screen billions of base pairs to find their unique target. And they have to do this twice, once for each chromosome that contains the site you want to change. Meanwhile, the cell, which does not like having its DNA chopped up and wants to fix things, to repair the things that have been broken, is also obliviously and ineluctably intent on turning into a foetus. The targets are moving in time and space.

Despite these major problems, this first use of CRISPR in a

primate had evident implications for the potential to change human genes. The day after Doudna learned of the experiment she mused to her husband over breakfast: 'How long will it be before someone tries this in humans?'[5] That vague questioning soon became a central focus of Doudna's life as she became increasingly drawn into ethical discussions around her invention:

> By mid-2014, I was concerned that CRISPR–Cas9 would be used in a way that was either dangerous, or perceived to be dangerous, before scientists had communicated enough about it to the wider world. I wouldn't have blamed my neighbours or friends for saying, 'All this was going on and you didn't tell us about it?'[6]

All that took time, and in the major review article on CRISPR by Doudna and Charpentier that appeared in *Science* in December 2014, germline editing was not mentioned.

*

Behind the scenes, events were moving fast. At the end of 2014, there were rumours that an article describing the use of CRISPR on human embryos was being submitted to leading journals. Both *Science* and *Nature* eventually rejected the paper, apparently on ethical grounds, and for a few months knowledge of the experiment was restricted to the charmed circle of editors, reviewers and their confidants. But in March 2015, even before the study that was causing all the disquiet was accepted by a journal, the rest of the world heard about it. First there was an article in *MIT Technology Review* by Antonio Regalado, who has a nose for a scoop.[7] According to Regalado, scientists in several groups around the world were working on human germline engineering. He also revealed that, at Doudna's initiative, a dozen or so Californian scientists and legal scholars had met in a hotel in Napa Valley at the end of January to discuss the apparently inevitable prospect of someone, somewhere, using CRISPR to change the human germline.[8]

Then, a week later, an opinion article appeared in *Nature*, signed by Edward Lanphier and Fyodor Urnov of Sangamo Biosciences, the company that had developed the therapeutic use of zinc finger nucleases.[9] The title was clear: 'Don't Edit the Human Germ Line'. Lanphier and Urnov's hostility to germline modification was focused on the problems associated with using CRISPR in humans. They swept aside the ethical arguments that had preoccupied the field for decades, highlighting the immediate issue – the safety and reliability of gene editing:

> Philosophical or ethical justifications of this technology – should any ever exist – are moot until it becomes possible to demonstrate safe outcomes and obtain reproducible data over multiple generations. ... At this early stage, scientists should agree not to modify the DNA of human reproductive cells. Should a truly compelling case ever arise for the therapeutic benefit of germline modification, we encourage an open discussion around the appropriate course of action.

A week after this, the Napa group (they described themselves as 'interested stakeholders') published a statement in *Science*.[10] With David Baltimore and Paul Berg as the fortuitous lead authors (they were listed in alphabetical order), the declaration – tellingly entitled 'A prudent path forward for germline engineering and germline gene modification' – inevitably invited comparisons with the 1974 letter signed by Berg and Baltimore that called for a moratorium on recombinant DNA research. But unlike the Berg letter, and unlike the Lanphier and Urnov article in *Nature*, the Napa statement preferred to merely 'strongly discourage' any clinical applications while international discussions were continuing to chart a 'prudent path' towards an outcome that was never seriously questioned. Germline editing was going to happen and the role of scientists and others was to outline the way towards that goal, it seemed.

On 1 April 2015, the much-whispered article on the use of CRISPR in human embryos finally appeared in the unprestigious journal *Protein & Cell*, a mere two days after it was submitted (the journal's editor claimed the decision to publish was made 'with extraordinary care, consideration and deliberation'[11]). The research had been carried out by a group of Chinese researchers led by Huang Junjiu and described the use of CRISPR to cleave the β-globin gene in single-cell human embryos.[12] Things were not as bad as some people had feared – the research was done on non-viable embryos that are inadvertently produced in routine IVF procedures when two sperm fertilise the same egg. There were no CRISPR babies. For the moment.

Huang's project had begun three months after the publication of the monkey study early in 2014 and had taken about six months to complete.[13] Given the widespread applicability of CRISPR, it was not surprising that it worked in human embryos, but the efficiency was alarmingly low for a potentially clinical procedure. Of eighty-six injected embryos, seventy-one survived the process, but only twenty-eight showed CRISPR cleavage of the target gene. Furthermore, many of these embryos were mosaic, there were off-target effects and sometimes new mutations were mistakenly introduced, affecting up to 63 per cent of the embryos. The article said not one word about the ethics of human embryo manipulation, merely emphasising the need for improved fidelity and specificity before any clinical application.

Despite these substantial problems, many researchers and commentators responded as though the era of human genetic engineering had finally dawned. In May, *Nature Biotechnology* solicited the views of leading figures in the world of gene editing, under the grandiose title 'CRISPR Germline Editing – The Community Speaks'.[14]* Most of the contributors agreed with Craig Venter, who idly blended techno-futurism with Nietzsche:

> I think that human germline engineering is inevitable and there will be basically no effective way to regulate or control the use of gene editing technology in human reproduction. Our species will stop at nothing to try to improve positive perceived traits and to eliminate disease risk or to remove perceived negative

* This phrase implied, first, that there is a 'community' of CRISPR scientists with some common, unique and important insights, and second, that the prominent individuals invited to give their views somehow represented that community. Both these suggestions are contestable.

traits from the future offspring, particularly by those with the means or access to editing and editing technology.[15]

Similar views were later expressed by Huang Junjiu himself, using steely Maoist rhetoric rather than Venter's free-market genebabble. For Huang, gene editing of the human genome was 'a necessity of history. … The development of science and technology is unavoidable.'[16] When scientists muse about the political and sociological consequences of their work, the results are often unedifying. This was no exception.

Other commentators claimed that the potential therapeutic benefits completely outweighed any other considerations. Psychologist Steven Pinker took to the pages of the *Boston Globe*, telling bioethicists to 'get out of the way' of biomedical research in general and gene editing in particular, warning hyperbolically that even a one-year delay in implementing an effective treatment 'could spell death, suffering, or disability for millions of people'.[17] Henry Miller – a one-time physician and long-standing and determined opponent of regulation on any genetic technology – presented a similar view in the letters page of *Science*, leading with a muscular clarion call: 'Germline Gene Therapy: We're Ready'. Miller dismissed the idea that Asilomar might provide a model for discussions over editing the germline – he described Asilomar's legacy as 'stultifying process-based approaches to regulation' that 'plagued genetic engineering research'.[18] Miller also took a sideswipe at Bob Pollack, the man who had set in train the debates leading to Asilomar through his telephone call to Paul Berg in 1971. Pollack had also written a letter to *Science*, opposing germline editing because of his hostility to eugenics.[19] Miller dismissed Pollack's worries as abstract concerns that counted for nothing when compared to the suffering of patients with genetic diseases. This kind of rhetorical move assumes that the immense safety issues in genetic manipulation can be easily resolved and above all fails to explain why germline modification for genetic diseases should be prioritised over finding a solution to any other form of suffering. Some more lucid physicians, well used to looking into the eyes of despairing parents, can resist the temptation to use their patients' misery to promote their pet therapy. French physician

Alain Fischer, who carried out gene therapy on 'bubble babies', is a firm opponent of germline modification under any circumstances.[20]

Opposition was also expressed by the Society for Developmental Biology, which highlighted that Huang's article reported alarmingly high rates of off-target mutations and called for 'a voluntary moratorium by members of the scientific community on all manipulations of pre-implantation human embryos by genome editing'.[21] At around the same time, the Comité d'éthique of the French medical research body, INSERM, insisted on maintaining the existing national ban on modification of the germline and opposed any change to the law.[22]

Other researchers teased apart the unstated implications of the growing acceptance of the prospect of germline editing. US bioethicist Annelien Bredenord rejected the idea that nothing could be done to prevent this outcome: 'I would prefer not to use the word "inevitable" because in the end it would be a consequence of human decision-making', while cell biologist Luigi Naldini pointed out that many embryos would have to be created in order to find the precise one that was correctly modified.[23] Any readers who have gone through IVF will flinch at the idea of harvesting hundreds of eggs. Away from the fevered dreams of the enthusiasts, gene editing in humans did not seem quite such an attractive prospect.

Doudna's view was resigned. She felt that it was impossible to control the technique she had helped create and that all that could be done was to prepare for the fallout. When I interviewed her at the beginning of 2016, she expressed her fears in terms that were remarkably prescient:

> I feel a little bit worried I'm going to wake up some morning and open up Google News or something and read about the first CRISPR baby being born somewhere in the world. And if that happens, or maybe *when* it happens, what is going to be the public perception of that, what are going to be the follow-on discussions that happen in government?[24]

The problem was that there was no international agreement over whether such a procedure would be legal, never mind ethical; there was not even a framework for discussing the issue.[25] Each country had its own regulatory approach, or lack thereof. For example, Japan and India had many IVF clinics but no enforceable rules on germline modification. In the United States, federal funds and facilities could not be used to manipulate embryos, but there was no federal law to prevent a private individual or institution from carrying out germline modification. And as to China, Qi Zhou of the Chinese Academy of Sciences said: 'The truth is, we have guidelines but some people never follow them.'[26]

*

The conclusion of this initial round of discovery and discussion was an International Summit on Human Gene Editing, held in Washington DC on 1–3 December 2015. Chaired by David Baltimore, the Summit was organised by the Royal Society, the US National Academies of Science, Engineering and Medicine (NASEM), and the Chinese Academy of Sciences. Although the organising committee was US-dominated, including Baltimore, Berg, Doudna and Lander, two Chinese researchers and three Europeans were also involved. Attended by over 500 people from around the world, with another 3,000 watching online, the Washington Summit was smoothly organised and featured few pointed arguments, in marked contrast to the chaotic and amateurish squabbling of Asilomar forty years earlier. And instead of ruling out ethical and political discussions, the Washington Summit encouraged debates on societal implications, including three sessions on governance and a presentation from a patients' group.

Far from the worries that dominated Asilomar – fears that flowed from the potentially disastrous consequences of the new technology and the threat of heavy-handed legislative regulation – parts of the Washington meeting had an almost evangelical tone. Opening the meeting, Baltimore declared:

> We could be on the cusp of a new era in human history. Today, we sense that we are close to being able to alter human heredity. Now we must face the questions that arise. How, if at all, do we as a society want to use this capability?[27]

The most straightforward discussions at the Summit revolved around somatic cell modification – gene therapy using CRISPR or other editing approaches. These techniques raised few ethical or safety issues, although even here there was the risk that the editing might not be as precise as expected. The real focus of the meeting was the clinical use of germline modification, examining its potential for treating various genetic diseases and debating under what circumstances it might be permissible. Nevertheless, some of the participants desired a very different kind of application. For example, my University of Manchester colleague, the bioethicist John Harris, gave full rein to his transhumanist dreams of genetic enhancement:

> What is clear is that we will at some point have to escape both beyond our fragile planet and beyond our fragile nature. One way to enhance our capacity to do both these things is by improving on human nature.[28]

There was little agreement at the meeting on whether enhancement of any kind was either desirable or feasible. Our current ignorance of human genetics suggests that such ideas will remain science fiction for centuries. Furthermore, if the long-term future of humanity is the issue that genetic enhancement is supposed to solve, we have far more pressing problems to deal with.

The presence of Chinese scientists at the Summit revealed significant differences in ethical stances, in particular with regard to the status of the embryo. While for decades the United States had been embroiled in a culture war over the legal status of stem cells and embryos, some Chinese researchers were bemused by the suggestion that any consideration should be paid to embryos or foetuses – the bioethicist Renzong Qiu explained that: 'According to Confucius, human being is only after birth.'[29] Even more profoundly, when sociologist Ruha Benjamin raised the growing awareness in the West of disability rights and explained that some congenitally deaf people would be extremely hostile to the prospect of their deafness being edited out of existence, Jinhua Cao, of the Chinese Academy of Science's Bureau of International Cooperation expressed surprise: 'That is too far away from a normal way of looking at things in

China. It is good for us to learn that perspective; nevertheless, that goes too far.'

The involvement of a patients' group in the Summit was a significant and positive move, a legacy of the activist campaigning of patients during the AIDS crisis. Patients have unique insights into their condition and can sometimes overcome institutional and intellectual inertia, encouraging – and even financing – underdeveloped research areas. But patient pressure can skew the debate towards the fate of individuals, which carries an emotional charge that cannot be matched by the cold language of population-level health statistics. This happened in Washington when Sarah Gray addressed the Summit from the floor. Her son Thomas had died of a genetic disease only six days after being born, his miserably brief life wracked by agonising seizures. Speaking through sobs she implored the scientists to act: 'If you have the skills and the knowledge to eliminate these diseases, then frickin' do it!'[30]

Emotion, while understandable and even essential in such debates – lives are at stake – makes for a poor policy guide. Some genetic conditions cannot be eliminated by germline gene editing because they are the product of new mutations that occur during the creation of egg and sperm, or in the case of recessive conditions because they lurk unsuspectedly within the parents' genome and are only detected when it is too late. Germline gene editing will never prevent all genetic disease.

Despite – or perhaps because of – the breadth of the debate at Washington in comparison to Asilomar, the outcome of the Summit was far less precise than at its predecessor. No stringent safety criteria were adopted and no call was made for a suspension of all clinical applications. Instead, the Summit simply reiterated the argument in the Napa letter that it would be irresponsible to go ahead with clinical application of germline editing until unspecified safety issues had been resolved and there was an undefined broad societal consensus on the question. Regulation was seen as a national issue, and there was a clear sense that those at the Summit felt they did not have the legal or moral authority to impose their views on a scientific community that was truly global and which involved tens of thousands of people.

Perhaps as a result of the recognition that there were many different national contexts that needed to be considered, the references to safety were vague, full of imprecise adjectives that could be interpreted in all sorts of ways, and lacking clear operational definitions of what safety might involve. The only binding proposal that emerged from the Summit was to meet again, with the aim being to create an 'international forum to discuss potential clinical uses of gene editing' that could inform governments, which alone could decide how to regulate their national affairs.

*

Many people drew the obvious parallels between the discussions of germline editing and the debates about recombinant DNA that took place around Asilomar. The Napa letter that effectively launched the Washington Summit suggested that the key feature of Asilomar had been the adoption of transparency and open public debate, while Jennifer Doudna, writing in *Nature* on the eve of the Summit, called on the genome-editing community 'to renew its commitment – which began more than 40 years ago – to wholeheartedly engage with the public'.[31] Although this rosy vision of Asilomar was widely shared by scientists, historians took the opportunity to revisit the events of the 1970s and concluded that the lessons of the past were rather different.[32]

For Shobita Parthasarathy of the University of Michigan, Asilomar 'was far too limited in terms of both its participants and its scope. As a result, it missed opportunities to anticipate and address emerging concerns related to the ethical, social, and economic concerns of biotechnology.'[33] Her critique continued:

> By failing to engage with the social, economic, and ethical issues surrounding recombinant DNA research and applications, the conference set a precedent for treating such issues as 'outside the scope of regulation'.

Despite Doudna's claim, the public became involved in debates over recombinant DNA because they were suspicious of the reassurances from the scientists that emerged from Asilomar, not because the scientists invited them in. As Parthasarathy pointed out, in contemporary debates over germline gene editing the contributions of non-scientists needed to take centre stage:

> To produce politically legitimate policymaking for CRISPR/Cas9, then, citizens must be engaged and treated as equals in the discussion. ... rather than serving as a model for governing emerging science and technology, the 1975 Asilomar conference offers us an important cautionary tale.

This view was echoed by Harvard academic Sheila Jasanoff and her colleagues, who explored the possibility that germline editing might increase inequality, for example in health access:

> The research community acknowledges the unfair distribution of health resources but tends to shrug it off as someone else's problem. ... Scientists at the frontiers of invention do not see it as their responsibility to address even the most obvious equity issues, such as whose illnesses are targeted for intervention or when money should be directed from high-cost individualised treatment to lower-cost public health interventions.[34]

For biologist and bioethicist Benjamin Hurlbut, the focus on safety that characterised Asilomar and afterwards had three major consequences. The public was effectively denied a role in complex and technical discussions, scientific participants presented themselves as the appropriate community to decide such matters, and political and social questions about future benefits and threats were transformed into technical issues.

This was all true, but in one respect the critics of Asilomar, like the scientists who invoked its legacy, were missing the fundamental point. They all seemed to assume that the gritty safety issues of using CRISPR had been overcome, or could easily be so, and that the fundamental questions relating to the technology were ethical, political or sociological. But whatever the restricted nature of the debate at Asilomar, the key outcome of 1975 – the adoption of clear safety protocols – was a significant step forward that was completely absent from the twenty-first-century debates around germline gene editing. This would turn out to have potentially tragic consequences, as would the failure to ask the deep question of what this technology could do that could not be done in a simpler way using other techniques.*

One reason why people did not grapple with these issues in 2015 may have been the general sense of inevitability that hovered around germline gene editing at the time. This was explored by Leah Ceccarelli of the University of Washington when she compared the language used in the Napa letter of 2015 with that of the Berg letter of 1974. In the Berg letter recombinant DNA was something that was *used by* scientists – verbs such as 'construct', 'create', 'link' and 'join' were employed, portraying researchers as the decisive component in the scientific process. In the 2015 Napa letter, CRISPR was often the *subject* of a sentence, suggesting that it was an autonomous agent of change rather than something that is under our control and that we can chose to employ or not.[35] Ceccarelli argued that this way of writing has profound implications – 'we are carried along by a linguistic and cognitive momentum that makes it less likely that we will orient toward this technology as a tool under the control of scientists and science regulators'. She concluded bleakly: 'scientists are powerless in CRISPR's world, carried along for the ride by a family of technologies that are revolutionising biomedicine'.[36]

This is a slight exaggeration and involves reading too much into a stylistic device. The 2015 Napa letter was not as passive as Ceccarelli implied, containing plenty of active verbs; furthermore, personifying a technology is not unique to gene editing – the same thing is done with nuclear power or smartphones. But Ceccareilli was right to highlight that the phraseology employed around CRISPR often gives the impression of an inexorable process, with the result that scientists are conveniently no longer responsible for the consequences of their experiments. Conceiving of a technology as an independent

*Reading the discussions from this period, I have repeatedly wished I could go back in time, grab people by the shoulders and yell that they are not looking at the right questions.

agent implicitly absolves those who deploy that technology, potentially excusing irresponsible behaviour.

*

While these somewhat abstract debates rumbled on, the genetic engineers pursued their apparently prescribed path, more or less prudently. In April 2016, a year after Huang Junjiu's breakthrough paper, another Chinese group, led by Fan Yong, reported that they, too, had altered the genes of non-viable embryos.[37] Like Huang's group, Fan's team had begun their work in early 2014, following the successful application of CRISPR to primates. And like Huang's group they first tried to publish in a well-known journal before being rejected and submitting their article to a much less prestigious title. There it was initially rejected and went through a total of eight months in revision before finally being accepted.[38]

Fan and his colleagues wanted to introduce Δ32 (pronounced 'delta 32'), an allele of *CCR5* implicated in resistance to HIV, into human embryos. Although they were generally able to alter the cells (the percentage of transformed embryos varied between 17 and 76 per cent), this was hardly an example of successful editing. In one phase of the experiment the desired Δ32 change was introduced in only four out of twenty-six embryos, and other mutations were detected in up to 38 per cent of cases. Most worryingly, once again all the embryos were mosaic. Somewhat sobered by the results, Fan and his colleagues concluded that there should be no genome editing of the human germline until 'a rigorous and thorough evaluation and discussion are undertaken by the global research and ethics communities'. How any attempt to bypass this injunction could be prevented was not explained. Four months earlier, Doudna had argued that a complete ban on germline engineering was impractical 'given the widespread accessibility and ease of use of CRISPR–Cas9'.[39] *

* Not all manipulation of embryos is aimed at editing the germline. For example, Kathy Niakan at the Francis Crick Institute in London has permission from the UK Human Fertilisation and Embryology Authority to use CRISPR on early human embryos as part of her work to understand the role of genes in miscarriage.

This feeling of inevitability was reinforced by an apparent breakthrough paper, which appeared in *Nature* in August 2017.[40] This study was made on viable embryos produced by fertilising donated eggs with sperm from a man with a dominant genetic condition, hypertrophic cardiomyopathy, and involved correcting the four-base mutation. This was the first time that viable embryos had been gene edited – the study was primarily performed in the United States, at a private institution (Oregon Health & Science University), using private funds, thereby bypassing the limited US federal regulation of embryonic manipulation. Investigating the background to this project, Canadian bioethicist Françoise Baylis discovered that part of the consent form signed by the donors implied that there might, at one point, have been the intention to implant these embryos.[41] The project administrators insisted that this was not the case and that the incriminating phrase had been included by mistake.

This study was led by Shoukhrat Mitalipov, who had previously cloned human cells and pioneered mitochondrial transfer therapy. It was remarkable not simply because the procedure worked, but also because of the apparent lack of mosaicism in the altered embryos. Furthermore, in a most surprising result, it appeared that introducing the CRISPR components at a particular point in the cell cycle caused the embryos to use the non-mutant version of the gene from the mother's chromosome as a template, instead of the synthetic DNA the researchers had introduced. The authors described their work as the 'correction' of the mutation, rather than 'editing'. It looked like Mitalipov's group had made the safety and efficiency breakthrough that many had hoped for. The world's press went crazy, with even the tabloids putting it on their front pages.

This optimism was reflected in two major reports on human genetic engineering and germline editing that were produced at the time. In 2017, NASEM published its contribution to the ethical debate, *Human Genome Editing: Science, Ethics, and Governance*.[42] Although the report recognised the uncertainty associated with the use of germline editing, the tone was resolutely positive: 'the technology is advancing rapidly, such that it may, in the not-so-distant future, become a realistic possibility that needs serious consideration'.[43] And there was no question of a moratorium:

THE Sun

20p LESS THAN THE DAILY MIRROR

WIN £40K A CARAVAN

YOUR ESSENTIAL FREE RACING GUIDE GLORIOUS GOODWOOD PLUS 8-PAGE ATHLETICS PULLOUT

Thursday, August 3, 2017 FOR A GREATER BRITAIN 50p thesun.co.uk

'DESIGNER BABY' WORLD FIRST

GENE GENIE

DNA 'fix' in embryos could wipe out inherited diseases

By NICK McDERMOTT, Health Editor

A MEDICAL breakthrough yesterday promised to end the curse of inherited disease.

US scientists revealed they successfully corrected faulty genes in human embryos by "snipping out" sections of defective DNA.

The revolutionary work, which could help end 10,000 hereditary diseases including cancer, may be continued in Britain.

But there were fears the controversial research could lead to "designer babies".

Full Story – Page Two

Cover of *The Sun*, August 2017.

> Heritable germline genome-editing trials must be approached with caution, but caution does not mean they must be prohibited. If the technical challenges are overcome and potential benefits are reasonable in light of the risks, clinical trials could be initiated.[44]

The call for public consensus, which had been clearly stated in the 2015 Washington Summit statement, had disappeared. As gene therapy scientist Edward Lanphier said of the report: 'It changes the tone to an affirmative position in the absence of the broad public debate this report calls for.'[45] Journalist Jocelyn Kaiser put it pithily in the headline to her article about the report in *Science*: 'A yellow light for embryo editing.'[46] *

Similar views were expressed in July 2018, in a report from the influential UK Nuffield Council on Bioethics. First, there was a large

* In 1975, a *Nature* editorial described the outcome of Asilomar as an 'Amber light for genetic manipulation' – *Nature* 253:295.

dose of optimism: 'it is likely that different CRISPR-Cas9 technologies will be clinically safe in the foreseeable future'. 'Safe' was not defined. Then, with suitable bet-hedging about genome editing only being appropriate for medical conditions, not enhancement, the idea of germline editing was given a stamp of moral approval:

> We can, indeed, envisage circumstances in which heritable genome editing interventions should be permitted. ... there are moral reasons to continue with the present lines of research and to secure the conditions under which heritable genome editing interventions would be possible.[47]

Both the key ethical guardians of heritable gene editing in the English-speaking world were, in Kaiser's words, giving a yellow light to the process. Yellow. Proceed with caution. Both organisations seemed more interested in exploring the exciting prospects of the new technology rather than detailing the dull technical challenges relating to safety or insisting on the once-key precondition of consensus. They could not even state what exactly the procedure would be uniquely useful for. Neither report defined safety in clear criteria nor stated that such experiments should not take place until those criteria were unambiguously met.

For some commentators, risk could even be argued away. In August 2018 an editorial in *Nature Medicine* highlighted the need to study off-target effects in CRISPR but also emphasised that 'a certain degree of risk is embedded in many promising and successful medical therapies'. The journal suggested that an acceptable safety criterion would be that 'gene editing tools that reach the market should not carry a risk of mutation rates much beyond the frequency of natural spontaneous mutations'.[48] As to global consensus, that could be dismissed. A year earlier, a group of ethicists argued that global regulation of germline editing was effectively impossible, certainly not if the public were to be involved – 'values from different cultural and religious backgrounds would make it nearly impossible to achieve any type of consensus', they said.[49] Instead, 'self-regulation seems reasonable, because scientists have expert knowledge pertaining to the potential risks and side effects of their research', they purred reassuringly, insisting that scientists have a common language and culture and are 'well qualified to judge the potential risks and side effects'. Their solution was what they called 'self-regulation *plus*' – the plus being national regulations that would differ from country to country.

These shifts in attitude by the leading ethical authorities were noted at the time. The German Ethics Council highlighted that the 2017 US Academies report contained a significant but unheralded change:

> It is clear that the US-American academies are no longer focusing on a partially fundamental, partially risk-related strong rejection of germline therapy by genome editing but on a fundamental permission guided by individual formal and material criteria.[50]

Despite such warnings, the world seemed set on a supposed prudent path towards genome editing. Another way of putting it is that germline editing was an accident waiting to happen.

*

On 22 November 2018, as Doudna was preparing to leave California for Hong Kong to attend the second human editing global summit,* she learned that all her fears had been realised. An email from a young Chinese researcher she knew vaguely, He Jiankui (his family name is pronounced 'Huh'), landed in her inbox. It had the subject line 'Babies born' and explained how CRISPR had been used to alter the DNA of two embryos; the outcome was the birth of two gene-edited baby girls.[51] Doudna recalled her stunned reaction:

> This is fake, right? This is a joke. 'Babies born'. Who puts that in

* For reasons that remain unclear, in 2017 the Chinese Academy of Sciences withdrew from the organisation of the Hong Kong summit; its place was taken by the Academy of Sciences of Hong Kong. Because of the pandemic, the Third International Summit on Human Genome Editing, due to take place in London, has been postponed until 2023.

> a subject line of an email of that kind of import? It just seemed shocking, in a crazy, almost comedic, way.[52]

But it was no joke. Three days later, the story was all over the internet, and phone alerts, newspaper websites and Twitter were all thrumming with the news. This was all due to some more expert sleuthing by Antonio Regalado, who had first sniffed something was up while talking to He off the record in October. Doggedly pursuing his hunch, in November Regalado found an online Chinese clinical trial registry that outlined He's intention to edit and then implant human embryos. The bombshell story was published on the *MIT Technology Review* website on 25 November (Regalado did not know that the babies had been born), immediately prompting Associated Press, which had been working confidentially with He Jiankui, to break a fuller version that included news of the birth of twin girls a few weeks earlier. Dr He himself rush-released YouTube videos describing what he had done as gene surgery.[53] In China, the *People's Daily* hailed 'a milestone accomplishment achieved by China in the area of gene-editing technologies'.[54] The prudent path had led to a brave new world.

He Jiankui had been scheduled to speak at the Hong Kong summit – rumours of his intentions to use gene editing on embryos reached the organisers in October and they decided to invite him, unaware that he had already turned his intentions into actions.[55] Some observers have speculated that He intended to pull a Steve Jobs-esque 'One more thing …' and announce the birth of the twins at the end of his talk, accompanied by the simultaneous release of the videos and exclusive coverage in friendly media. Whatever the case, the careful planning had been thrown into disarray by Regalado following his nose and the Summit organisers insisted that He present his data in full. By the time that He walked onto the stage on the second day of the Hong Kong summit, on 28 November 2018, everyone knew what he was going to say. What was not clear was what exactly he had done and above all why he had done it.

As He haltingly explained his experiment, experts around the world tried to make sense of the data presented on the video feed, reporting their interpretations on Twitter in real time using the hashtag #CRISPRbabies while the rest of us watched, aghast. I had been explaining CRISPR to the general public and to my students since 2015; in my talks I would raise the probability that someone would edit an embryo and bring it to term. Focusing abstractly on the big picture, I glibly emphasised that whatever the ethical considerations, this would not lead to any massive change in society – germline editing would be too cumbersome to become anything more than a fringe activity. The worst that could happen, I would coldly say, was that the babies might die. Watching my flickering Twitter feed as the horror unfolded, I found that I was outraged and absolutely furious. Two real mutated babies carried far more weight that any number of hypothetical gene-edited individuals.

Dr He explained he had used CRISPR to alter the *CCR5* gene of two embryos to prevent them from getting HIV in the future. This made no medical sense, as there are simple, well-known and very effective ways of avoiding HIV infection. Furthermore, despite his claims that there were no off-target effects and that the babies were perfectly normal, it was clear from the limited and confused data he presented that not only had he introduced two completely different mutations into the two embryos, but also that both girls were mosaic. This was not gene surgery, it was gene butchery. The two embryos had been normal and healthy before He got his clumsy hands on them. Now they were certainly not normal; no one knew if their health would be affected. We still do not know.

After the presentation came the questions, although none were as astute and pertinent as those being asked on Twitter at the same time – for example, the issue of mosaicism was not raised. Nevertheless, two important points were made. In response to a question, He revealed that there was a third germline-edited pregnancy, which had yet to come to term.* Above all, He's pretensions to have any ethical understanding of what he had done were erased by a simple question from Harvard's David Liu: 'What was the unmet medical

---

*There were persistent doubts about the reality of this third pregnancy, because there was no evidence it existed. In December 2021, journalist Vivien Marx confirmed the birth of the third child, who she called Amy. Marx, V. (2021), *Nature Biotechnology* 39:1486–90. There may well be more edited embryos in deep freeze storage somewhere.

need?' Liu asked. He Jiankui blinked and had no answer. He then left the Summit and left Hong Kong, returning immediately to China.

Over the next few days the CRISPR babies affair turned into a scandal as the full scale of what He had done became apparent. On *The Atlantic* website, science journalist Ed Yong summarised the '15 worrying things' about the experiment and how it had been received, while one of the main commentators on Twitter, Sean Ryder of the University of Massachusetts Medical School, published a scathing article in the *CRISPR Journal*.[56] Subsequent analysis and discussion of both the science and the ethics of the affair confirmed and expanded this critique, in particular some brilliant investigations by Kevin Davies in his book *Editing Humanity*, a precise and chilling dissection of the article He had submitted to *Nature* by human geneticist Kiran Musunuru* and a sober but relentless account by legal ethicist Henry Greely of Stanford, first in a long academic article, then in the best book on the whole business, *CRISPR People*.[57] Greely pulled no punches: this was a 'fiasco', 'a reckless ethical disaster', he stated.[58] The key issues highlighted by the critics included:

- The babies were not facing an unmet medical need – there was no need to do the experiment to prevent the children getting HIV later in life.
- The editing went horribly wrong – neither baby has the desired Δ32 mutation. One child has a deletion of fifteen base pairs in one copy of the gene, the other copy is normal; the other child has two different mutations, one a deletion of four bases, the other an insert of a single base (neither of these variants have ever been seen before).
- Both children are mosaic.
- It is not clear that the parents understood what they were

* He's article was full of errors, lies and disturbing ethical missteps. Musunuru was mailed an embargoed copy of the manuscript a week before the Hong Kong summit. In his memoir describing how he became unwittingly tangled in He's web, Musunuru takes apart the paper in a 'forensic analysis of an experiment gone wrong'. Musunuru says he let out a guttural scream when he realised that He Jiankui had implanted mosaic embryos. Musunuru, K., *The CRISPR Generation: The Story of the World's First Gene-Edited Babies* (2019).

signing up to; according to the consent forms the project involved the development of an AIDS vaccine.

- The procedure was not open or transparent; it was done in secret.
- Ethical advice was sought but then ignored.
- The procedure breached a series of Chinese regulations.

In 2021, Fyodor Urnov gave me this summary of He's experiment:

> It is breathtakingly low quality science. He did not do elementary things that I would have instructed a first year graduate student in my lab to do, and it did not work in the way he wanted it to work. And he still went ahead and had the embryos implanted. ... As a scientist, you need to meet the basic standards of quality control, and he did not. It breaks my heart, it really does.[59]

Above all, remember that this was an experiment conducted not only on these healthy embryos, but also *on their children*. The consequences of this botched experiment may carry on down the generations.

In the days that followed He's announcement there was a small minority of scientists and commentators who suggested that the experiment was not so bad, but as the full horror and pointlessness of He's handiwork became clear, the mood rapidly changed. The article He had submitted to *Nature* was rejected (it was immediately resubmitted to the *Journal of the American Medical Association*, which also rejected it), and a bland piece by He on the ethics of germline editing which had been accepted by the *CRISPR Journal* was retracted by the journal (the article had presumably been written as an attempt at ethical cover). He Jiankui was simultaneously notorious and becoming an unperson.

Most significant was the shift in opinion in China, where the trumpeting of the *People's Daily* about a milestone for Chinese science was rapidly deleted (but not forgotten). Both of the institutions implicated – He's university and the hospital that had given ethical approval – quickly denied knowledge or involvement. He Jiankui himself was apparently placed under house arrest and condemned by his university, which issued a statement saying that he had contravened China's 2003 Regulations on Human Assisted Reproduction, which prohibit the implantation of a genetically modified human embryo.[60] And he was attacked by his peers – on the evening of He's presentation a group of 122 Chinese scientists posted a statement on Weibo, the Chinese equivalent of Twitter, attacking the experiment as 'crazy', deserving of 'strong condemnation!!!' and 'a huge blow to the global reputation and development of Chinese science'.[61]

Opposition from more official sources came in *The Lancet* in January 2019, which contained highly critical letters from the Chinese Academy of Medical Sciences, from the Chinese Academy of Engineering and from 149 Chinese HIV professionals.[62] Although the Guandong Health Commission launched an immediate investigation, the outcome has not been revealed, nor has any more systematic inquiry taken place. In May 2019 Qiu Renzong of the Chinese Academy of Social Sciences was reported in *Science* as saying: 'Nothing suggests another investigation is underway. ... There has been no transparency regarding who the members of the investigating team are, or what they have found.'[63]

In December 2019 it was announced that He Jiankui and two of his colleagues had been convicted after a secret trial; He was sentenced to three years in jail (he is apparently now free) and a substantial fine, while his colleagues received suspended sentences.[64] They were all forbidden from ever working again on assisted reproduction. The man who dreamed of a Nobel Prize is now a universal pariah. The condition of the CRISPR babies, and above all their future health, are both unknown.

– THIRTEEN –

AFTERMATH

As the news of He Jiankui's dreadful experiment echoed around the world, Western journalists and researchers traced his path, trying to glean information about his backstory and character, as though those factors would explain what he had done. It soon became clear that this would only tell part of the story.

In carrying out a terrible procedure for his own ends, potentially ruining the lives of three girls, He Jiankui revealed himself to be a vain and foolish man. But he was not merely a rogue medic, an isolated case. His reckless behaviour was born of China's desire for rapid technical advance and success at all costs.[1] After a PhD in the United States, He gained fame and a relative fortune in China as an entrepreneur developing DNA sequencing services (one of his companies was allegedly worth around $200 million) and in 2017 Chinese television profiled him as 'The new top shot in the gene world'.[2] Recruited to the newly created Shenzhen Technical University in 2011 under its Peacock Programme, which was designed to attract foreign and Chinese scientific talent, He received major funding, inevitably with the approval of the regional Communist Party apparatus. He was also selected to be part of the Thousand Talents Plan, run by an office of the Central Committee of the Communist Party. This was a period when China was encouraging the development of 'Strategic Emerging Industries', which included $9.2 billion funding for the application of genetics to medicine.[3]

Chinese scientists were pressured to focus on rapid success, with an emphasis on short-term gain – publishing exciting results in leading journals. Inevitably there were several cases of apparent misconduct, including a claim to have developed a new form of gene editing which was published in *Nature Biotechnology* in 2016 before being retracted nearly eighteen months later.[4] This was He Jiankui's world, and it explains an awful lot. As Benjamin Hurlbut has put it: 'Far from "going rogue" and rejecting the norms and expectations of his professional community, JK [He] was guided by them.'[5] In a series of discussions with Hurlbut after the scandal broke, He Jiankui – then under house arrest – explained that he had the impression that US colleagues saw Chinese science as second class. He said he felt 'disgust' in 2015 when US scientists criticised Huang's report of the first gene editing of human embryos and it was rejected by leading journals; this then turned to a feeling that there was 'scientific racism against the Chinese' when in 2017 Mitalipov's experiment was not only published in *Nature* but was widely praised.[6]

What occurred in 2018 was almost inevitable. In 2015 the *New York Times* reported the words of Dr Yi Raoi of Beijing University regarding the lack of ethical oversight in China:

> 'The more technology we have, the more dangerous we are to ourselves and entire humankind. ... Right now, human gene editing is the main thing' Mr Yi said. Geneticists in China 'don't want to be guided by Western people.' The mind-set among Chinese researchers, according to Mr Yi: 'We're going to do it, then see what's wrong, then fix it. But the conceptual discussion may be missing.'[7]

As four Chinese bioethicists wrote in *Nature* after the scandal broke, the country was at a crossroads and needed to make 'substantial changes to protect others from the potential effects of reckless human experimentation'. In particular, they detected an unpleasant undertow in the attitudes of sectors of the population, and in particular in academia:

> China should step up its efforts to counter prejudice against people with disabilities and the eugenic thinking that has persisted among a small proportion of Chinese scholars.[8]

But although the situation in China meant that none of He's peers told him to stop, there were perhaps sixty people in the West who knew or suspected what he was doing, either before he did it or after the embryos were implanted, and who should have behaved differently. Not one of these people revealed what they knew before the scandal broke. They formed what the science journalist Jon Cohen called He's 'circle of trust' and included Michael Deem, He's former PhD supervisor at Rice University, who was in Shenzhen when at least one pair of prospective parents signed the dubious consent forms. Others, such as Stephen Quake of Stanford, were not as involved, but were still positive – on hearing of the first pregnancy Quake emailed He: 'Wow, that's quite an achievement! Hopefully she will carry to term.'[9]

Some were more circumspect: when Nobel Prize winner Craig Mello heard of the pregnancy he replied 'I'm glad for you', but immediately distanced himself ('I'd rather not be kept in the loop on this') and correctly highlighted the fundamental problem: 'I just don't see why you are doing this.'[10] In 2018, He Jiankui visited various US universities to discuss ethical questions around germline editing, and again met dissent. For example, accompanied by a colleague, he visited stem cell scientist Matthew Porteus at Stanford; Porteus explained in no uncertain terms that He's project – there was no indication that it was a reality – would be a Very Bad Idea:

> I spent the next half-hour, 45 minutes telling them about all the reasons that it was wrong, that there was no medical justification. ... I left it at that, assuming that I had dissuaded him.[11]

But the criticism had no effect, and He continued on his reckless path, obsessed by his delusional vision of fame, glory and a Nobel Prize.

To those outside his circle of trust, He dissimulated. In February 2017, He gave a talk at a meeting co-organised by Doudna in which he described his rather unoriginal work editing *CCR5* in non-viable human embryos, closing with the consensual statement that until safety issues were resolved it would be 'extremely irresponsible' to edit the human germline; seven months later he presented the same work at Cold Spring Harbor, closing with a slide on Jesse Gelsinger. We now know that in between those two talks he submitted a medical ethics approval form to the Shenzhen HarMoniCare Women and Children's Hospital outlining his experiment, claiming: 'This is going to be a great science and medicine achievement ever since the IVF technology which was awarded the Nobel Prize in 2010, and will also bring hope to numerous genetic disease patients.'[12]

He also dissimulated to the parents, although they have insisted that they understood the procedure and that they were simply trying to escape the widespread prejudice against HIV-positive people in China: 'At a certain level our participation in the experiment was indeed forced, but we were not coerced by any person in particular. We were coerced by society', they said.[13]

Whatever the parents may say, the reality is that the genetic manipulation of their daughters – and of any offspring the girls may have – was not necessary to achieve the aim of an HIV-free child. The IVF procedure itself ensured that outcome, by removing all traces of the virus. That could have been performed in any IVF clinic, but the catch was that the parents did not have the means to pay for such treatment. This was the true nature of the coercion they were subjected to and reveals the profoundly unethical approach adopted by He and his colleagues. In reality, the parents had no choice.

The final, fundamental failure that led to He's experiment was collective. It was the failure of the international scientific and bioethics community, which over the previous three years had pondered the possibility of germline editing in reports and meetings but had failed to speak clearly or to act decisively. Not everyone agrees with this. At the close of the 2018 Hong Kong summit, David Baltimore emphasised that self-regulation could work only if everyone complied: 'I think there has been a failure of self-regulation by the scientific community because of a lack of transparency.' Alta Charo, a bioethicist who had been involved in the discussions since the Napa meeting of 2015, put the blame solely on He Jiankui: 'the failure was his, not the failure of the scientific community', she said.[14]

He's responsibility is obvious, but an immense part of the blame for what happened surely lies with those who – unlike Lanphier and Urnov in their letter to *Nature* at the beginning of 2015, unlike the Society for Developmental Biology – did not clearly say 'do not do this'. Although the various declarations by the 'scientific community' stated their disapproval of irresponsible germline editing, from 2017 they gave what was widely perceived as a yellow light to the procedure. They should have displayed a stark red signal and also clearly argued for a moratorium. By focusing on general ethical issues rather than dealing first with precise safety measures, and above all addressing the question of why one might wish to carry out such editing, by abandoning the emphasis on consensus as a prerequisite and by treating the advent of germline editing as something that appeared to be inevitable, the 'gene editing community' played an important role in the He catastrophe. They inadvertently created an atmosphere in which the simple metaphors of molecular scissors and editing that go hand in hand with CRISPR began to be seen as literally true.

In 2021, I interviewed David Baltimore and put it to him that the scientific community had been collectively seduced by the simplicity of those metaphors and that this had played a role in He's behaviour. Baltimore agreed and said that he had been thinking along these lines for some time. Fyodor Urnov, who back in 2005 had come up with the phrase 'genome editing', went even further, telling me he now wished that he had used some other term, one which downplayed the apparent simplicity of changing genetic sequences.

These difficulties had been noted in 2015, when a group of researchers explored the CRISPR metaphors used in newspapers and popular science publications and concluded that two of the main terms – 'editing' and 'targeting' – were misleading:

> We see a pattern of reduced complexity and exaggerated control of outcomes that has troubling implications. … 'Editing' does not convey a sense of risk or a need for caution … the metaphors that are gaining traction obscure and mislead in important ways.[15]

Nobody paid any attention, and it was probably too late anyway. The metaphors were already too deeply fixed in the collective language of both scientists and the public, surreptitiously shaping the way the technology was seen, even by those most aware of its technical details.

It was not only the metaphors that were misunderstood. The basic positions that the 'scientific community' thought it had clarified were not at all obvious to He Jiankui and his colleagues. On the eve of the Hong Kong summit, after the news had broken, Doudna, together with Alta Charo and others, had dinner with He to find out exactly what he had done and why. Charo asked He if he had understood the significance of the NASEM and Nuffield reports. Dr He replied: 'I absolutely feel like I complied with all the criteria.'[16] Because those criteria were not clear, nor rooted in precise details, He imagined that he could go ahead – the light was yellow, after all. The reports did advise against germline editing under existing conditions but this was ignored by He – that was his ethical crime. But in his ethics proposal He claimed that the 2017 NASEM report had 'for the first time' approved germline editing to treat disease, which, in a way, it did. As Ed Yong put it: 'It's as if he took the absence of a red light as a green one.'[17] Kevin Davies also saw the problem, in retrospect: 'the refusal to issue a blanket prohibition of germline editing from either of these two prestigious bodies [NASEM and Nuffield] was tantamount to lifting a velvet rope'.[18]

He Jiankui apparently took the decision to duck under that rope because, like many others, he was entranced by the fantasy that an individual physician could make a major clinical breakthrough, by will alone and almost irrespective of the costs. In a remarkably prescient contribution to a 1983 Cold Spring Harbor Laboratory meeting, David Baltimore had predicted exactly what might happen if this attitude was applied to human germline gene editing:

> Many, many medical advances have come from an individual who vaguely sees a way to do something and although there are lots of problems goes on and does it. I am not sure this field can afford this type of behaviour. I would look for a terrible backlash, because it is so publicly sensitive.[19]

Nevertheless, like previous ambitious and arrogant gene therapy researchers, such as Stanfield Rogers, who injected those German girls with Shope virus in 1970, or Martin Cline who sidestepped ethical controls to introduce recombinant plasmids carrying β-haemoglobin into two women patients in 1980, He allowed his ambition to blind his judgement. This was quite conscious on his part. As he explained to Benjamin Hurlbut after the scandal broke, his model was the work of Bob Edwards and Patrick Steptoe on IVF, which led to Edwards winning the Nobel Prize in 2010 (Steptoe died in 1988). He clearly did not understand how slowly their work progressed nor the persistent and profound ethical control and inquiry it was subjected to by both the relevant authorities and by the media. For He, it all seemed to come down to the action of an individual researcher, who by sheer power of will, could change the whole of society. As he told Hurlbut:

> If we are waiting for society to reach a consensus ... it's never going to happen. ... But once one or a couple of scientists make first kid, it's safe, healthy, then the entire society including science, ethics, law, will be accelerated. Speed up and make new rules. ... So, I break the glass.[20]

*

Within weeks of the glass being broken, in January 2019, a group of eighteen scientists led by Eric Lander, and including Emmanuelle Charpentier, Paul Berg, Feng Zhang and bioethicist Françoise Baylis, published a letter in *Nature* calling for a five-year global moratorium on all clinical uses of human germline editing (this would not apply to academic research into the early embryo, or to work on non-viable embryos).[21] This position subsequently gained the significant support of Francis Collins, the head of the NIH. While research was voluntarily suspended, the signatories argued, there should be 'discussions about the technical, scientific, medical, societal, ethical and moral issues that must be considered before germline editing is permitted'. Above all, the time should be used to establish an international governance framework which individual countries would then sign up to, if they so wished.

Doudna turned down an invitation to sign the letter, initially arguing that a moratorium would be too strong a response. As the *Washington Post* reported:

> 'To me that word implies enforcement,' Doudna said. 'I don't want to drive others underground with this. I would rather they feel that they can discuss it openly. Gene editing, it's not gone, it's not going away, it's not going to end.'[22]

Others, like Baltimore, just did not get it – 'I don't see the need for, or rationale for, a moratorium', he said.[23] To those who argued that a moratorium would not prevent an individual, or a country, from carrying on regardless, Lander insisted that was not the point. Instead, he wanted to focus governments' attention on their responsibility for the actions of those within their jurisdictions.[24] That seemed to be the view of the Director-General of the World Health Organization, Tedros Ghebreyesus, who in July 2019 stated: 'regulatory authorities in all countries should not allow any further work in this area until its implications have been properly considered'.[25] Other researchers were more proactive, proposing that there should be a grassroots academic boycott of those involved in editing viable embryos and of journals and institutions that supported such research.[26]

Alongside the widespread condemnation of He's actions there were also subtle signs of the acceptance of a new reality, just as He had hoped. Amazingly, despite He's botched experiment, the statement issued at the end of the Hong Kong summit dropped the references to the need for societal consensus that had characterised the discussions in 2015 and boldly claimed that it was now 'time to define a rigorous, responsible translational pathway toward clinical trials'.[27] A couple of weeks later, in a joint editorial in *Science*, the presidents of the US National Academy of Medicine, the US National Academy of Sciences and the Chinese Academy of Science called for the rapid production of a report that would outline 'criteria and standards to which all genome editing in human embryos for reproductive purposes must conform'.[28]

The world had stepped over the threshold of germline manipulation and there was no going back, so we had better find a way of living with it, these people were saying.

In the following months a whole series of proposed international structures that might regulate or discuss germline editing were imagined. The WHO set up a committee to develop an international framework and eventually created an international registry of clinical research using human gene editing. The US National Academy of Sciences, the US National Academy of Medicine and the Royal Society set up a Joint Commission to prepare a framework to guide clinical research in germline gene editing. Other proposals included an international governance framework and a global citizen's assembly, set up as much to inform the public as to involve them.[29] This latter suggestion, published in *Science* in September 2020 by a group of twenty-five researchers, led by John Dryzek of the University of Canberra and including George Church, had the virtue of recognising that experts in science and ethics are not the only people who need to be involved:

> Though we might expect scientists to be good at reflecting on scientific values, their role gives them no specific insight on the public interest. Ethicists are professionally capable when it comes to moral principles – but these are not necessarily the same as public values.[30]

Despite all this imagination and effort, little concrete progress was made before the world was blown off-course by the COVID-19 pandemic. At the time of writing there is no agreement on anything much, beyond that He's experiment was wrong and that there should be no clinical application of germline gene editing for the foreseeable future.

Not even that is accepted by everyone. As Jennifer Doudna has argued from the outset, part of the problem of regulating CRISPR lies in its relative ease of use which means unprincipled researchers can simply ignore and evade regulation. Although editing embryos requires complex IVF facilities, which can be regulated and indeed are in many countries, the case of He Jiankui shows that this does not preclude a disastrous attempt to edit the human germline.

There are those, egotistical and avid for the limelight, who have apparently refused to learn the lessons of the CRISPR babies scandal.

In Russia, geneticist Denis Rebrikov announced in spring 2019 that he was intending to follow in He's footsteps and edit *CCR5* in human embryos. Rebrikov claimed he would only go ahead if the procedure was safe but added unreassuringly 'I think I'm crazy enough to do it'.[31] He later changed his proposed target to *GJB2*, a gene involved in deafness, before seeming to accept the Russian Health Ministry's endorsement of the World Health Organization's view that it was too soon to do such experiments. Keen to have the last word, in 2019 Rebrikov ominously told *Nature*: 'What does it mean, too soon? Lenin said, "yesterday was too early, tomorrow it will be too late".'[32]*

To the excitement of the media there are even biohackers – hobbyists who think they can easily master complex techniques and ignore ethical guidelines – who say they are going to edit the germline.[33] There is one gang of fantasists who mix cryptocurrency funding and transhumanist nonsense in a toxic, nauseating nightmare, claiming that they will use CRISPR germline editing to produce babies who will live to be 'super-centenarians' or will 'grow muscle without weightlifting'. The cunning plan of these attention-seekers involves injecting CRISPR components into the testicles of a male volunteer and then finding a woman to carry the baby. Good luck with that. Lacking the necessary intellectual or technical resources, they outsourced the project to a laboratory in Ukraine, which said it needed the funding but was clearly more in need of training in ethical behaviour.[34] Nothing will come of this except that some people will lose a lot of money (and perhaps a testicle or two) while a lucky few will enjoy their fleeting notoriety and will presumably laugh all the way to the bank.

Even previously enthusiastic scientists began to back-pedal as it gradually became clear that gene editing in human embryos is fraught with currently intractable problems. In 2018, the seemingly more reliable 'correcting' technique described by Mitalipov in 2017, with its claimed zero level of mosaicism, came under fire as scientists suggested that the procedure might in fact have led to

---

*In 2022, Rebrikov revealed that as soon as he could find a willing deaf couple he would carry out the gene editing and then freeze the resultant embryos, waiting for changes to the regulatory frameworks that would make it possible to implant them. Mallapaty, S. (2022), *Nature* 603:213–4.

undetected deletions of substantial stretches of DNA.[35] Mitalipov's team responded to these criticisms, but subsequent studies have found deeply worrying unintended on-target editing outcomes in a wide range of CRISPR applications in various mammalian tissues, including human embryos. Bits of DNA in and around the target gene have been chewed away, sometimes involving deletions of up to 20,000 base pairs, or even whole parts of chromosomes. There are now many reports confirming these findings.[36] In 2022, an article in the *CRISPR Journal* warned of 'more chaotic than expected' on-target adverse effects from CRISPR.[37] This is all profoundly disturbing – although there were differences between how these experiments were done and the protocol employed by He Jiankui, it is possible that the babies mutated by He have even more genetic damage than was initially suspected. We are told that CRISPR involves 'molecular scissors'; in some cases, it would appear to involve something more like a chainsaw gone amok.

One of the perplexing things about these disconcerting results is the variability that is observed even from the same laboratory. Depending on the precise protocol, the proportion of cells with deletions can vary by over an order of magnitude and we do not know why. Our understanding of the early embryo and of its DNA repair mechanisms remains extremely rudimentary; even without the profound ethical issues, it will be many, many years before this technology is ready for a clinical trial, never mind for clinical application.[38]

Nonetheless, the enthusiasm continues. In 2019 a Chinese group used a CRISPR offshoot technique called base editing (which alters a single base in a sequence) in two-cell embryos with apparent 80 per cent success.[39] Although the lead researcher, Yang Hui, cautioned that 'an almost 100 per cent efficiency is required before the technology can be used on humans', he also went on to cast his work in terms that suggested that the atmosphere in Chinese gene-editing circles had not changed much following the He scandal: 'Base-editing technology with a high safety standard should be ready for clinical use in one or two years ... we are ahead of the competition in the United States', he boasted, going on to use an alarming analogy: 'We are working with the same spirit as building the first nuclear bomb.'[40]

We have been warned.

*

One positive development is that many people are finally asking the fundamental question of what exactly germline editing might be used for. For a whole series of reasons, there are now few serious people advocating that it could be used for the development of superhumans with great strength or resistance to pain. The debate has thankfully shifted against such views – the inaccuracy of our current methods of editing mean that such an approach would be likely to fail, even if simple edits could produce such effects (they could not). There will be no CRISPRed X-Men. Even apparently simple 'improvements' intended to reduce the risk of disease are more complex than appears. For example, it would be possible to introduce a variant of the *SLC39A8* gene which is associated with a lower risk of hypertension and Parkinson's disease. But this same variant increases the risk of schizophrenia, Crohn's disease and obesity.[41] The genetics of most human characteristics, including disease, is horrendously complicated.

Many of the leading agencies and organisations involved in discussing gene editing have recognised that even if such developments were possible, they would encourage the growth of inequalities and should therefore be forbidden. In the world of 2020, where COVID-19 was amplifying the differences between rich and poor, both within and between countries, and where social justice had become increasingly significant in rhetoric if not reality following the explosion of Black Lives Matter, the idea of advocating that a tiny minority should profit from some spurious upgrade seemed completely off-message. This could be seen in the way that the Joint Commission of the National Academies of Science and of Medicine and the Royal Society used the COVID-19 pandemic and the struggle against racial injustice to frame their 2020 report on what they called heritable human genome editing (HHGE):*

* The Joint Commission abandoned the traditional technical term 'human germline editing' for the more accessible 'heritable human genome editing', which was also used as the title of the report.

> These twin upheavals have underscored that we live in an interconnected world, where what happens in one country touches all countries, and that science occurs in a societal context. Although of a very different nature, the potential use of HHGE is an issue that transcends individual countries, deserves wide-ranging global discussions, and entails important issues of equity.[42]

This was obvious long before the world was turned upside down in 2020 – even to scientists, who can be a bit dim about such matters. As Jennifer Doudna told *Le Monde* back in 2016:

> The ethical question is who wants to apply these techniques, who has access to them, who decides to use them, to what end.[43]

Three years later, Feng Zhang highlighted the obscenity of unequal access to healthcare and how it could be exacerbated by genetic engineering of our germline:

> In a world in which there are people who don't get access to eyeglasses, it's hard to imagine how we will find a way to have equal access to gene enhancements. Think of what that will do to our species.

Even Jim Watson could see clearly on this point, as he explained to Walter Isaacson in 2019:

> If it's only used to solve the problems and desires of the top 10 per cent, that will be horrible. We have evolved more and more in the past few decades into an inequitable society, and this would make it much worse.[44]*

* Watson appears to have changed his tune. At a 1998 conference on 'Engineering the Human Germ Line' he argued that any international regulation of germline engineering would be 'a complete disaster'. Wadman, M. (1998), *Nature* 392:317.

As well as clearly situating the question of germline editing in a social context, the March 2020 Joint Commission report was the first to be honest about why the procedure might be carried out, assuming that far-fetched and divisive 'enhancements' were ruled out on both ethical and practical grounds. Unlike gene therapy, heritable human genome editing will not cure anyone with a genetic illness. Instead, it allows the creation of a modified human in the context of assisted reproduction. As Peter Mills of the Nuffield Council put it in 2020:

> HHGE does not treat a child; it brings a certain kind of child into existence ... HHGE is about expanding reproductive choice and not about therapy.[45]

The Joint Commission accepted this view and recognised that only a tiny number of people could potentially benefit from the process:

> Since there are already viable alternatives for prospective parents to have genetically-related, unaffected offspring in the vast majority of cases, the benefits will accrue to very few prospective parents.[46]

As Harvard scientist David Liu put it in 2020:

> I continue to struggle to imagine plausible situations in which clinical germline editing provides a path forward to address an unmet medical need that cannot be provided by other options such as preimplantation embryo testing.[47]

The only cases in which the desire to have a genetically related child free of a particular genetic disease could be met only by germline gene therapy would be either where one prospective parent has two copies of a dominant affected gene (they are described as being homozygous), or in the case of a recessive disease, where both parents are homozygous for the disease gene. [48] The number of such couples is vanishingly small. In the case of the recessive disease cystic fibrosis, one of the most common genetic diseases, there are perhaps one or two US couples where both potential parents are homozygous for the mutated gene, while only a few dozen homozygous patients with the dominant Huntington's disease have ever been described, anywhere in the world.[49] Unbelievable as it may seem, once the excitement and hype have died down, the only conceivable ethical justification for germline gene editing is to meet the potential reproductive desires of a hundred or so couples around the world.

In June 2021 the cumbersomely titled WHO Expert Advisory Committee on Developing Global Standards for the Governance and Oversight of Human Genome Editing released a Framework for Governance which, in keeping with the global mood, put ethical issues at the centre of its reflection, expressing its concern about the possible effects of HHGE on 'fairness, social justice and non-discrimination, as well as potential disregard for the individual dignity of persons with disabilities'.[50] As to whether anybody, anywhere, should edit the germline of an embryo, the best the WHO could do was support the 2019 statement of its Director-General which opposed future work until the implications were clear.[51] The Framework notably left open the role of the global public in determining policy with regard to germline editing. There are a number of reasons behind this shift in policymakers' attitudes away from considering societal consensus to be a precondition. As Françoise Baylis has highlighted, consensus implies the distribution of power from professionals to the public, something not all scientists and physicians are prepared to accept.[52]

In reality, there are already legal controls on germline editing in many countries. The 1997 Oviedo Convention, which has been ratified by twenty-eight countries and binds them legally, states clearly that 'an intervention seeking to modify the human genome may be undertaken for preventive, diagnostic or therapeutic purposes and only if its aim is not to introduce any modification in the genome of any descendants'.[53] However, of the leading industrialised countries, only France, Spain and Switzerland have signed this Convention. The United Kingdom, Germany, the United States, China, Russia, Japan, Australia, Canada and many other countries have declined the opportunity. Even so, in many of these countries there are clear laws and regulations governing the use of assisted reproduction and/or genetic manipulation – over fifty countries now restrict or forbid germline modification.[54] But without a global agreement, there would always be the potential for gene-editing tourism to develop, as rich and desperate clients took their egg and sperm to clinics in countries that would allow the practice.

The situation in China remains unclear. In 2019 the Standing Committee of the National People's Congress passed a Biosafety Law and drafted new regulations for biomedical and biotechnology research, but as three leading Chinese bioethicists highlighted:

> There is no direct regulation on gene therapy at the legal (narrow sense) level. The current regulatory norms are mainly technical management methods or ethical guidelines. China's legal (broad) framework for gene therapy mainly includes administrative regulations.[55]

In this context, the enthusiasm of some Chinese scientists to race with colleagues in the United States to be first to use new editing techniques on embryos is worrying to say the least. It is not clear that the full lessons of the He affair have been learned by anybody, anywhere.

*

The eugenic fantasies tapped into by the *Gattaca* poster and more recently copied by some US clinics specialising in assisted reproduction and pre-implantation screening are just that, fantasies. But if it were possible to tweak babies' genes in some desirable and safe manner, one unintended consequence of making individual choices along these lines would inevitably be to amplify inequality. We can already see this at work in non-germline editing genetic contexts. In many Western countries, improvements in the diagnosis of Down syndrome in the foetus, by testing the blood of the pregnant mother, have led to increasing numbers of abortions of affected foetuses and a consequent decline in the number of Down syndrome babies being born (Down syndrome is not inherited and people with Down syndrome are virtually always sterile). Those individual, understandable and legitimate choices may eventually lead to increased prejudice against people with Down syndrome as they become less common in the community. Support networks for families and cultural acceptance of people with Down syndrome will gradually be weakened. We will all be impoverished by the unintended consequences of individual choices.

The same argument can be applied to those with various forms of disability (many would contest that term), such as deafness. Some deaf people fear for the future of their culture if genetic cures, be they heritable or somatic, are developed. Others wish to use reproductive technology to ensure that their offspring carry the same genes as they do and that they are deaf too. In 2018, Krishanu Saha, Benjamin Hurlbut, Sheila Jasanoff and their colleagues highlighted these problems in a plea for a broader debate:

> We need to make room for voices and concerns that have gone largely unheard when debates are driven by the imperative of speed at the frontiers of biological research. These neglected voices are no less important for shaping the human future than the voices of those who are already positioned to radically remake it.[56]

These are complex issues to which there are no simple answers – I certainly have none – but they reveal that uncritical advocacy of germline editing is wrong. Caution, suspicion and even fear are all far more reasonable.

In 1989, at the beginning of debates about gene therapy, David Suzuki and Peter Knudtson argued in their book *Genethics* that 'Germ-cell therapy, without the consent of all members of society, ought to be explicitly forbidden.'[57] While unanimous consent on a planetary scale about anything is unfeasible, the central argument remains valid over forty years later.

For years the point of germline genome editing was never systematically questioned by those who should have known better. We have now arrived at a situation in which dreadful procedures have been undertaken, for the weakest of reasons. In her 2019 book *Altered Inheritance*, Françoise Baylis was acerbic:

> Not to put too fine a point on it, why should scientists, governments, philanthropists, and other investigators direct resources to address the desire for genetically healthy and genetically related children using genome-editing technology when, in almost all cases, there are safer, simpler, and cheaper ways of achieving this goal and building loving families?[58]

Put that way, there really does not seem to be much point to germline editing at all, no matter how enthusiastic some boneheaded scientists, ambitious physicians and tedious transhumanists might be about the prospect. In the words of Fyodor Urnov, the bathetic reality is that human heritable genome editing is a solution in search of a problem.[59]

Excitable books about future humans might sell by the shedload and enable readers to dream about weird people living on other planets, but all that is not only literally fantastic, it is a distraction from the problems we have now, on planet Earth.[60] If we want to transcend our current situation, then solving our social problems, along with the existential threat posed by climate change, the emergence of new diseases and the re-emergence of the threat of nuclear conflict, is more than enough challenge for our ingenuity.

And as that broader focus implies, heritable human gene editing is not the only worrying application of our understanding of the double helix – scientists also have their eyes on nature itself.

– FOURTEEN –

ECOCIDE

From the ancient use of sulphur to the catastrophically effective deployment of DDT in the twentieth century, we have killed gazillions of arthropods and damaged the environment to secure food supplies and prevent disease. That could soon be behind us. Genetic engineering promises to fix the problem by altering whole species and ecosystems through artificial genetic elements called gene drives. These are designed to rapidly sweep through a wild population, ignoring the ordinary laws of inheritance, copying themselves madly from generation to generation in alarming exponential growth. This explosive potential could be used to make mosquitoes unable to transmit malaria, or it could eliminate them entirely from a region.

You do not need to be a Luddite or a technothriller writer to imagine how this could all go horribly wrong. True, those worries echo the fears that have been conjured up by each phase of genetic engineering, but that does not mean they are unjustified. As scientists working on gene drives have recognised, sometimes it is right to be frightened by technology.

*

Ever since the 1960s, the ecologically intelligent solution to insect pests and disease vectors has involved the restricted release of clouds of sterile males, which mate with females but leave no descendants, leading to a temporary crash in the local population. First imagined in the 1930s, the technique was tested in 1954 when US researchers irradiated millions of male screwworm flies, rendering them sterile, and then released them on Curacao off the Venezuelan coast. Within a few months this voracious farm animal pest was eradicated from the island; by 1966 sterile male releases had eliminated the screwworm from the whole of the United States.[1]

Because this approach involves the continual release of millions of sterile males, insects which have to be bred and sterilised by X-rays or chemicals, from the 1960s onwards researchers sought to employ natural genetic examples where the laws of heredity are twisted and crosses between members of the same species produce only males or only a few offspring.[2] These effects can be caused by a variety of factors – chromosome abnormalities, transposons (mobile pieces of selfish DNA) or microbes.[3] Progress was slow, however, and in 1985 Chris Curtis, a British pioneer in the field, noted ruefully that despite all the promises, the old way of creating sterile insects by chemicals or X-rays was still dominant:

> It would perhaps be premature to write off all the excitement about inherited factors for genetic control as a 'lead balloon', but we have not seen lift-off yet and, meanwhile, it is the conventional approach to sterile insect technique which has matured into a cost-effective and growing industry.[4]

At the beginning of the 1990s the WHO adopted a twenty-year plan to beat malaria using genetics, focusing on three aims: developing genetic tools to engineer mosquitoes, discovering how to make the insects genetically immune to malaria and developing ways of spreading that genetic immunity in natural populations.[5] This last target was the most challenging – after ten years the WHO could report important advances towards its first two objectives, but had to recognise that 'there has been no significant progress in developing methods for driving desirable genes into wild populations'.[6]

Few of the studies on genetic control of insect pests that were carried out in the closing decades of the twentieth century mentioned the ethical or political issues associated with releasing self-sustaining genetic elements affecting fertility. The researchers involved did not ask if it was the right thing to do, or if there was a better solution, and few seemed curious about potential damage to the ecosystem and even less about the need to consult the local population. And yet there was good evidence that such releases could cause protest and misunderstanding.

In 1967, a WHO project in a village near Rangoon released male mosquitoes carrying a genetic factor that massively reduced the number of viable offspring, leading to the temporary local eradication of the insects. However, the researchers warned that there could be problematic consequences:

> If in the future a population in an extended area is completely suppressed, the problem of filling the vacuum created could be serious. ... Another species of mosquito could come to occupy the niche. It might be harmless or a vector of the same importance.[7]

In 1975 the WHO had to call off a release of sterilised male mosquitoes in twelve villages near New Delhi because the Indian press claimed that the mosquitoes had been weaponised and would introduce yellow fever to the subcontinent (male mosquitoes do not bite). In a pompous but accurate editorial, *Nature* chided the WHO that it 'should have pursued a policy of being open with the Indian press – open about failures as well as successes'.[8] Three decades later, the row blew up again following the US admission that it had indeed carried out biological warfare tests using mosquitoes, including a 1965 release of yellow-fever-transmitting mosquitoes on uninhabited Baker Island in the South Pacific.[9] In reality, both releases were merely intended to track the dispersal of marked sterile males – the first, harmless, step to understanding how modified mosquitoes behave. The fact that an incident provoked by an anodyne trial still resonated nearly thirty years later shows the depth of public disquiet when it comes to the genetic manipulation of wild populations.

*

In 1992, the dream – or nightmare – of self-sustaining genetic control of pest species took a significant step towards becoming a reality when researchers discovered strains of the flour beetle *Tribolium* which, if crossed, led to the death of all offspring. The system was given a fancy acronym, 'maternal-effect dominant embryonic arrest' (Medea – in Euripides' Greek tragedy Medea kills her own children), and it was soon shown that the character could spread rapidly through a population.[10] The precise molecular basis of Medea was obscure, but it involved two components, a maternal factor that produced a toxin and a paternal element that provided an antidote. If the antidote component was not present, the insect died in the egg. In 2007, an artificial Medea system was built in *Drosophila*; within twelve generations, the construct had spread to every fly in the laboratory cages.[11] One entomologist predicted that 'in 5–10 years, fully functional gene drive systems with effector genes could be available in target species'.[12]

That turned out to be pretty accurate, although things did not proceed as anyone expected. At the beginning of the century, evolutionary biologist Austin Burt of Imperial College London was studying an example of biased inheritance seen in bacteria, fungi and viruses – but not animals – caused by homing endonuclease genes.[13] These genes produce a DNA-cutting enzyme (an endonuclease) that is precisely targeted on the location of the gene that encodes it. When one parent has the homing endonuclease gene and the other does not, any offspring initially has just one copy of the gene. But soon after fertilisation the nuclease cuts the DNA sequence on the chromosome that does not carry the gene, and the cell's DNA repair mechanisms then use the intact chromosome as a template to reconstruct the apparent gap in the DNA sequence. Where there was only one copy of the gene, there are now two, one on each chromosome. The same thing will happen in the next generation and the gene's frequency in the population will grow exponentially. This is in complete contrast to what normally happens with new mutations – unless they confer an advantage to the organism that carries them, they generally disappear from the population, through simple dilution.

In 2003 Burt realised that by hitching one of these endonuclease genes to a gene that would alter the target animal – for example, by inducing sterility or making a mosquito immune to malaria – it would theoretically be possible to drive that character into the population. One approach imagined by Burt was the use of multiple homing endonuclease genes, each with a slightly different target. If carried on the male's Y chromosome they would attack and destroy the X chromosome in his spermatozoa (most male insects, like us, have one X and one Y chromosome), thereby preventing any females from being produced (most female insects have two X chromosomes). This system was later given the alarming name X-shredder.[14]

Such a construct could theoretically drive an entire species to destruction. As critics put it back in the 1970s during the recombinant DNA debate around Asilomar, once an organism is released into the wild, you cannot recall it. Burt – who did not use the term gene drive, which had yet to be coined* – recognised this, warning that such a self-sustaining system could not be controlled and would continue to drive until it affected the whole species. But Burt also pointed out the cost of *not* acting – despite all the insecticide that is sprayed, the use of protective bed nets and so on, the diseases transmitted by mosquitoes cause over 700,000 deaths per year; malaria alone kills a child under five years old every two minutes.[15] He wrote:

> Finally, wide-ranging discussions are needed on the criteria for deciding whether to eradicate or genetically engineer an entire species. Clearly, the technology described here is not to be used lightly. Given the suffering caused by some species, neither is it obviously one to be ignored.[16]

When a new and potentially problematic technology appears, advocates routinely warn that restriction or regulation will lead to the public being deprived of future benefits.[17] The difference with gene drives is that they claim to be able to save millions of human lives, not make millions of dollars. Furthermore, in this case, researchers

---

*The earliest use I have found of the term was by Anthony James of the University of California, Irvine. James, A. (2005), *Trends in Parasitology* 21:64–7.

doing the work were calling attention to the dangers so that, if possible, lives could be safely saved. As at Asilomar, the cheerleaders and the Cassandras were the same people.

It would be another eight years before Burt's alarming idea could be tested. In 2011, Burt and colleagues from Imperial College, the United States and Italy introduced a yeast mitochondrial homing endonuclease gene into a malaria mosquito, where it rapidly invaded laboratory cage populations. It did no harm but it was everywhere. This was simply a proof of principle, but the principle it was proving had dramatic implications. As the researchers put it, this technique 'could be used to take the step from the genetic engineering of individuals to the genetic engineering of populations'.[18] Three years later, the Imperial College group realised another of Burt's ideas, when they created a synthetic X-shredder in mosquitoes. This led to over 95 per cent male offspring and rapidly suppressed a caged population of the insects.[19]

Even before such exotic genetic constructs were available, some scientists were thinking about the potential consequences of eliminating all mosquitoes. In 2010, *Nature* quizzed a variety of entomologists, ecologists and molecular geneticists about the possibility.[20] Although some of the ecologists pointed out that mosquitoes are important pollinators and only a very small minority of the 3,500 mosquito species transmit disease (for example, only forty species transmit malaria), most of those questioned were surprisingly cheery about the prospect. 'It's difficult to see what the downside would be to removal', said one researcher. Another dismissed the problem – 'If we eradicated them tomorrow, the ecosystems where they are active will hiccup and then get on with life' – before adding somewhat ominously: 'Something better or worse would take over.'

To be clear: no one is planning to inflict massive biocide, like Thanos in the Marvel Avengers films. The avowed aims of gene-drive researchers are precise, localised in time and space, and laudably humanitarian. But, as the saying goes, the road to Hell is paved with good intentions.

With the advent of CRISPR in 2013, scientists could now see how to create gene drives in pest organisms. In summer 2014, Kevin Esvelt and some of his Harvard colleagues, including George Church,

published a theoretical article on gene drives in which they let their imaginations run free. RNA-guided gene drives, they claimed, 'would represent an entirely new approach to ecological engineering with many potential applications relevant to human health, agriculture, biodiversity, and ecological science'.[21] The team described the complex molecular biology involved in creating such drives and outlined different types of system that would hypothetically either suppress or alter a population, including drives that would make a species sensitive to a particular molecule or restore susceptibility in a population that had become resistant to a particular pesticide.[22] They even saw the potential to use gene drives to aid conservation, for example, by targeting invasive species that cause untold damage to ecosystems across the planet. But 'invasive' is a relative term – these species are generally benign in their original location. Any gene drive targeting an invasive species might end up affecting its parent population too.

In 2021, Esvelt told me of his feelings when he realised what he could do:

> When I first thought of CRISPR-based gene drives, I was pretty elated on the first day, thinking this is the answer to malaria and to so many other ecological problems ... And then the next day I woke up and I was absolutely terrified, because I had realised that individual researchers in their labs, if they knew

> how to edit the genome of a relevant organism with CRISPR, they would have the power to do so, and let it go, and it would spread to affect everyone. Which from a security standpoint sounds pretty terrifying. So I actually didn't tell anyone – even my supervisor – for over a month.

Esvelt's foreboding was shared by his colleagues, so their 2014 paper emphasised the need to find solutions to problems before implementation, not afterwards. They tried to imagine ways of halting a drive once released, or of undoing any genetic damage that might occur if it all went wrong. They called for robust safeguards and methods of control, which included the inevitable technical solutions that might or might not work, but most significantly focused on the need for societal review and consent. Gene drives were too important to be left to the scientists.

Other researchers were less impressed by the idea of using gene drives in conservation. In 2015 a group of Australian scientists pointed out what might happen if an invasive species were suddenly removed:

> As with critical food resources or apex predators filling gaps after earlier human-driven extinctions, there remains a risk that removing species with gene drive technology could produce unintended cascades that may represent a greater net threat than that of the target species.[23]

Without a clear regulatory framework, they concluded, 'this putative silver bullet technology could become a global conservation threat'.

*

While these debates were taking place, the first CRISPR-based gene drive was being created, almost by accident. In 2014, Valentino Gantz of the University of California San Diego was finishing his PhD on the development of the *Drosophila* wing; unable to create enough flies carrying two copies of the mutation he was studying, Gantz thought about using CRISPR to multiply the number of flies to meet his needs. Neither Gantz nor his supervisor, Ethan Bier, had any inkling of the substantial literature and discussion of gene drives that had accumulated over previous years (as Kevin Esvelt put it, 'they hadn't read any of the earlier publications, seen any news coverage, or heard the warnings from other scientists'[24]). Indeed, they assumed their pilot experiment, using the well-known *yellow* body colour mutant, would not work. But it did, to their great excitement and alarm. According to the title of their article they had created a mutagenic chain reaction. To put it another way, they had made a genetic atom bomb.

At the end of 2014, as Gantz and Bier began writing up their discovery, the disturbing implications of what they had done began to dawn on them. At one point, Gantz recalled, 'we were thinking, should we even publish this?'[25] In the end, they went ahead but took care to emphasise the dangers:

> We are keenly aware of the substantial risks associated with this highly invasive method. … a dialogue on this topic should become an immediate high-priority issue. Perhaps, by analogy to the famous Asilomar meeting of 1975 that assessed the risks of recombinant DNA technology, a similar conference could be convened.[26]

Those risks did not stop them from immediately linking up with Anthony James, the mosquito biologist who had first used the term 'gene drive' in print. Within a few months they had extended their technique to mosquitoes, easily driving two genes that confer resistance to malaria.[27] However, although the experiments worked well, they were not carried out in population cages over multiple generations – a more naturalistic test than laboratory conditions. A few weeks later, Andrea Crisanti and Tony Nolan of Imperial College published a study that took that extra step: a CRISPR-based gene drive produced female sterility in cage populations of mosquitoes and grew in frequency over four generations.[28]

Even so, this was still not proof that a gene drive would work in the long run. The results of that test appeared in 2017, when the Imperial College group described a gene drive in two cage

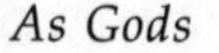

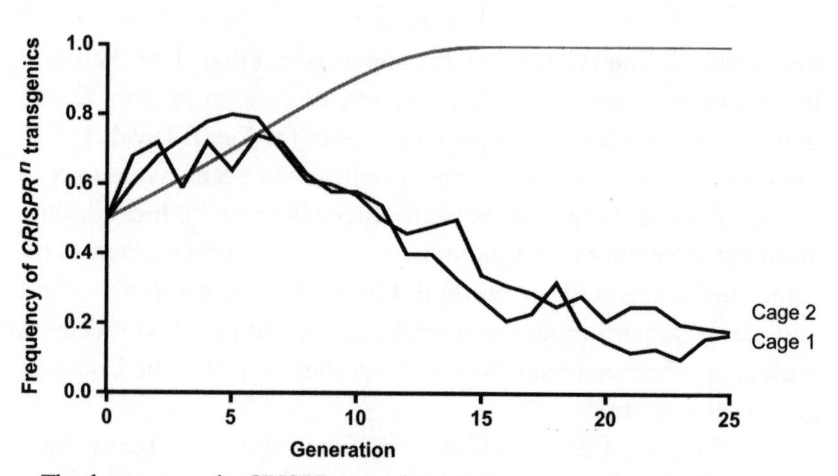

The frequency of a CRISPR gene drive in two cages of mosquitoes over twenty-five generations. The upper light grey line shows the expected result. From Hammond et al. (2017), *PLoS Genetics* 13:e1007039.

populations that spanned twenty-five generations. The results were initially exactly as expected, but after six generations the frequency of the drive suddenly dropped.[29] Nolan and Crisanti eventually concluded that the nuclease was causing mutations in its target site. The construct no longer recognised where it was supposed to cut and so it gradually declined in frequency. Similar results were observed by other researchers, studying two different gene drives in *Drosophila*. Their conclusion was gloomy for some but reassuring for others:

> Our results demonstrate that the evolution of resistance will likely impose a severe limitation to the effectiveness of current CRISPR gene drive approaches, especially when applied to diverse natural populations.[30]

All this had been predicted by mathematical models of gene drives, with researchers pointing out that the appearance of resistance might actually be an advantage as it would mean that the effectiveness of a drive would be limited in time, acting like a safety valve were things to go wrong.[31] Further problems were revealed when genomic and experimental studies of natural populations of three different insects – flour beetles, African malaria mosquitoes and the invasive fruit pest *Drosophila suzukii* – all reported genetic variation in target sites that reduced drive efficiency and led to the appearance of resistance. As one group concluded: 'standing genetic variation in a wild target population has the potential to severely compromise CRISPR/Cas9-based gene drives'.[32] In other words, despite all the alarm about the damage they might do, it was not evident that gene drives would work in the real world.

Scientists love a challenge, and within a year Crisanti and Nolan had used a highly conserved sex determination gene as the target for the drive – this made it unlikely that there would be genetic variation in natural populations, or that viable resistance mutants could arise. The construct was able to drive the gene into a cage population, producing a collapse in egg production and no sign of resistance.[33] This was strong evidence that gene drives, if properly designed, can make populations extinct. The group then replicated the result in larger population cages, which allow the flies to show more natural mating behaviour – in all the cages, the population disappeared in less than 320 days.[34] There appears to be no major technical obstacle to the release of such a genetic bomb.

Some people have argued for gene drives to be used to suppress weed populations, or to make resistant weed strains sensitive to herbicides once more by changing the resistance genes. This would be extremely slow (many plants reproduce only once a year), some plants are self-fertilising, limiting the effectiveness of the gene drive, while the way that plant cells respond to double strand breaks in their DNA means that endonuclease-based gene drives will probably be of such low efficiency that they would not work in practice. For the moment there are no plant gene drives and it is quite possible that none will ever be developed.[35]

The use of gene drives in conservation campaigns targeting invasive mammals, in particular rodents, may also be more difficult than expected. In 2019, Kim Cooper of the University of California San Diego reported an attempt to create a gene drive in mice using eight different strategies, but the DNA repair mechanisms at different stages of the mouse's life cycle meant that the drives were not always effective.[36] Furthermore, the constructs did not self-propagate and were not studied over multiple generations. Other researchers have also found creating a rodent gene drive to be difficult, concluding that it will be many years before one is made, if ever.[37] As Cooper told *Nature*: 'There's still so much work to be done to show that something like this is even feasible.'[38] For the moment, Elizabeth Kolbert's grim joke in the *New Yorker* remains fiction:

> Gene-drive technology has been compared to Kurt Vonnegut's *ice-nine*, a single shard of which is enough to freeze all the water in the world. A single X-shredder mouse on the loose could, it's feared, have a similarly chilling effect – a sort of *mice-nine*.[39]

*

In response to the growing concern about gene drives, National Academies of Science, Engineering and Medicine (NASEM) set up a Committee on Gene Drive Research in Non-Human Organisms to report on the ethical and technical issues associated with the new technology. The Committee's work was jointly funded and framed by the NIH, the Foundation for the National Health, the Bill and Melinda Gates Foundation and the Defence Advanced Research Projects Agency (DARPA) – by far the biggest global funder of gene-drive research.* As with the debates that were taking place on heritable human gene editing at around the same time, the main institutional players were not interested in asking if the technology *should* be applied, but instead wanted to identify 'the key scientific techniques for reducing ecological and other risks' prior to deployment.[40] At the first meeting of the Committee, DARPA representative Colonel Daniel Wattendorf warned that the panel's work might simply be ignored if it did not come up with the answer DARPA wanted:

---

*DARPA's interest in gene drives flows from the 2016 declaration by James Clapper, Obama's National Security Administrator, that gene editing is a weapon of mass destruction. DARPA immediately invested millions of dollars in gene-drive research, in particular the 2017 Safe Genes program, designed to find ways of stopping a drive once it has been released and to 'protect warfighters and the homeland against intentional or accidental misuses of genome editing technologies'. A condition of Safe Genes funding is that researchers will not release any gene drives.

> We may not have the time in this case to actually wait for, and make calls for, certain scientific actions and communities to deliberate. We may actually need to be working on technology solutions right now.[41]

DARPA knew they wanted to use gene drives and merely required a green – or yellow – light.

To their immense credit, the members of the Committee refused to accept the remit given by the funders, complaining that 'the tone of the statement of task presumes a certain level of technological development of gene drives is inevitable'. Instead, they proceeded as they saw fit. The Committee's report – *Gene Drives on the Horizon* – appeared in March 2016. Its fundamental conclusion, printed in bold type, was unequivocal:

> **There is insufficient evidence available at this time to support the release of gene-drive modified organisms into the environment.**[42]

Because of the potential benefits to human health, the Committee supported further research in both laboratory and highly controlled field trials, but placed a number of conditions on that approval, many of which would not have impressed the shadowy defence agency. Research groups were told to cooperate and produce open-access data repositories, public debate was needed to highlight the potential irreversibility of the environmental effects (and not simply the claimed health benefits) and it was essential that any proposed use of a gene drive should be governed by a strict decision-making process involving complex ecological models and public involvement.

None of this had the force of law, nor even the financial clout of the NIH in 1975, but the sober report carried weight around the world because of the panel's thoroughness. Despite the desires of DARPA and the other sponsors, the Committee injected a massive dose of caution and responsibility into gene drive research. In a letter to *Science* the Committee members explained that they were concerned that the thrill of discovery might push scientists to move from the laboratory to the field; the report aimed to counteract that impulse, they said.[43] Their explanation of this 'innovation thrill' was deliberately chilling. They cited the words of Robert Oppenheimer, leader of the Manhattan Project:

> When you see something that is technically sweet, you go ahead and do it and you argue about what to do about it only after you have had your technical success.

This was what the NASEM panel wanted to avoid at all costs. Their position was in dramatic contrast to the yellow light given to heritable human gene editing in 2017 by NASEM and the following year by the UK Nuffield Council on Bioethics. If only the scientists, physicians and bioethicists involved in those discussions had been as clear and cautious as the NASEM gene-drive committee.

As in the debates around recombinant DNA in the 1970s, worried scientists were more focused on technical issues associated with physical and biological containment than on the deeper question of whether the thing they had invented should ever be used. For example, in August 2015, all of the leading gene-drive researchers published an article in *Science* arguing for a combination of stringent physical, molecular, ecological and reproductive confinement strategies during experimental phases. So far, so good, but the stated aim was to build public trust for future applications when confinement could be safely abandoned.[44] This was effectively another prudent path to an application that was seen as inevitable. Later that same year, Esvelt and Church described methods of 'molecular confinement' that would reduce the risks associated with accidental release.[45] They even developed a way of returning the introduced gene form back to its original version, and then removing all the drive elements – Church semi-jokingly described this procedure as 'undo' or 'control-z'. But while this method might return the genome of the target organism to its original form, it would be unlikely to repair any damage to the ecosystem.

2017 marked a significant shift in scientific attitudes to gene drives, as the complexities of the question became clearer to many researchers. The Harvard group around Esvelt and Church begin to think more carefully about how to develop self-limiting, rather than self-sustaining, gene drives. They realised they needed to develop genetic constructs that would run out of steam after a set amount of time – or, to be more accurate, after a certain number of cycles of reproduction. One of their more interesting ideas was a 'daisy-chain drive', which would have several different components on different chromosomes or in different populations, and which would not work if any one element was absent.[46] That would inevitably happen, either as chromosomes were shuffled during sexual reproduction, or as different populations no longer exchanged their genes when the number of organisms declined because of the initial effects of the drive. Meanwhile, other researchers – funded by DARPA – developed an anti-gene-drive system that could stop CRISPR gene drives by using a protein derived from bacteriophages, which have been engaged in a struggle with bacteria and their CRISPR system for billions of years.[47]

These ideas for making gene drives safe are all very clever and clearly echo Sydney Brenner's 1975 aim of using genetics to 'disarm the bug' in order to proceed safely and to calm legitimate fears. Apart from the fact that many of these methods remain hypothetical (for example, no daisy-chain drive has even been built, never mind shown to work over the long term in a population cage), they all suffer from the overall problem that something might go horribly wrong, either through accidental release or some unforeseen genetic or ecological event.

These concerns were shared by some of the researchers. In 2016, Esvelt predicted that by around 2030 there would be an unfortunate event involving gene drives. 'It's not going to be bioterror, it's going to be bioerror', he said.[48] Like many other scientists he feared that, as with previous catastrophic incidents such as the death of Jesse Gelsinger, public support for genetic engineering could be undermined for many years as a consequence.[49] The solution was to work out what could go wrong before it actually did, to build in regulatory safeguards and to nurture public understanding of the key issues:

> History suggests that safety engineering becomes a primary concern only after a well-publicised disaster. That is not a pattern we care to perpetuate. Now is the time to be bold in our caution.[50]

Thinking about things in this way led Esvelt to admit that it had been 'an embarrassing mistake' to suggest that self-sustaining gene drives could be used for conservation. 'The kind of gene drive that is invasive and self-propagating is in many ways the equivalent of an invasive species', he realised.[51]

*

While scientists have focused on technical solutions to contain and control gene drives, scholars from other disciplines have adopted an organisational, regulatory approach. For example, Stanford researchers, including the political scientist Francis Fukuyama, argued for the creation of 'permanent, credible national-level oversight and standard-setting bodies that work in partnership through international entities'.[52] Their model was the International Civil Aviation Organization (ICAO), a voluntary but authoritative international regulatory body which enables us to fly in relative safety. Something similar could be set up to govern work around gene drives. This is an excellent proposal, but like the International Atomic Energy Agency, another non-statutory international organisation that regulates a very dangerous technology on a voluntary basis, the ICAO was set up after the Second World War when all countries, including the United States, had an appetite for international regulation. That is no longer the case. Bigger geopolitical issues than gene drives would need to be resolved before this could become a reality, but nevertheless, this is the route the world needs to take to ensure that, if it were ever to be used, gene-drive technology would be safe.

In 2018, Yale University researcher Natalie Kofler and a group of colleagues put forward a slightly different version of the same idea, calling for a 'global coordination task force' that could deal with 'the potentially ramifying global implications of environmental release of gene-edited organisms ... an approach that places great weight on local perspectives within a larger global vision'.[53] Their ambitions for this hypothetical body, while laudable, were sky-high:

> Through the collective creation of this new governance model, the first that proposes a connection between local needs and

> global frameworks and expertise, our world may realise this technology's most profound benefit – the opportunity to inspire a more healthy and just future for all who share our planet.

Part of the dilemma for US researchers is that their country is out of step with the rest of the world on the question of how to regulate developments in biology. The main international instrument that relates to gene drives is the United Nations Convention on Biological Diversity (CBD) and its implementation through the Cartagena and Nagoya Protocols.[54] Nearly 200 countries have signed up to the CBD, which involves a strong precautionary approach. Alone out of all the UN member states, the United States has not signed the Convention, nor is it likely to.* As a result, US researchers exploring the need for regulation of all aspects of genetic engineering are obliged to search for alternatives, even if some of these proposals seem like reinventing the wheel or are worryingly parochial, focusing solely on US regulation when faced with a global issue.[55]

Because the CBD is the only recognised international body that has even a chance of regulating gene drives, opponents of the technology have focused much of their activity around CBD meetings, which are held every two years, most recently in Cancun (2016) and Sharm el-Sheikh (2018) (the 2020 meeting was postponed because of the pandemic, taking place in two parts over 2021 and 2022). Opponents claim that the true aim of gene-drive research is to transform agriculture, as revealed by patent applications, which routinely include hundreds of agricultural weed species and insect pests. But patent applications are intentionally cast as wide as possible, such as including plant endonuclease gene drives that will almost certainly never work. Patents can also be used to control how gene-drive technology is employed – the Broad Institute licensed its CRISPR patents to Monsanto, but explicitly prohibited their use for creating gene drives or terminator seed or for use with tobacco.[56]

---

* The United States disagrees with the content of the Convention; it is also generally opposed to international control over its activities – it has not, for example, signed up to the International Criminal Court.

In the run-up to the 2016 Cancun meeting of the CBD, thirty leading conservationists called for 'a halt to all proposals for the use of gene drive technologies, but especially in conservation'. The idea of a moratorium was rejected by the meeting and little progress was made, leaving nobody satisfied.[57] Two years later, shortly before the 2018 meeting at Sharm el-Sheikh in Egypt, a major row erupted when a US freedom-of-information (FoI) request led to the publication of 1,200 emails between gene-drive researchers in which they discussed how they would organise around the forthcoming CBD meeting.[58] The ETC group of environmental activists, which was responsible for the FoI request, claimed the emails revealed that the researchers were trying to manipulate the process. On one level this was not so different from the behaviour of gene-drive opponents themselves, except with far better funding. Although this was a storm in a teacup, one part of the email haul did attract wider media coverage – the US military's interest in research on deploying and stopping gene drives. The emails revealed that DARPA funding of gene drives was even higher than the $65 million for the Safe Genes project, hitting $100 million in total.[59]

At Sharm el-Sheikh there was a renewed call for a global moratorium on gene drives, signed by over 100 organisations, including many small conservation and anti-GM groups from around the world, as well as José Bové's Confédération Paysanne.[60] In general, the demands raised by gene-drive opponents were things that many gene-drive scientists would agree with – the use of strict containment procedures, the need for reversibility to be engineered into drives from the outset, the importance of obtaining 'free, prior, informed consent' from the local population and opposition to 'dual-use', whereby a scientific invention may have a military application.[61] There was little sign of the claimed attempt at coordinated activity by gene-drive advocates.

The meeting itself was tense – Natalie Kofler described the atmosphere as 'pretty raw'– as activists drew comparisons between gene drives and nuclear weapons and claimed that malaria research was a cover for the development of profitable agricultural gene drives. In the end, the meeting again rejected calls for a moratorium but nevertheless insisted on the need for detailed environmental risk assessment procedures and above all argued for the active involvement of communities before any release took place.[62]

This focus on obtaining free, prior and informed consent from local communities was also seen in a significant WHO document which appeared in June 2021 – the updated Guidance Framework for GM mosquitoes.[63] All GM insects released so far have been engineered only to be sterile or to produce no female offspring and have been used simply to temporarily suppress local populations. Nevertheless, the bad reputation of GM and deep fears of altering the ecosystem, perhaps by the transfer of genes to local populations, have combined to provoke opposition from communities where commercial companies have carried out such releases.[64] The revised WHO Guidelines emphasised the need for gene-drive projects to be co-developed with local communities, including a series of checkpoints that could lead to the project being abandoned.

The WHO's serious and detailed approach to environmental risk assessment will carry weight, even if the organisation does not have any direct regulatory authority. Some idea of what will be involved in meeting these standards can be seen in the work of Imperial College researchers funded by Target Malaria, a not-for-profit research consortium funded by the Bill and Melinda Gates Foundation. Target Malaria organised a series of workshops in Africa that explored the kinds of ecological problems that should be studied, and they identified eight major effects that could occur through forty-six plausible pathways.[65] Each of these possibilities would have to be tested in the field before any decision could be taken about deployment, even if the local community were in favour. If ecology is complicated, so too is ecological risk assessment, if done properly.

*

Although it is now widely accepted (but not legally required) that free, prior and informed consent is needed from the community before any gene-drive release could take place, obtaining such consent turns out to be quite difficult. With the approval of the Burkina Faso government, in July 2019 Target Malaria released male flies that had been sterilised and dusted with fluorescent powder in a small number of villages with high rates of malaria.[66] Like the aborted WHO release in India in 1975, the aim was simply to see how far marked flies dispersed, and whether they could be captured.* Unlike the 1975 release, the project embraced the need to fully inform the population beforehand. Which is where the problems began.

The local language has no word for 'gene' or 'genetically modified', so terms had to be invented by the Target Malaria workers.[67] However, the words spontaneously used by the villagers revealed the long-term negative impact of Burkina Faso's failed experiment with Bt cotton at the beginning of the century – the popular term used to describe the mosquitoes (*soussou*) was *soussou-BT*.[68] Explaining the Target Malaria plans to the population was challenging – some people did not know that only females bite, or that bed nets could protect from malaria.[69] Unfounded fears expressed by villagers included concerns that modified mosquitoes would transmit HIV, or that the insects might grow to the size of a helicopter, while others asked how it was possible to have something that was half human, half mosquito.[70] One part of the campaign therefore involved the use of theatre to explain the project, ensuring that illiteracy would not be a barrier to understanding and decision-making.[71]

Some residents expressed enthusiasm – including those paid by Target Malaria to catch mosquitoes as part of the study – and felt the project was of great significance. 'We are searching for a solution to malaria to save tomorrow's people', said one young mosquito collector.[72] Nevertheless, when subsequently questioned by journalists, some locals claimed they had not been properly informed. 'They didn't tell us about the risks, only the advantages', said one farmer, while a woman resident complained that there was not enough opportunity for discussion: 'We were able to ask all the questions we wanted, but after they had gone new questions arose. If I had known, I would have asked them "And if it doesn't work, what are the risks for our village?"'[73]

The gulf in knowledge – in fact, a power asymmetry – between the researchers and the population was immense and led to some

*Not far, and yes, were the answers. The experimental mosquitoes also tended to die earlier than the controls.

villagers feeling impotent. 'They tell us they are going to eradicate malaria, but because we aren't scientists, we believe them, but we still have questions about future risks', one farmer told *Le Monde*.[74] Anti-GM campaigner Ali Tapsoba highlighted this problem, claiming that there was 'no real free and informed consent but rather an abuse of the ignorance and illiteracy of local communities'. In response to Target Malaria's statement that it was not possible to obtain consent from everyone, journalist and film-maker Zahra Moloo turned the argument on its head:

> But the reason it is difficult to acquire informed consent from all people affected by gene-drive experiments is the same reason that doing so is absolutely critical. ... Securing the consent of only a handful of local residents simply is not good enough.[75]

Before the release there was a 1,000-strong protest march in Ouagadougou, the capital. Ali Tapsoba warned that any eradicated species might be replaced by another species that could become more dangerous and claimed that although Target Malaria had opened a dialogue, the organisation had refused to meet opponents or respond to questions for over a year.[76]

As the example of Burkina Faso shows, ensuring free, prior and informed consent turns out to be remarkably hard. Communication needs to go beyond merely informing, to a style that involves the community in questioning the most fundamental aspects of the project, including exploring alternative means to achieve the desired end. This is particularly important where people might be rightly wary of scientists in white coats, given a past history of colonialist medical interventions.[77] Furthermore, although existing leaders need to be respectfully consulted, in the words of one group of US academics, 'researchers should be cautious not to cement these actors as permanent gatekeepers to information and participation'.[78] Communities are not homogeneous, and it is possible that some sections might be difficult to access, due to power structures within the society.[79] As one woman in Burkina Faso told *Le Monde*: 'In any case, we won't have any say in it, it's the men who take all the decisions here.'[80]

All these points, including the need to engage with opponents and take their views into consideration, have recently been recognised by Target Malaria following a workshop in Kenya.[81] Putting these principles into practice will be far harder than developing a safe gene drive.

*

Maybe there is nothing to worry about. Maybe gene drives will not work because of the amount of genetic variation in natural populations, or because populations are fragmented without sufficient gene flow, or because ecological factors will prove more challenging than expected, or for some other reason that we have not thought of. Or maybe they will work brilliantly, precisely affecting only a small local population of disease vectors. It is also possible that the disappearance of a species from a particular locality will have no untoward ecological effect. Gene-drive advocates point out that none of the species currently being targeted is the sole food source for any other species, implying that if they were to disappear, no harm would be done. But the opposite might be true. For example, the malaria mosquito *Anopheles gambiae* is eaten by a whole range of flies, true bugs, dragonflies and damselflies, shrimp, spiders, flatworms, amphibians, fish, birds, bats and aquatic parasites.[82] If even some of the species within these large groups go just slightly hungry, an unforeseen cascade of ecological problems might arise. Simply saying no single predatory species would be heavily affected does not mean that a release would have no deleterious effects on the mind-bogglingly complex webs of interactions that make up an ecosystem.

It is also depressingly true that even if the worst were to happen, this would be minor compared with the environmental havoc that we have long been wreaking through intensive agriculture, habitat destruction and now climate change. But it is precisely because things are already pretty bad that we need to do all we can to avoid adding to the planet's problems. The real possibility that mating between species might in some cases allow gene drives to jump from one species to another represents an additional reason to be extremely cautious and to choose targets carefully.[83]

Maybe gene-drive technology will not be used to tackle

insect-transmitted diseases because simpler solutions emerge. After years of stagnation, malaria vaccines are now making important progress, while if *Aedes aegypti* mosquitoes are infected with naturally occurring *Wolbachia* bacteria (these are one of the causes of the natural biased inheritance patterns in insects that have so intrigued evolutionary biologists), they are less able to be infected with the dengue and Chikungunya viruses or the malaria-causing *Plasmodium* parasite.[84] A field trial in Indonesia showed that deliberately infecting mosquitoes with *Wolbachia* led to a significant decline in dengue infection rates in humans. In some localities this may represent a stable solution to this unpleasant tropical disease, and perhaps to others, including malaria.[85]

Ultimately, the promises and perils of gene drives do not relate simply to tropical diseases, or only to ecology. US academic Maywa Montenegro de Wit has explored how we can practically balance the possibility of saving millions of lives with the difficulty of identifying potential dangers and the need to obtain free, prior and informed consent from the local community.[86] Providing local communities with a veto is essential, but self-sustaining gene drives challenge our notions of what 'local' is. As Kevin Esvelt has put it, 'a release anywhere, is likely a release everywhere'.[87] A malaria-ridden village might understandably want to be rid of mosquitoes and might perhaps be prepared to do anything to save their children's lives. But it is not clear that they should have the right to determine the future for the rest of the region, country, continent or even planet. This is why some kind of international oversight body with the power of regulation, like the International Civil Aviation Organization, is essential. Ultimately, gene drives are simply one piece of a vast puzzle. As Montenegro de Wit has put it, the issues raised by gene drives are much more profound than fancy genetics:

> How do we solve the root problems – of poverty, inequality, and late-modern globalisation – that propagate the crises that gene drives are designed to address?[88]

– FIFTEEN –

WEAPONS

When genetic engineering became a reality, there were fears that scientists might create deadly and infectious life forms that could escape from the laboratory. It was because of these concerns that the Asilomar process was launched, culminating in the 1975 conference. As it became clear that recombinant bacteria and viruses could be manipulated safely, the nightmares faded and scientists openly applied genetic engineering in the most ingenious and productive ways. That is the official history, which has been more or less followed in the preceding pages.

But alongside this version of events there was also a secret history of perverted science. It played a role in some of the major conflicts of the last half-century, forming an ever-present lurking threat to humanity that we were not even aware of. Hidden in the shadows for most of the last fifty years, this parallel world involved using genetic technology to create terrible biological weapons.

It was there from the very beginning, hiding in plain sight.

*

On the evening of the third day of the Asilomar meeting a photocopied telegram was handed out to delegates. It was from a dissident

Soviet molecular biologist, Alexander Goldfarb, whose recent request to emigrate to Israel had been denied because his work on *E. coli* was deemed 'important for the state security of the USSR'.[1] Goldfarb pleaded with the meeting to discuss whether such research might be used for biological warfare. His call was ignored: from the outset, the Asilomar committee had decided that there would be no discussion of ethical issues, human genetic engineering or biological warfare. The overwhelming majority of the scientists at the meeting agreed – they considered that their expertise lay in simply developing containment protocols that would enable benign research to go ahead safely. There was no point in speculating how malicious forces might use the new science.

To some, it appeared that the threat of biological warfare had already been neutralised. In November 1969, President Nixon unexpectedly and unilaterally discontinued the US biological warfare programme; within two years US bioweapon stockpiles were destroyed.[2] These weapons all used existing microbes to infect humans or animals and were of the kind that had been developed during and after the Second World War (the Japanese used them and Churchill was keen on the idea, ordering 500,000 anthrax bombs from the United States). A month after Asilomar, in March 1975, the 1972 Biological Weapons Convention (BWC), signed by eighty-five nations including the United States and the Soviet Union, was due to take effect. The Asilomar conference recognised the potential danger and the meeting proposed, and the NIH accepted, that research involving extremely dangerous pathogens should be forbidden, but that was as far as it went. Although David Baltimore stated at the opening of Asilomar that recombinant DNA 'is possibly the most potent technology for biological warfare', delegates could be excused for thinking that the nightmare of recombinant DNA bioweapons could be forgotten.[3] International diplomacy had confronted the threat of weaponised genetic engineering and it would now never be developed.

That turned out to be wishful thinking. Close examination of the list of Asilomar attendees reveals that the five-man Soviet delegation included three leading molecular biologists from the Institute of Molecular Biology in Moscow – Alexander Bayev, Vladimir

Engelhardt and Andrej Mirzabekov* – all of whom had been involved in the recent decision by the USSR to launch a top-secret offensive biological warfare programme.[4] This programme was composed of two branches, one aimed against humans, code name Ferment (this means 'enzyme' in Russian), the other, which is still poorly understood, was targeted against agricultural plants and animals and was called Ekology.[5] In 1971 Bayev had co-authored a letter to Soviet premier Leonid Brezhnev proposing a new biological weapons programme. Together with his collaborator Mirzabekov, Bayev was a member of the secret Interagency Scientific and Technical Council for Molecular Biology and Genetics. Its role was to oversee Ferment and to ensure that the technical issues associated with the development of biological weapons ran smoothly.[6] The third Asilomar attendee, Soviet Academy of Sciences member Engelhardt, was the director of the Institute of Molecular Biology, one of the main contractors for Ferment research.[7]

Ferment soon mutated into a huge network of laboratories, institutes and manufacturing facilities, known as Biopreparat, that sprawled across the Soviet Union. It began operation in April 1974, just as Berg, Baltimore and the others were writing their open letter drawing attention to the dangers of genetic engineering. Within fifteen years, the Biopreparat archipelago employed 30,000 people in nearly forty institutions, carrying out some of the deadliest research imaginable, involving the use of genetics to create highly infectious and lethal microbes.

The Soviet delegation at Asilomar gave no hint of their country's interest in using genetic engineering to create bioweapons. Instead, they behaved exactly as their Western hosts expected. During the meeting, one senior US microbiologist chuckled over dinner:

---

*Mirzabekov was well-known to Western molecular biologists. In 1971 he worked on sequencing a transfer RNA molecule in Cambridge with the support of Francis Crick, Frederick Sanger and Max Perutz. He remained in contact with Jim Watson and other leaders of molecular biology until the early 1990s. After attending Asilomar, Mirzabekov travelled to Harvard to visit Wally Gilbert; Gilbert later said that some of Mirzabekov's remarks over dinner prompted him to come up with a method for sequencing DNA, which five years later led to Gilbert's Nobel Prize.

> These Russians, they just send over the old guys from the academy who don't know *anything*. You ask them something and they hedge around – and it's not that they're hiding things … they just don't *know* in the first place.[8]

Rolling Stone journalist Michael Rogers was more curious about what he described as the somewhat puzzling Soviet presence at Asilomar. He even went so far as to ask one of the delegates whether the USSR considered molecular biology to be militarily significant. He got no direct answer but did not pursue the matter any further.

The Soviet bioweapons programme was the brainchild of Yury Ovchinnikov, a young chemist and a member of the Central Committee of the Communist Party of the Soviet Union, as well as a confidant of Brezhnev.[9] The head of the Soviet delegation to Asilomar, Alexander Bayev, was a colleague of Ovchinnikov and had closely followed developments in genetic engineering after Beckwith's isolation of the *lac* gene in 1969.[10] Although much of Biopreparat's work revolved around 'improving' traditional bioweapons such as anthrax and plague, the key promise that had sold the project to the Soviet leadership was that genetic engineering would lead to radical new lethal agents. That required funding, buildings and personnel.

To hide the growth of Biopreparat, a civilian cover was created in the shape of a new research programme into physicochemical biology and biotechnology, which soon provided both civilian and military breakthoughs.[11] In 1977 the *Los Angeles Times* described the growth in Soviet genetic engineering research and gave hints of a military appetite for the technology.[12] But the newspaper obscured these insights with claims that Soviet scientists such as Bayev shared the concerns of their US counterparts about the safety of recombinant DNA – the NIH regulations were adopted by Soviet laboratories and were prominently displayed when Western journalists or academics visited.[13] Bayev was quoted as saying that biological warfare would not only contradict the BWC, but also 'the moral code of science'. He was the thoughtful civilian frontman for a murderous secret project.*

---

*In 1995, Bayev wrote a memoir, 'The Paths of My Life', in which he recounted his absurd prosecution as a 'Bukharinite terrorist' in the 1930s, as a result

The first research programme developed by Ferment was called Koster (Bonfire); it used genetic engineering to create three classes of novel bioweapons: antibiotic-resistant bacteria, microbes with altered antigenic structures that would avoid detection and treatment, and bacteria with 'wholly new and unexpected properties'.[14] Other research, commissioned from laboratories outside of Biopreparat, included creating recombinant antibiotic-resistant versions of the plague bacterium *Yersina pestis* and of anthrax. While the latter was finally achieved in the late 1980s, the former never succeeded, although Soviet scientists were able to alter the antigen profile of the plague bacillus, thereby making it unidentifiable, potentially confusing enemy medical personnel and delaying treatment.

Biopreparat suffered from the counterproductive Soviet bureaucratic regime of command and control, as well as its own fragmented structure.[15] It was also crippled by a lack of resources, despite the vast funding the programme enjoyed. For example, the Soviet system restricted access to restriction enzymes, which eventually forced Biopreparat to set up its own supply chain. Perhaps because of these problems, success was a long time coming. According to Biopreparat scientist Kanatzhan Alibekov, a Kazakh who later Westernised his name to Ken Alibek, by 1989 'most of us had given up hope of ever obtaining results'.[16] In the end, however, they were able to make the breakthroughs they desired, which were greeted with a mixture of pride and horror. Alibek recalled how in 1989 a researcher on the Bonfire programme announced the conclusion of a long attempt to engineer a close relative of the plague bacillus to produce a toxin against myelin, a key component of nerves, effectively producing an artificial version of multiple sclerosis that would rapidly incapacitate the victim's senses and their motor systems:

> The test was a success. A single genetically engineered agent had produced symptoms of two different diseases, one of which could not be traced. The room was absolutely silent.

---

of which he spent a decade in the Gulag. Bayev described his work with Mirzabekov but said not one word about Biopreparat. Bayev, A. (1999), *Comprehensive Biochemistry* 38:439–79.

> We all recognised the implications of what the scientist had achieved. A new class of weapons had been found.[17]

*

It was not only the Asilomar scientists who had no idea of the real motivations behind the new Soviet interest in genetic engineering.[18]* Western intelligence was equally in the dark.

In 1976 the US Defense Intelligence Agency produced its first report on Soviet genetic research.[19] It gave no indication that the USSR might be considering any military application of the technology. Instead, the US intelligence community complacently considered that the sudden increase in Soviet genetic engineering capacity was driven by embarrassment at the US lead in the field. Following an anthrax outbreak in Sverdlovsk (now Yekaterinburg) which killed sixty-four people in 1979, the United States informed the Soviet Union that it considered there had been a breach of the BWC.[20] The Soviets denied that the incident had anything to do with bioweapons, claiming that the victims had eaten anthrax-contaminated meat.† A series of independent US scientific visits in the late 1980s and early 1990s, led by sociologist Jeanne Guillemin and her husband the molecular geneticist Matthew Meselson, showed that the geography and prevailing winds at the site strongly suggested that the source of the infection was a nearby Biopreparat facility. According to Meselson, the amount of anthrax released may have been tiny – perhaps merely milligrams. The exact cause of the release – explosion, accident or experiment – remains unknown.

Despite the Sverdlovsk outbreak, lack of data continued to

---

* The presence at Asilomar of a sixth, non-scientific, member of the Soviet delegation, described by Michael Rogers as 'a charming, dapper, San Francisco vice-consul' who could speak perfect English, might have rung a few alarm bells.

† In June 1985, Bayev urged US microbiologist Joshua Lederberg, on a visit to Moscow, to ignore claims from 'dissidents' that there had been a bioweapons incident at Sverdlovsk, insisting that it was a public health matter. Ovchinnikov told Lederberg he knew nothing about the incident at all. http://resource.nlm.nih.gov/101584906X18644

hamper Western intelligence. As late as February 1982, the US Defense Intelligence Agency remained open to the idea that the Sverdlovsk anthrax infections were a natural event and viewed a breach of the BWC as something that might happen in the future.[21] But by the end of 1983 this attitude had changed and intelligence reports began to speculate – quite accurately, as it turned out – about the kind of work that the Soviets might be doing. However, not only did the United States have no firm evidence, relying instead on hearsay and suspicion, it also continually got lost in an intelligence hall of mirrors as confirmation bias led it to believe false stories.

In the fevered atmosphere of the new Cold War, one of the United States' prime propaganda claims was that the Soviet Union had developed a fungal toxin weapon known as Yellow Rain, which it supposedly used in Afghanistan and also supplied to Vietnam, Cambodia and Laos.[22] But there was no evidence to support the US government's claim that these yellow blobs on leaves in the forests of south-east Asia were a bioweapon. The actual explanation was found by the indefatigable Matthew Meselson in 1983 – the blobs were bee crap.[23] In response to this debunking, in 1984 the *Wall Street Journal* went on the offensive and published a series of eight informed but occasionally lurid articles by one of the newspaper's editorialists, William Kucewicz, claiming that the Soviet Union had now gone 'beyond Yellow Rain' and was using genetic engineering 'to recombine the venom-producing genes from cobra snakes with ordinary viruses'.[24]

Although the US intelligence sources who presumably briefed Kucewicz sometimes furnished him with accurate information – for example, the articles fingered Ovchinnikov as the brains behind the Soviet bioweapons programme – there was still no evidence that the programme existed, for the simple reason that Western intelligence had no proof of what was happening in the USSR beyond what they heard from exiles of varying degrees of credibility.[25] At the same time, whether out of curiosity, fear of missing out or malicious intent, the number of Pentagon-funded projects using recombinant DNA technology increased from zero in 1980 to more than forty in 1984.[26] This did not go unnoticed in the Soviet Union, where politicians and generals had always been convinced that the United States, too, was surreptitiously breaching the BWC.[27]

BEYOND
'YELLOW
RAIN'
THE THREAT
OF SOVIET
GENETIC
ENGINEERING
FIFTH OF A SERIES

Logo for a series of articles on genetic engineering in the Soviet Union that appeared in 1984 in the *Wall Street Journal*.

In the summer of 1989, Western intelligence learned the full extent of what had been happening in the Soviet Union when a leading Biopreparat scientist defected. Vladimir Pasechnik's bosses had allowed him to travel to Toulouse to complete the purchase of chemical equipment; once he got there he telephoned the British Embassy in Paris and announced his desire to defect. He was rapidly spirited to London on a commercial flight using a false passport.

The information Pasechnik provided during his long debriefing alarmed both British intelligence and the CIA (the American interrogators included Nobel Prize winner and Asilomar attendee Joshua Lederberg).[28] Pasechnik's evidence, which everyone agreed was reliable, revealed both the full extent of Soviet genetic engineering of new weapons and the West's utter failure to have the slightest idea of what was going on. As a British debriefer explained:

> The information was so stunning. A whole ministry exposed, billions of roubles spent, a complete organisation shown to be a front. ... It just went on and on.[29]

Armed with Pasechnik's evidence, Margaret Thatcher and George Bush pressured Soviet premier Gorbachev into accepting a visit by US–UK weapons inspectors. The inspection team was met with denials and Biopreparat laboratories that had been hastily emptied. One UK inspector said: 'This was clearly the most successful biological weapons programme on earth. These people just sat there and lied to us, and lied, and lied.'[30]

A year or so later, after the collapse of the Soviet Union, other Biopreparat scientists defected, were intensively debriefed by US and UK security agencies, and were eventually allowed to reveal some of their alarming secrets to a startled global public. Kanatzhan Alibekov (a.k.a. Ken Alibek) had once been First Deputy Director of Biopreparat; in the West he became the most widely known Biopreparat defector. In March 1998, once his intelligence handlers were satisfied they had wrung all useful information out of him and that he could be relied upon to talk only about what he was supposed to talk about, Alibek was interviewed on US television, six years after his defection.[31] He subsequently published a disturbing and powerful book, *Biohazard*, and became a key public source for information about Biopreparat.[32]

With the public admission by Russian president Boris Yeltsin that the Soviet Union had indeed developed an offensive biological weapons programme, there was a joint UK–US–Russian agreement on dismantling the system and destroying the weapons. This agreement eventually ceased to function in 1996.[33] In the new Russia, Biopreparat was privatised, and carried out secret contract work for the Ministry of Defence before eventually turning to civilian biomedical research and rapidly becoming embroiled in the kind of financial misappropriation that was widespread in this period.[34] For a while, potential bioweapons research continued more or less openly – at the end of 1997, Russian scientists from an ex-Biopreparat facility published an article in a Western scientific journal describing how they had genetically engineered the anthrax bacterium so that it could infect vaccinated hamsters.[35] Then everything went dark. It is not clear whether any of the ex-Biopreparat facilities still retain a biological warfare function, or if new institutions have emerged in Putin's Russia that are fulfilling this role.[36] It would surely be naive to imagine that Russia has completely abandoned all interest in bioweapons, given its use of chemical nerve agents in Salisbury in 2018.

*

Of the major powers, only the Soviet Union appears to have used genetic engineering to develop new biological weapons during the twentieth century. The US programme was discontinued before the advent of recombinant DNA, there was apparently no interest from the British, while the French were quite content with their nuclear weapons.[37] China, which was attacked with biological weapons by the Japanese during the Second World War and eventually signed up to the BWC in 1984, repeatedly insisted that the People's Liberation Army had no offensive biological warfare capacity, but (like the United States) it did have an anti-biological warfare unit. In the late 1980s, Soviet analysts concluded that an outbreak of haemorrhagic fever near the Chinese Lop Nor nuclear testing facility indicated that bioweapons research was taking place there.[38]

Effective regulation depends on openness, inspection, trust and potential sanctions. If any organisation is prepared to betray trust and hide information, regulation collapses. This is particularly true if, as in the case of the BWC, the regulatory body has no provisions for monitoring and verifying compliance, or of imposing sanctions in the case of a breach.[39] The BWC signatories are aware of this, and in the 1990s they sought to adopt a legally binding inspection regime. This protocol was rejected by the United States, which claimed that inspections on US soil would threaten the country's biodefence and pharmaceutical sectors. The Convention remains toothless.

The relative weakness of the BWC is shown both by the Soviet example and by South Africa, which developed a programme called Project Coast that was revealed only in 1996, two years after Nelson Mandela became president.[40] From 1981 onwards, the apartheid government, which had signed up to the BWC from the very beginning, sought to develop weapons that could be used against individuals or crowds, under the leadership of Wouter Basson, later known in South Africa as 'Doctor Death'. The Project Coast production facility, Roodeplaat Research Laboratories, was privatised in 1991 but went bust in 1994 under a cloud of rumour and claims of misappropriation. At the turn of the century, individuals attempted to sell the collection of microbes created by Project Coast to the US government for $200 million – this included an engineered strain of *E. coli* that had an additional gene that would produce epsilon toxin, 'the third most potent bacterial toxin known'.[41] As part of the attempted sale, samples of the microbes were taken to the United States on a commercial flight, in a glass container concealed in a tube of toothpaste. Despite verifying that the material was indeed as claimed, the US security agencies decided not to go through with the deal and informed the South African authorities.

Other major countries with an apparent interest in developing biological weapons have simply refused to engage with the BWC, provoking speculation and suspicion. For example, Israel is widely suspected to have developed biological weapons, in particular through the activities of the Israel Institute of Biological Research (IIBR).[42] Set up in 1952, the IIBR admits only to research that is consistent with defence against bioweapons – detection, diagnosis and treatment. But the difference between defensive and offensive research is largely a matter of motivation rather than some absolute distinction. For example, it was revealed in 1998 that the US Naval Research Laboratory in Washington DC had developed recombinant fungi that could degrade anti-radar coatings on aircraft, thereby enabling them to be detected and destroyed. Other fungi were developed that targeted crops such as coca plants and opium poppies as part of the United States' 'war on drugs'.[43] All this was done in the name of defence.

Although Iraq signed the BWC in 1972, under Saddam Hussein it consistently refused to engage with the required processes, with tragic consequences. During the first Gulf War in 1991, there were repeated concerns that Iraq would use chemical weapons (in 1988 thousands of Kurds had been killed in Halabja by an Iraqi chemical weapon attack). After the defeat of Iraq, UN weapons inspectors discovered that, unknown to Western intelligence, the Saddam Hussein regime had developed a bioweapons capacity. Iraq repeatedly denied this before eventually admitting that it had both produced and weaponised anthrax and botulinum toxins.[44] Although the Iraqi bioweapons programme was ended and its stockpiles were destroyed in the early 1990s, US intelligence either could not or would not recognise this. Having failed to spot the existence of the programme when it was in progress, the spies now assumed it still existed after it had been stopped.

From 2001 the CIA, the Bush administration and the UK Blair government were seduced by stories from defectors and were determined to press the case – any case – for war. In the run-up to the US–UK invasion of 2003, Iraq was repeatedly accused of developing biological weapons of mass destruction alongside its supposed nuclear programme. There is no evidence for any of these claims. Western intelligence sources claimed that Iraq established a genetic engineering unit with the intention of creating antibiotic-resistant anthrax, but no evidence has ever emerged. Despite all the intensive searches that took place in the chaotic wake of the 2003 Iraq War, no trace of genetic engineering of bioweapons was ever found.[45]

*

When the Cold War ended, a new focus for bioweapon fears appeared as Western governments became increasingly concerned about the potential use of such agents by terrorists. In 1989, Commander Stephen Rose of the US Navy warned that advances in genetic engineering had transformed biological warfare 'from a major undertaking into a cottage industry – simple, cheap, quick, precise'.[46] In the future, he claimed, 'power may come from the mouth of a test-tube as well as from the barrel of a gun'. This was not just rhetoric. In 1984 a religious cult in Oregon released *Salmonella* bacteria in local restaurants and caused over 750 infections in a bizarre attempt to rig a local election, while from 1990 to 1995 the Aum Shinriko cult in Japan, which had a molecular biologist in its leadership, repeatedly attempted to use both anthrax and botulinum toxin on the Tokyo subway (it failed, but had a murderous success with the chemical weapon sarin).[47]

Although none of these crazed and dangerous schemes involved any genetic manipulation, there was a widespread anxiety that this would eventually happen. Such concerns were vulnerable to manipulation, as politicians sought to mobilise legitimate fears of bioweapons in support of their foreign policy objectives. In 1999, the US Secretary of State for Defence, William Cohen, brandished a now-familiar amalgam of bogeymen states and vaguely sketched bad guys:

> In North Africa and the Middle East, rogue states – Libya, Syria, Iran, and Iraq – are poised to use all means at their disposal to threaten US and allied interests in the region and beyond. Many of these countries have ties to terrorists, religious zealots, or organised crime groups who are also seeking to use these weapons. ... As the new millennium approaches, the United States faces a heightened prospect that regional aggressors, third-rate armies, terrorist cells, and even religious cults will wield disproportionate power by using – or threatening to use – biological weapons against our troops in the field or our citizens at home.[48]

This kind of rhetoric both reflected and reinforced US government policy – by the end of the Clinton administration in 2000, funding for biodefence had increased fourfold to over $400 million a year.[49] President Clinton's awareness of biological weapons was partly fuelled by a massively successful 1998 technothriller, *The Cobra Event*, by *New Yorker* journalist Richard Preston, who had extensively interviewed Ken Alibek and was one of the first to bring the reality of Biopreparat to the public's attention. The book's plot focused on a disgruntled scientist who genetically engineered a lethal viral weapon called Cobra which he then released in various US cities with terrible effects – it attacked the human brain, making people intensely aggressive and even driving them to eat their own bodies. Shortly after Clinton read the book, the US Department of Defense set up Project Jefferson (Jefferson is Clinton's middle name), which was charged with trying to replicate the creation of a genetically engineered anthrax bacterium that had been published the previous year by Russian researchers. This was 'defensive' research, of course.

The fear that terrorists or states could develop bioweapons simply by reading the scientific literature gained traction in

December 2000 after Australian researchers described how they had accidentally increased the virulence of mousepox, a virus similar to smallpox.[50] They were trying to develop a vaccine that would prevent pregnancy in invasive wild mice, which are a menace to native Australian wildlife. As part of this experiment, they introduced a gene for interleukin-4, a molecule involved in suppressing the immune response, into the mousepox virus. The virus was to act as a vector to deliver the *Il-4* gene to the mouse, so they used mice that were genetically resistant to mousepox. To their surprise, all the mice died after they were injected. The same happened if the recombinant virus was injected into mice that had recently been vaccinated. The similarity of mousepox and smallpox implied that introducing *Il-4* into the smallpox virus would make vaccinated people susceptible to the engineered virus and would probably render all existing smallpox vaccines useless. This was confirmed three years later by a study of cowpox, a virus that is even more closely related to smallpox.[51]

Alarmed by their unexpected discovery, the Australian researchers spent eighteen months mulling over whether to publish their findings – including a period of consultation with the Australian Department of Defence – before eventually deciding that openness was essential. The researchers framed their work in the most alarming way possible: 'This is the public's worst fears about GMOs come true', said one, while another claimed: 'It would be safe to assume that if some idiot did put human *Il-4* into human smallpox, they'd increase the lethality quite dramatically.' A UK biological warfare expert who had helped debrief Biopreparat defectors warned that clandestine manipulations of this kind might have already been carried out in the Soviet Union.[52] When a leading Russian virologist working in an ex-Biopreparat facility was told of the experiment, he remarked disconcertingly: 'Of course, this is not a surprise.'[53]

The Australian mousepox study marked the beginning of a long-running conflict between governments' desires to limit the circulation of knowledge for security reasons and scientists' need to exchange information as openly as possible. The central problem was summed up a few years later by two researchers: 'the understanding of how to make viruses more pathogenic further blurs the distinction between fundamental academic research and bioweapons development'.[54]

This debate was soon supercharged by one of the defining events of the first part of the twenty-first century.

*

The horrific al-Qaeda terrorist attacks of 11 September 2001 achieved one of their desired ends – they shifted global politics and created a frenzied atmosphere of anxiety and suspicion in Western countries and in particular in the United States. This mood was further ramped up a few weeks later when a series of letters containing anthrax spores were delivered to US addresses, including to congressional offices and various media. At least twenty-two people fell ill, and five died.* Although the anthrax letters had no apparent political motive and did not involve any form of genetic engineering, they amplified existing worries about the ability of scientists to produce biological weapons and the risk of describing potentially dangerous experiments in scientific journals where they might be read by malicious forces. Within weeks, Anthony Fauci and his colleagues at the Centers for Disease Control (CDC) were highlighting bioterrorism as 'a clear and present danger' and warning of a massive potential threat to the public.[55] The overwhelming focus was on the menace of Islamist terrorist organisations, in supposed liaison with a hotchpotch of baddies such as Iraq, Iran and various ex-Soviet states. In response to the grim mood in the United States, the NIH budget for 'anti-bioterrorism' research, which had been steadily growing since 1992, almost doubled between 2001 and 2002, to around $90 million.

US investment in understanding potential bioweapons soon bore fruit. What Biopreparat researchers had done in secret was now carried out in the open and published in leading journals, as scientists were insidiously conscripted into the war on terror. There was no

* The spores were accompanied by letters using Islamist rhetoric, but US authorities soon discounted this – the motivation for the attacks has never been made clear. In 2008 the key suspect, Bruce Ivins, a microbiologist at the US Army Medical Research Institute of Infectious Diseases, committed suicide shortly before he was due to be charged. Part of the investigation revealed that much of the United States' extensive stocks of anthrax was not properly secured or documented.

overall plan, no central scientific scheme for the development of new treatments or vaccines, nor – as far as we know – a coherent project to create Western bioweapons. Instead, there was a growing, chaotic and very well-funded focus on understanding viral and microbial diseases by a variety of means, including increasing their virulence through genetic engineering in what are known as gain-of-function studies. This term refers to any process that increases the activity of a gene. It was initially used by geneticists to describe anodyne aspects of their research but has since become identified with this particular kind of study of pathogenic organisms.[56]

In 2001 and 2002, as the US population reeled from the consequences of 9/11 and unquestioning patriotism teetered into near paranoia, laws were passed to regulate work on infectious agents and in 2003 a US National Research Council report, *Biotechnology Research in an Age of Terrorism: Confronting the Dual-Use Dilemma*, highlighted seven types of 'experiment of concern'.[57] Similar laws were rapidly adopted in the United Kingdom and France, and then in the European Union.[58] Dual-use refers to legitimate research that might easily be transformed into an offensive capability, or simply entails a massive risk. The suspect topics that would require special approval before being funded included all kinds of pathogen manipulation, in particular 'gain-of-function' studies of virulence and host range, which create disease agents that can evade detection or resist treatment, and so on. In response to the report the Bush administration set up the National Science Advisory Board for Biosecurity (NSABB), a largely toothless oversight body. 'The NSABB was set up not to do anything', one security expert said later.[59] The Bush administration was more concerned that the research should go ahead, rather than that it should be shackled by administrative requirements.

Legal and administrative oversight of research, driven by security concerns, went hand in hand with a massive increase in funding, creating a contradictory situation in which more and more researchers began to work in the area, thereby increasing the security risks. By 2006 it was estimated that more than 300 US institutions had access to biological warfare agents, with over 16,000 people approved to handle them. The level of expertise in handling dangerous pathogens possessed by those researchers was hugely variable – for example, virtually all of the scientists who received National Institute of Allergy and Infectious Diseases funding to study bioweapon agents had never worked on the topic before. Bacteriologist Richard Ebright of Rutgers University ridiculed the background checks on these new researchers, saying that the 9/11 terrorist Mohammed Atta 'would have passed those tests without difficulty'. Ebright even claimed that because of the sudden, massive expansion in US research in the area, 'the NIH was funding a research and development arm of al-Qaeda'.[60]

The underlying political context of this research changed over time as the face of the enemy perpetually shifted – from Iraq and its non-existent links to al-Qaeda in 2001–2003, through the fear of Islamist terrorist groups flourishing in the ashes of Iraq, Afghanistan and then Syria, to ever-present and legitimate concerns about the near-inevitability of a natural global pandemic. The only thing that was constant was the research itself and the growing likelihood that something might go terribly wrong, either deliberately or by accident. A whole generation of researchers matured and flourished in this atmosphere, using their brilliance and mastery of genetic engineering to carry out experiments that would have horrified the delegates at Asilomar, but which were now an accepted part of the scientific landscape.

In 2002 researchers at the University of Pennsylvania reconstructed a gene called *SPICE* from the smallpox virus and showed that it increased the virulence of a related virus by inhibiting the immune system.[61] A couple of months later, researchers from Stony Brook University on Long Island described how they had ordered DNA through the post and used it to reconstruct the polio virus,[62] while at the end of 2003, researchers from the University of California Berkeley mutated the tuberculosis bacterium to make it hypervirulent and Craig Venter and his colleagues synthesised a bacteriophage virus from scratch in only two weeks.[63] There was a renewed debate about whether to destroy what are supposed to be the two remaining stocks of smallpox – one in Russia, the other at the CDC in Atlanta – with the WHO repeatedly arguing that they should be retained for research purposes (they still exist).*

---

*Those are the only publicly declared stocks, but the reality is much more alarming. As I was writing this, over a dozen vials labelled 'smallpox' were

Worse was to come. In 2005 researchers sequenced and then recreated the deadly 1918 influenza virus, which nearly a century earlier had killed up to fifty million people around the world. The sequencing work, which involved extracting viral DNA from body parts kept in formalin and from an Alaskan victim who had been buried in the permafrost in November 1918, was carried out at the US Armed Forces Institute of Pathology. Scientific and public awareness of the possibility of a natural global pandemic was high, given that the world had dodged a bullet in 2002–2003 when a SARS coronavirus epidemic emerged in southern China. The origin of this coronavirus was a spillover event from bats (it took fifteen years to identify the source) – the outbreak resulted in over 8,000 infections with an alarming 10 per cent mortality rate before being stifled by basic public health measures.[64]

Whereas sequencing the influenza virus was dangerous but could be justified, resurrecting it seemed dangerous but pointless. The researchers who did the work at the CDC in Atlanta claimed that it would show how the genetic sequence of this influenza virus gave it its exceptional virulence.[65] But all they learned was that the virus had a unique combination of genes, all of which had to be present for it to wreak havoc. Resurrecting the virus provoked anxieties about safety and the possibility of misuse – one researcher said it was a recipe for disaster, while another called it the creation of perhaps the most effective bioweapon known and argued that the risk of accidental release verged on inevitability.[66] This concern was quite legitimate; in 2003 and 2004 there had been three escapes of SARS from containment facilities like that used in the CDC experiment – one in Singapore and two in China.

These studies provoked alarm because the very thing that

---

found in a Pennsylvania laboratory freezer (*The Guardian*, 18 November 2021). In 2014, smallpox samples were discovered in an NIH laboratory in Maryland, while a few months earlier smallpox DNA was found in a South African laboratory. In 1991, Russian researchers tried to recover the virus from corpses buried in the Siberian permafrost; they failed, but with global warming the prospect of zombie smallpox particles emerging from the melting ground cannot be discounted. Reardon, S. (2014), *Nature* 514:544; Enserink, M. and Stone, R. (2002), *Science* 295:2001–5.

science thrives on – the exchange of information – might now have lethal consequences, in particular if someone with malicious intent and a well-funded laboratory were to use that information. Scientists were divided between those who wished to restrict information about potentially dangerous reagents and those who argued that the free circulation of data was required to respond to future threats, be they natural or malicious.[67] Both the American Society for Microbiology and the National Academy of Sciences set up procedures for the editors of the journals they published to vet submitted articles for worrying content; they were joined by thirty other journals in agreeing voluntary guidelines to ensure that nothing untoward was published.[68] The limits of this approach were soon felt when in 2005 two Stanford mathematicians submitted an article to the *Proceedings of the National Academy of Sciences* describing how to contaminate the US milk supply with botulinum toxin. Despite requests from the US government and the National Academy of Sciences' own internal processes, the article was published.[69]

The final element of this febrile mix was the advent of synthetic biology, a kind of souped-up genetic engineering, which often involves manipulating many genes and biochemical steps to achieve the desired end.[70] Synthetic biology views biological processes as circuits, which can be created at will in microbes and refined by successive cycles of design → build → test → design → build → test and makes use of increased computer power both to design molecular components and to predict their behaviour. As early as 1981, MIT researcher Eric Drexler seized on the idea of using genetic engineering to create new proteins that 'could lead to great advances in computational devices and in the ability to manipulate biological materials' through the construction of 'molecular machines'.[71] Although Drexler's molecular engineering did not come about, his vision was finally realised – in the laboratory at least – two decades later.

In 2000, researchers produced a biochemical 'latch' whereby the physiological state of the cell altered gene function like a switch, while another team created an even more complex biochemical oscillator.[72] By synthesising and distributing modular biological components such as genes for enzymes and so on, synthetic biology

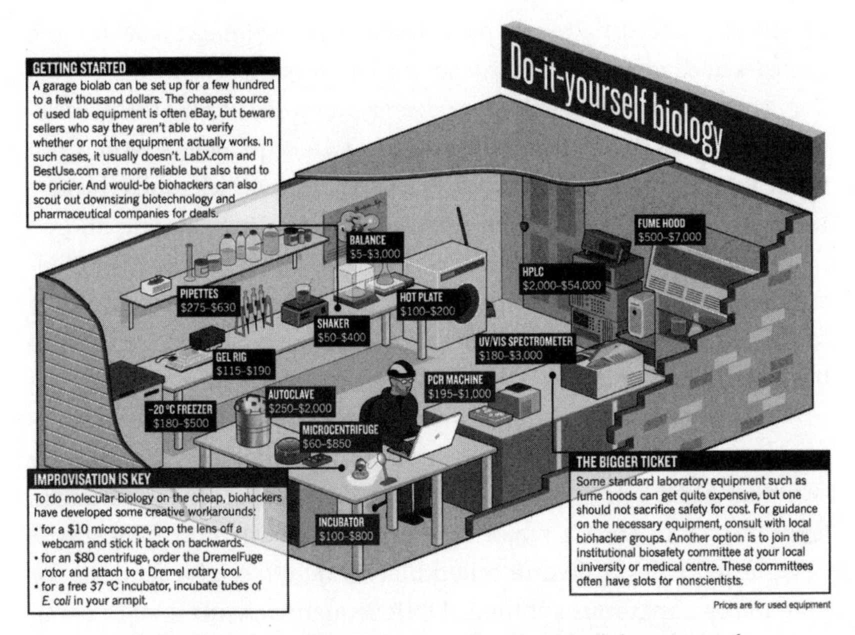

Illustration from *Nature* in 2010 showing the laboratory of a biohacker. It could equally be a bioterrorist's lair.

initially promised to democratise the scientific process by enabling hackers and hobbyists to join the game, contributing to and refining those circuits. This vision was partly realised through an immensely successful MIT student competition to design new organisms using such components – the international Genetically Engineered Machines (iGEM) competition, which uses genetic 'BioBricks' from a Registry of Standard Biological Parts (there are currently over 20,000 such genetic components in the registry).[73]

By claiming that virtually anybody could genetically engineer microbes using this modular approach, highlighting the advantages of 'open-source biology' to go along with open-source computing, the synthetic biology advocates attracted the attention of the security services. Democratising the genetic manipulation of organisms inevitably raised the spectre of terrorists creating bioweapons in a basement. The FBI's Weapons of Mass Destruction Directorate took a particular interest in the new approach and meetings took place between US and UK scientists and security experts in a variety of locations – according to one participant 'some were like a set for *Yes, Minister* and others were like a set for *Dr Strangelove*'.[74]

There were worries that the decreasing price of DNA synthesisers – machines that can create bespoke DNA sequences – would put them into the hands of hobbyists, who might accidentally misuse them, or terrorists, for whom misuse would be the sole intention.[75] Synthetic biology advocates such as George Church took the potential dangers of the new approach very seriously and there were calls for DNA synthesisers to be internationally registered and for users to have an obligation to report what they were studying.

To cope with the supposed threat from terrorists unable or unwilling to shell out $10,000 for a used synthesiser, other approaches were adopted. Most synthetic DNA is sold to researchers by commercial laboratories; in 2002 a major US supplier, Blue Heron Biotechnology, started using software called Blackwatch to screen requests for potentially dangerous synthetic DNA sequences, with the FBI being informed if there was no legitimate justification. Different approaches by various trade organisations and by the US government led to three different standards being adopted, each of which involved the same principles of surveillance of requested sequences, full records of the person making the request and contact with the security agencies in case of concern.[76] Nevertheless, in 2006, science journalist James Randerson was able to order the DNA sequence of part of the smallpox virus and have it delivered to a residential address in London, for a mere £40.08, including postage. The company that produced the DNA said that it was not aware of what exactly it had synthesised and dispatched.[77]

During this period there were repeated attempts to define the object of all the worries – dual-use research – but no satisfactory definition could be arrived at. Under criteria eventually adopted by the NSABB, the Australian mousepox experiment, the polio virus study and others would all have escaped regulation.[78]

Although in 2012 the World Economic Forum declared synthetic biology as a key twenty-first-century technology, second only to informatics, the hype eventually subsided. This was partly because it was replaced by a new genetic engineering kid on the block – CRISPR – and partly because much of the biology turned out to be far more complicated than suggested by some of the more enthusiastic evangelists. The science fiction promises that synthetic biology would lead to the development of smart therapies 'where the therapeutic agent can perform computational and logic operations and make complex decisions' turned out to be just that – science fiction.[79] Leading researchers in the field gradually became more measured in their claims about synthetic biology's supposed revolutionary potential and ease of application, although these are still regularly repeated.[80]

Many scientists – me included – are not entirely convinced that synthetic biology even exists as a specific discipline, beyond a modularly minded version of genetic engineering. However, when I stated this on Twitter at the beginning of 2022, Fyodor Urnov replied that 'in my field, genomic therapies, synthetic biology is very much a thing. It refers to design and deployment in the clinic of entirely engineered molecular circuits. This requires a distinct mindset and skillset.'[81] Furthermore, as microbiologist Carlos Acevedo-Rocha accurately predicted in 2016, 'synthetic genomics promises to revolutionise the medical sector by reducing the time needed for the production of synthetic flu vaccines in case of pandemics'.[82]

Whatever the case, the field has produced some amazing technical achievements. For over two decades, Craig Venter's research group has been radically tinkering with the genome of a microbe, *Mycoplasma mycoides*, with the aim of producing an organism with a minimal genome – the smallest amount of DNA compatible with life, supposedly allowing the molecular and biological function of every gene to be understood. In 2010 an artificial version of the genome of *M. mycoides* was introduced into the cell of a closely related species, from which the chromosome had been removed.[83] The genes functioned correctly; a kind of new life had been created, which Venter modestly called *M. mycoides* JCVI-syn1.0 (JCVI stands for J. Craig Venter Institute).

Over subsequent years the genome of this semi-artificial life form was repeatedly refined, leading to a reduction of around 50 per cent in the size of the genome.[84] Even these simplified cells have hidden complexities: seven of the genes eliminated in the reduction cycles may not be necessary for survival in the laboratory, but they do appear to be required for the bacterium to have the correct shape.[85] These genes are found in most bacteria, suggesting a great deal remains to be understood about what is a 'minimal' genome. Which genes are essential will depend on environmental conditions, and a gene that might not be essential for survival and reproduction in one condition could have a fundamental role to play in another.

Researchers have also been altering the very structure of the DNA molecule. As long ago as 1990, Steven Benner – one of the founding figures of synthetic biology – created two new complementary bases that could slot into the double helix, extending the genetic code from sixty-four possible combinations to 216.[86] Since then, many other researchers have introduced yet more bases, and have altered part of the backbone of the DNA molecule – its ribose sugar structure – in a range of ways. All these bits of amazing biochemistry are variously called xeno nucleic acids (XNAs) or, more prosaically, non-canonical genetic material. With suitable tweaking they can even be expressed and replicated in bacteria.[87] Most recently, in an astonishing feat of genome engineering, Jason Chin's group at Cambridge rewrote the genetic code in a bacterium, changing the function of three codons and producing an organism that both resisted bacteriophage infection and used those three new codons to assemble non-canonical amino acids into a completely novel protein.[88] As well as being unbelievably clever, this kind of work may eventually help us develop new nucleic acid therapies with precisely designed molecules, or enable the production of novel biopolymers with uses in both medicine and material science.[89] Our imaginations may soon become the limiting factor.

As these extraordinary developments demonstrate, despite suggestions that the democratisation of synthetic biology would lead to a start-up revolution, a great deal of skill, knowledge, equipment and teamwork is required to do anything really interesting. After an initial period of enthusiasm, the hacker side of synthetic biology faded and their online communities dwindled.[90] Similarly, security fears about terrorists acquiring biological weapons declined as tragic events around the world demonstrated that the terrorists' murderous purpose could be served quite simply by vehicles, kitchen knives and home-made explosives. Although synthetic biology is still often used as a catch-all term to describe dual-use research of concern, far more tangible threats to international biosecurity emerged exactly where you might expect – in fancy high-tech biology laboratories carrying out government-funded genetic engineering research into terrifying pathogens.[91]

*

In September 2011, Rotterdam virologist Ron Fouchier spoke at a European meeting on influenza, held in Malta. He confessed to the audience that while working on an NIH-funded project on the highly pathogenic H5N1 bird flu virus, he had done something 'really, really, stupid'. Fouchier said he had 'mutated the hell out of H5N1', enabling it to be spread through the air from one of his laboratory animals – ferrets – to another. A few weeks later, Yoshihiro Kawaoka of the University of Wisconsin, also funded by the NIH, announced that his group had carried out a similar experiment.[92] This virus, which can cause sickness and death in humans, is normally transmitted only by physical contact – this has enabled repeated outbreaks around the world to be contained with little evidence of community transmission (in the first decade of the century there were only about 600 cases, but with a 60 per cent mortality rate – around 100 times higher than for COVID). Both these mind-bogglingly alarming projects altered the virus so it could now be transmitted by aerosol or respiratory droplets.

Within a few months, both groups of researchers submitted papers describing their work – one to *Science*, the other to *Nature*.[93] These gain-of-function studies, which were supported by the US National Institute of Allergy and Infectious Diseases as part of a programme to predict future pandemics, understandably caused tremendous concern, both because of the danger of accidental release and the potential for the weaponisation of the discovery. Had this work been attempted in a Soviet laboratory in the 1980s it would undoubtedly have been seen as a breach of the BWC. Before one of the papers could be published, the Dutch government intervened (some of the research was done in the Netherlands) and insisted that under EU regulations controlling the export of dual-use items, a licence was required before publication. In the United States, the

NSABB stirred itself and recommended that both articles should be censored, with key methodological details removed.[94]

At the beginning of 2012, forty virologists – led by Ron Fouchier – expressed their deep alarm at the situation and adopted an immediate voluntary sixty-day 'pause' on gain-of-function research in H5N1 viruses (the Asilomar 'm word' was not mentioned). This was necessary, they argued, because 'organizations and governments around the world need time to find the best solutions for opportunities and challenges that stem from the work'.[95] But, as at Asilomar, the scientists' concerns were purely technical and safety-oriented – there was no challenge to the legitimacy of such studies. Nevertheless, despite its severe limitations, this initiative was the first time since 1974 that geneticists decided to stop doing potentially dangerous experiments.

Two months later, after long discussions including interviews with the researchers involved, a WHO meeting recommended that the articles be published in full, but that the research pause be continued until a system of biosafety and biosecurity review had been established.[96] Virologists and others directly involved continued to be divided over the issue, with some arguing that gain-of-function research helped understand the potential evolution of the virus, while others considered that such apparent insights were either non-existent or were vastly outweighed by the inherent danger of a potential leak.

The debate was still unresolved when H5N1 gain-of-function research resumed at the beginning of 2013, with the researchers arguing they had 'a public-health responsibility to resume this important work because the risk exists in nature that an H5N1 virus capable of transmission in mammals may emerge'. The WHO appeared to agree and issued well-meaning but vague and toothless guidelines:

> Given the potential of these newly developed laboratory-modified H5N1 strains to start a pandemic, it is important that facilities that are NOT able to identify and appropriately control the risks associated with these agents REFRAIN from working with them.[97]

In March 2013, the Chinese authorities reported the existence of a new form of bird flu, H7N9, which over the next year infected over 400 people, causing over 100 deaths. The reaction of Fouchier, Kawaoka and their colleagues was to call in *Science* and *Nature* for a new round of gain-of-function studies, to render the virus even more transmissible.[98] Their justification was that this would give us added time to develop therapies, as the wild version of the virus had yet to evolve this capability.

At the same time, the issue of safety would not go away. In 2014 it was revealed that eighty-four CDC workers had been potentially exposed to deadly anthrax when live samples were accidentally distributed to three different laboratories, while in a separate incident another CDC laboratory contaminated a normal flu sample with H5N1 and then shipped it to a government facility.[99] These breaches were not one-offs. In 2012 it was estimated that there were more than two possible release or loss events *every week* in US laboratories working with the most dangerous pathogens. Between 2004 and 2010 there were eleven instances of infection through inadvertent pathogen exposure.[100] None of those incidents led to a fatality, nor were there any identified cases of secondary transmission, but that was a matter of luck.

Then, in June 2014, US and Japanese scientists reported that they had created a new pathogen composed of bits of H5N1 avian flu virus with high homology to the 1918 influenza pandemic virus.[101] The resultant entity was both functional and virulent and could be made to transmit by respiratory droplets. The outcry was immediate. Lord May, a former president of the Royal Society, described the study as 'absolutely crazy', while virologist Simon Wain-Hobson of the Institut Pasteur in Paris said:

> It's madness, folly. It shows profound lack of respect for the collective decision-making process we've always shown in fighting infections. If society, the intelligent layperson, understood what was going on, they would say 'What the F are you doing?'[102]

Four months later, after the CDC anthrax biosecurity failure became known, the US government imposed a moratorium on the funding of eighteen gain-of-function experiments designed to increase viral pathogenicity in influenza, SARS or MERS viruses. (MERS was another coronavirus early warning we did not properly heed – the outbreak began in the Middle East in 2012 and has rumbled on since, killing over 800 people.) At the time there was a feverish atmosphere in the United States – in a completely unrelated incident, nursing staff treating a patient with Ebola in Los Angeles had recently contracted the disease, causing a nationwide wave of panic that was completely out of proportion to the actual threat.

In January 2017 the gain-of-function funding ban was rescinded but future research proposals had to be independently vetted by a panel whose membership remained confidential. In 2019 the panel approved two new gain-of-function experiments on H5N1, leading to more arguments about the process and the anonymity of the panel.[103] At the moment, only 5 per cent of countries regulate dual-use research; nowhere – including in the United States – is such regulation clearly effective.[104]

In 2016, the smallpox concerns kicked off again after Canadian scientists announced that they had reconstructed the related horsepox virus using bits of DNA that had been ordered over the internet.[105] The security procedures used by the suppliers of synthetic DNA evidently did not flag any potential danger in the requested jigsaw of sequences. Although the researchers justified their work in terms of the insight it might provide into creating a new vaccine against smallpox, and the fact that horsepox is not infectious to humans, the study immediately raised concerns that the smallpox virus could be assembled with equal ease. According to one estimate, all that would be required would be a handful of scientists with a little specialised knowledge and $100,000 to spare.[106] In response to such criticisms, the Canadian group argued that this kind of research enabled public health agencies to protect humanity even from extinct diseases. As they wrote ominously: 'The advance of technology means that no disease-causing organism can forever be eradicated.'[107]

Even non-bioweapon research could be terrifying. In 2014, Jennifer Doudna became alarmed when she heard a researcher describe using CRISPR to give a mouse a form of human lung cancer – the researchers used an airborne virus to carry the CRISPR components into mice, simply by breathing. It was not clear from the presentation (or from the subsequent paper) if the virus could potentially infect people.[108] Whatever the case, Doudna was not happy. As she said to a reporter:

> It seemed incredibly scary that you might have students who were working with such a thing. It's important for people to appreciate what this technology can do.[109]

The endless bubbling anxieties of this period, largely a product of poor laboratory biosecurity and audacious but disturbing gain-of-function studies, shed light on later suspicions about the origin of the SARS-CoV-2 virus responsible for the COVID-19 pandemic. Just as fears of bioterrorists replaced fears of rogue states in the early years of the century, fears of bioterrorists subsequently faded and were replaced by a generalised fear of China, rooted in its growing economic power and political influence. This has undoubtedly fuelled the idea that the COVID-19 pandemic was not a natural spillover event. Supporters of this hypothesis often suspect the Wuhan Institute of Virology of being at the origin of the pandemic through an accidental leak of samples that were either legitimately cultured (this is possible) or, in more fevered versions, were the product of reckless gain-of-function research (there is no evidence of this).[110]

Anxiety and suspicion on all sides, coupled with missteps by US researchers and by funders of research in Wuhan and elsewhere, as well as a Chinese governmental appetite for secrecy, have prevented openness and clarity and provided fuel for conspiracy theorists and opportunist politicians.[111] As a result, it is quite possible that the true origin of the pandemic will never be known. At the time of writing there is no evidence that the virus had anything other than a natural source, although concerns about laboratory security should never be dismissed. The data point to a natural spillover event such as we have seen in the past and will almost certainly see again.[112] The long, painstaking research that was required before the bat origin of SARS was identified explains why there was no immediate agreement on which animal species was the original host of SARS-CoV-2 – such things take a long time even in the absence of a global pandemic.[113]

One possible solution to concerns about identifying manipulated pathogens, and indeed a potential resolution to some of the more outlandish speculation about the origin of SARS-CoV-2, may lie in the use of genetic engineering forensics – complex bioinformatic analyses – to determine whether an organism involved in a disease outbreak has been genetically modified and, if so, to infer its likely origin. This work is in its infancy, but a network of laboratories, RefBio, has recently been set up under the auspices of the United Nations to gather sequence data from future events.[114]

*

Throughout the half-century history of genetic engineering there have been persistent concerns that the apparent simplicity of the methods involved might enable terrorists or biohackers to replicate experiments with potentially disastrous results. Janet Mertz told me that her 1972 breakthrough in assembling recombinant DNA molecules put the technique within the grasp of 'a bright high school student'; in 1989, Commander Stephen Rose wrote of genetic engineering becoming a 'cottage industry', while in 2011 US Secretary of State Hillary Clinton warned that a weapon could be made 'by using a small sample of any number of widely available pathogens, inexpensive equipment, and college-level chemistry and biology'.[115] In September 2021, marking the twentieth anniversary of the 9/11 attacks, Tony Blair gave a speech in which he carried on beating his old drum, warning of the supposed threat of Islamist terrorists wielding bioweapons:

> Bio-terror possibilities may seem like the realm of science fiction; but we would be wise now to prepare for their potential use by non-state actors.[116]

And yet, despite the alarming nature of much of the research done into genetically engineered bioweapons over the last half-century, one salient fact stands out: no weapons of this kind have ever been deployed, nor is there any evidence that the knowledge upon which they are based has been taken up and applied by terrorists or biohackers. As a number of scholars such as Katherine Vogel, Sonia Ben Ouagrham-Gormley and Claire Marris have emphasised, one major explanation of this apparent paradox is that it is one thing to understand a piece of research but quite another to be able to replicate or apply it.[117] This is perhaps particularly the case in molecular biology, where experiments are difficult to do and often fail (anyone who has tried to do a supposedly straightforward PCR experiment will know this).

As well as an academic understanding of the subject, actually doing the experiment requires tacit knowledge, often built up in a laboratory team over years, which enables scientists to know which techniques work best and how to perform them. Vogel has described this as 'the unarticulated, personally held knowledge that one acquires through a practical, hands-on process, through either "learning by doing" or "learning by example"'.[118] Even when you have acquired such expert skills, experiments that have worked regularly can stop giving the expected result for a number of reasons, including seasonal variations in the reagents being used.[119]

A scientific article presents a set of selected facts in a constructed narrative and does not describe the underlying social connections that were required for the whole thing to happen, nor will it mention the myriad occasions when experiments failed for unknown or tedious reasons. The 2005 synthesis of the polio virus built upon decades of knowledge and tradition in the laboratory and could not simply be replicated by any old basement biohacker. The same applies to today's concerns about bioterrorists using CRISPR – molecular biology is much more complicated than it may appear. Furthermore, even if an amateur were to synthesise a pathogen, they would then find it hard, if not impossible, to weaponise it. In 2008 the virologist Jens Kuhn pointed out: 'The methods to stabilise, coat, store, and disperse a biological agent are highly complicated, known only to a few people, and rarely published.'[120]

That does not mean that there is nothing to worry about, but it does put things into context. Furthermore, all the talk of simplicity and easy availability, so prevalent in alarm-inducing security discourse over the last thirty years, may have been counterproductive. After the invasion of Afghanistan in 2001, US forces found a memo written in 1999 by al-Qaeda deputy head Ayman Al Zawahiri, in which the terrorist leader explained the organisation's interest in producing bioweapons (they failed): 'the enemy drew our attention to them by repeatedly expressing concerns that they can be produced simply with easily available materials'.[121]

Rather than worrying about threats from terrorists or biohackers, the real biosecurity danger comes from states secretly seeking to create weapons, and from the possibility of an inadvertent leak from one of the many laboratories carrying out gain-of-function research on potential pandemic pathogens. The fundamental justification for such studies – that they will help us prepare for future pandemics – took a severe knock in 2020 when the world's response to COVID-19 was woefully inadequate. Despite some kick-back from gain-of-function researchers highlighting the importance of their research for the development of treatments for COVID-19, the fundamental way we have responded to the pandemic was not guided by these studies.[122] Furthermore, these same researchers have highlighted the danger from the sudden influx of laboratories into the field of coronavirus research, much as happened in the early years of the century. As viral gain-of-function pioneer Ralph Baric has warned, some of these new researchers may have 'less respect for the inherent risk posed by this group of pathogens'.[123]

In 2004, George Poste, a bioterrorism expert at Arizona State University, referenced what happened to physics after the creation of the atomic bomb and predicted: 'Biology is poised to lose its innocence.'[124] Thankfully, that has not yet happened and we can still stop it.

– SIXTEEN –

GODS?

As soon as genetic engineering became a reality over half a century ago, critics said scientists were 'playing God', implying that changing genes in a directed, precise way is unnatural and that it involves tampering with forces beyond our understanding.[1] This view can be applied to the three areas of concern that prompted me to write this book – heritable human gene editing, gene drives and pathogen manipulation. In each of these fields our ability to alter genes is both extraordinarily far-reaching and profoundly threatening, for ourselves and for the rest of the planet. As mythology teaches us, gods are not always benevolent.

But there is another side to this view of scientists 'playing God'. In 1968, in the heart of what would soon be called Silicon Valley, Stanford biology graduate Stewart Brand published the first issue of the *Whole Earth Catalog.** Part hippie mail-order listing, part instruction manual for the Age of Aquarius, Brand called the *Catalog* 'a guide to resources'. In the Introduction to the 1969 edition, written shortly before genetic engineering became a reality and at a time of growing

*Brand appears in the opening paragraphs of another iconic late-1960s book, Tom Wolfe's *The Electric Kool-Aid Acid Test* (1968). Brand was one of Ken Kesey's band of Merry Pranksters.

doubts about science, Brand made a striking declaration that served as a manifesto for his view of the modern world: 'We are as gods and might as well get good at it.'

Brand's brand of hippie modernist techno-optimism runs like a glowing thread through the history of genetic engineering and its commercial application, much of which was pioneered in the same region of California, created by people of Brand's generation and outlook.[2] Down the decades that optimism, together with the entrepreneurial spirit it often encouraged, has motivated both pure research and overheated predictions of future applications. It can be seen today in solid medium-term prospects in three fields where our genetic ingenuity might soon improve lives – conservation, medicine and agriculture.[3] However, in each case, the optimism is not unalloyed. Dreams have to be turned into reality, and scaling a brilliant idea into production, guarding against misuse, ensuring that this is truly the right thing to do, is often hard. In genetic terms at least, being a god is relatively straightforward these days; getting good at it is another matter.

*

The most truly god-like ability provided by genetic engineering is the apparent possibility of resurrecting extinct species – 'de-extinction'. The most attention-grabbing proposal, which has been around for over a decade, is the idea of recreating the woolly mammoth, using our knowledge of the mammoth genome, obtained from tissue frozen in the tundra.[4] It is claimed that de-extincting mammoths and other large tundra-dwelling mammals would not only enable us to encounter fantastic beasts but would also help mitigate some of the effects of climate change – as these creatures trampled and disturbed the ground, they would supposedly allow the winter frost to penetrate deeper, keeping carbon trapped for longer.[5]

Leaving aside the obvious lessons of *Jurassic Park*, the technical issues involved in recreating a mammoth appear insurmountable. It would require taking the genome of the mammoth's closest living relative – an Asian elephant – and introducing all the relevant changes (synthesising a whole chromosome is an incredibly arduous process

Stewart Brand
@stewartbrand
Want a full time biotech job to revive the Woolly Mammoth and eventually restore the species to the northern wild?
We're hiring. Details at the link and below.
reviverestore.org/funding-opport...
5:49 PM · Mar 3, 2021 · Twitter Web App
903 Retweets 782 Quote Tweets 2,667 Likes

Tweet by Stewart Brand, March 2021.

and has not yet been performed in an animal; a woolly mammoth had twenty-nine pairs of chromosomes). These two species split between 2.5 and 5 million years ago; there are millions of base pairs that differ between them.[6] Not all of those differences are significant, and we do not know which are the ones that count, but it is technically possible that all the relevant differences could be identified and then introduced into the elephant genome (the Asian elephant has one fewer pair of chromosomes than a mammoth, so the whole process might prove problematic). Assuming these difficulties can be overcome, the modified chromosomes would then have to be introduced into an elephant cell and safely interact there with all of the cell's organelles and molecules. But this cellular environment would be different in so many unknown ways to the cells that the mammoth genome co-evolved with that there is no guarantee that this would work. As an indication of the challenge involved, it took Venter's group twenty years to master this procedure in a bacterial cell, which has only one chromosome and does not have a nucleus or any of the complex structures found in eukaryotes such as elephants and mammoths and does not develop into a large, hairy, intelligent animal.

Even if all this went smoothly, the resultant embryo would have to be implanted into a surrogate – again, an Asian elephant would be best – and the myriad interactions between embryo and mother would have to function appropriately for months and not kill either or both. This is much trickier than you might think – a few years ago a Spanish-led team of researchers tried to clone an extinct subspecies of Pyrenean ibex; hundreds of embryos were created and implanted in a series of surrogate mothers of the same species, but only one animal was born and she died a few minutes after birth.[7] Things could so easily go awry using a surrogate from a different species. Perhaps for this reason, one version of the project involves artificial wombs (these are currently hypothetical), like those in *Brave New World*, but which would presumably have to be the size of a small car. Finally, mammoths were not simply bags of cells and DNA, but complex social animals that lived in an environment that has now disappeared along with the cultural aspects of their social organisation. Even if every one of those unbelievably complex steps could be overcome, there would be profound ethical questions about the well-being of these extraordinary animals in the alien world in which they would find themselves.

As a result of these very real problems, the woolly mammoth de-extinction scheme, initially the brainchild of George Church, has recently become less mammoth in scope. The current version of the project, partly funded by Revive & Restore (R&R), a US foundation co-founded by Stewart Brand, focuses on identifying those alleles that adapted the mammoth to cold temperatures, with the dream of maybe some day creating an elephant with enough mammothy characteristics to enable it to live in northern climes. By 2022, a biotech start-up co-founded by Church called Colossal Biosciences had raised $75 million* to try to create a 'cold-resistant elephant', which

*Among the investors were the billionaire executive producer of *Jurassic World*

it was claimed would be 'functionally equivalent' to a mammoth. So not a mammoth.

All these nuances were lost in the inevitable media excitement at the Colossal announcement, which yet again led to headlines about de-extincting mammoths (these have recurred every few years).[8] Tori Herridge, a mammoth expert at the Natural History Museum in London, was invited to be involved but declined. She explained why in a thoughtful article in *Nature*:

> Colossal has committed to 'radical' transparency, inclusion and community engagement, but has the chance to set the bar even higher, by empowering the public as part of its de-extinction journey. ... The ethical road to de-extinction has to include informed citizen voices, alongside experts and activists. This might mean that the process takes longer than five years, but private enterprises working for the common good shouldn't shy away from the views of those they seek to serve. Let the people decide the future world they want to build.[9]

As evolutionary biologist and R&R board member Beth Shapiro explains in her best-selling if somewhat misleadingly titled *How to Clone a Mammoth*, it is extremely unlikely that we will ever truly recreate any extinct species, not only because it is technically extremely difficult, but also because the ecological relationships that made up their world have all been lost.[10] In that respect, extinction is indeed forever.

Even more disturbing and god-like are the semi-serious, profoundly problematic proposals involving the recreation of Neanderthals – extinct humans who lived in Europe, Asia and the Middle East before us.[11] Leaving aside some absurd internet fantasists, no scientist is actually considering such an experiment – cloning a

---

and the celebrity Paris Hilton. That might sound a lot of money but given the scale of what is being attempted it is peanuts. Being a mammoth involved so much more than simply resisting cold – all the foraging behaviours, responses to the smell of food deep beneath the snow, and so on, would also have to be in place. We do not have the foggiest idea about the genes involved in these other essential aspects of mammothness.

Neanderthal and implanting the embryo in a surrogate human mother would be nauseatingly unethical on every level. However, researchers have studied the function of Neanderthal genes in human brain organoids – lentil-size blobs of tissue that can be grown in the laboratory from stem cells. A Neanderthal allele in a gene that plays a role in neural development and function altered the organisation and activity of otherwise human tissue.[12] Brain organoid research is slightly creepy (they spontaneously grow eye-like structures that respond to light) as well as being a potential ethical minefield (could such a structure become even dimly conscious? How would we know?[13]), but the researchers involved claim that this kind of study might shed light on how the functions of some of our genes may have differed in our close relatives.

Despite all the substantial technical and ethical issues relating to de-extinction, in 2014 the International Union for the Conservation of Nature set up a De-extinction Task Force which produced a set of guiding principles outlining the kind of detailed ecological and financial risk assessments that would be necessary before genetic engineering was applied to endangered or extinct species.[14] In 2019, a subgroup of the Task Force published a positive exploration of the potential impact of genetic technology on conservation.[15] At around the same time, ecologists from the University of California Santa Barbara and Imperial College took a far more cautious view, arguing that any de-extinction programme should focus on recently extinct species (their ecology would be more likely to be intact) that could be restored in sufficient numbers to enable the recreation of their lost ecological function.[16] That would rule out the mammoth, or a mammothified elephant.

What looks like a spiffy technofix might provoke excitement and attract funding from the public or mega-rich celebrities but is unlikely to be a solution to extinction for any but a tiny handful of cases. If molecular biology really must be employed, then scientists should try cloning dead members of an endangered species to increase genetic diversity, as has been done in the case of the US black-footed ferret, with funding from the R&R foundation.[17] In 2022, an attempt to recover the genome of the extinct Christmas Island rat, using different extant rat genomes as models, showed that it was impossible to recover around 5 per cent of the genome, with over twenty genes being completely absent. Key genes involved in the sense of smell and the immune system were particularly affected. De-extinction, if it is ever possible, will not be simple. The primary focus of conservation efforts needs to be on the prevention of species loss, not on de-extinction. In the vast majority of cases, that will involve basic work to prevent habitat degradation and to keep humans away from the endangered organisms. That might be less sexy, but it is a lot more sensible.

*

The use of CRISPR to alter genes in somatic cells is creating growing excitement among physicians and patients, in particular the possibility of a genetic treatment for sickle cell disease (SCD), which affects millions of people worldwide, leading to chronic pain, anaemia, substantially reduced lifespan and an increased risk of stroke and organ damage. SCD is caused by a malformed haemoglobin protein, the result of a single amino acid change produced by a single nucleotide base mutation that alters a single DNA codon from GAG to GTG.[18] In July 2019, Victoria Gray, a 33-year-old African American woman with SCD, received an experimental somatic CRISPR therapy for her disease. This targeted her bone marrow cells, which produce red blood cells, focusing on a gene that boosts production of a kind of fetal haemoglobin which is not affected by the sickle cell mutation.

In 2021, the results of the experiment revealed that Gray was pain-free for the first time in her life, while the levels of her edited gene were stable in the twelve months following treatment.[19] For Fyodor Urnov, pioneer of gene editing, this result was '*magnificent*' and so unexpectedly good as to be 'borderline utopian'.[20] Seven similar therapies are being developed; if all goes well some or all of these treatments may be approved in five years or so.* Because the same treatment would also be appropriate for β-thalassaemia, which

* As you may have noticed in previous examples, when a new application of genetic engineering is proposed it is often said to be about five years away. This is close enough to seem exciting and enticing, but not so close that people expect instant results or so far that it seems like a pipedream, even if it often is.

has similar effects to SCD but a different genetic basis and geographical distribution, the global impact could be substantial. Meanwhile, US regulators have approved a CRISPR-based clinical trial of a direct correction of the SCD mutation with the hope of a cure.[21] If these therapies become affordable and widely available in decades to come, this could make a major difference to millions of people around the world. You may have noticed that 'if' conceals a morass of social and financial problems relating to global healthcare inequalities.

Fundamental and applied discoveries can be expected to come from new forms of gene editing, which are being developed to avoid the off-and-on targets of CRISPR that are so worrying in clinical applications. In 2016, Alexis Komor and Nicole Gaudelli, two young researchers in David Liu's laboratory at Harvard, described a technique called base editing, which uses elements of CRISPR to precisely change a single base of DNA in a sequence.[22] The process is much safer than traditional CRISPR because it cuts only one DNA strand (the altered Cas enzyme 'nicks' the DNA molecule, so is called a nickase) and uses a guide sequence to replace one base of interest on the other strand. The cell then recognises the cut, and simply inserts the new base's complementary nucleotide into the cut strand. In 2021, Liu's start-up, Beam Therapeutics, successfully used base editing to target the sickle cell mutation directly in human cell lines, turning the disease-causing GTG codon into the normal GAG version. If safe and approved (this will take years) this could provide another way of targeting SCD, as well as other single nucleotide diseases.[23]

Base editing has provided another source of optimism by targeting one of the world's biggest killers – heart disease. In 2021 researchers led by Kiran Musunuru targeted *PCSK9*, a gene involved in the production of cholesterol – people with a natural mutation in this gene produce low levels of 'bad' cholesterol and are at reduced risk of atherosclerotic cardiovascular disease.[24] A single transfusion of CRISPR base editors in lipid nanoparticles (the things that deliver the mRNA COVID vaccines) produced a near-complete knockdown of *PCSK9* in the liver of a monkey, leading to a 60 per cent reduction in the animal's blood cholesterol levels that remained stable for eight months. This extraordinary treatment, still many years from an experimental trial in humans, could offer an additional solution to the global problem of cardiovascular disease and the heart attacks and strokes it produces.

Liu's powerhouse group has recently developed a new version of this approach called prime editing, which in the laboratory allows any base to be turned into any other; to prove its effectiveness Liu's group immediately used it in human cells to correct two genetic diseases.[25] As I was completing this chapter, Liu's laboratory announced the development of twin prime editing (twinPE), which is able to delete or insert large stretches of DNA without inducing a double-strand break. Liu and his colleagues suggest this might eventually be used to correct complex human genetic diseases.[26] Other new editing systems will undoubtedly appear in the coming years. With such a range of editing options, somatic gene therapy, which has promised so much since the 1980s, may eventually not only offer bespoke therapies for a few hundred people, but life-changing treatments for millions.

Gene editing may also transform our lives through the use of animals as a source of organs for transplants – xenotransplants. One of the most popular donor candidates, at least among humans, is the pig, whose kidneys would make a suitable replacement for a human organ, assuming the recipient accepted the procedure. However, the pig genome contains sixty-two retroviruses that could pose a problem were a pig organ to be transplanted into the human body. In 2017, Chinese researchers from George Church's eGenesis start-up, along with scientists in China and elsewhere, inactivated the retroviruses using CRISPR and were able to clone the resultant cells, producing cute retrovirus-free piglets that appeared on the front cover of *Science*.[27]

Three years later Church's group teamed up with researchers at Qihan Bio in China to further edit the pig genome, removing three pig genes that would induce a strong immune response in humans and adding nine human genes that would enhance the immunological and physiological compatibility of any transplanted organ.[28] Together with researchers from US company Revivcor, the team then began to transplant organs from the pigs into primates as a first test of feasibility for a possible human transplantation. In autumn 2021, one of these edited pig kidneys was connected to the blood vessels of a deceased woman for two days to see if it would be able to filter the blood without signs of rejection.[29] Everything went as expected and two similar experiments soon followed, but it will be several years before approval is given for transplanting one of these kidneys into a living human.

Or will it? The day after I wrote those lines, news broke that 57-year-old heart patient David Bennett, who was in a critical condition and was otherwise doomed to die, had received a transgenic pig heart. The pig had been edited using TALENs to make ten genetic modifications, four that targeted pig genes and six additions of human genes to make the transplanted organ more acceptable to the recipient.[30] Of course, fiction had got there first. In 1997, children's author Malorie Blackman published *Pig Heart Boy*, a novel centred on the life of a boy who receives a transplanted pig heart. Bennett died two months after the operation.

Although these pioneering studies will bring hope to millions of patients around the world, none of these potential therapies will be a magic bullet. The treatments are still highly complex as well as being incredibly expensive – the tragic healthcare inequalities around the world, including within countries, will limit their availability. The way that COVID-19 vaccines are being distributed around the world shows that unless our global system changes profoundly, breakthroughs in medical treatment will be restricted to the countries that can pay and, in many cases, to privileged layers within those countries. Medicine, like science, is deeply rooted in society; finding a sweet technical solution is far from the most difficult challenge involved in changing and saving lives.

*

The final area of genetic engineering's exciting promises is that of the creation of better, stronger and faster-growing plants and cunningly engineered microbes.[31] Using TALENs and CRISPR, researchers have created wheat that is resistant to powdery mildew (that would be great for my honeysuckle) and maize that shows improved drought resistance.[32] Changes in regulations mean that putting these plants in fields may be more straightforward than for GM crops. US regulation of CRISPR gene-edited plants is leaning towards the flexible, and

Japan has recommended that there should be no regulation of gene-edited plants if no DNA from another organism is present, while the United Kingdom post-Brexit is reviewing its regulation of gene-edited food. China, still in the vanguard of the application of genetic engineering, has approved the sale of gene-edited crops.[33] There is even the possibility that European legislation might change – despite a 2018 ruling by the European Court of Justice that CRISPR-edited crops were no different to classic GM plants, in April 2021 the European Commission argued that new, less-restrictive regulation should cover CRISPRed crops.[34] For the moment, there is no sign of popular unease about the possibility of the use and sale of gene-edited crops in any country. In September 2021, the first commercial CRISPRed crop, a tomato which contains high levels of the amino acid GABA, went on sale in Japan. With claims for apparent health benefits, this variant of the Sicilian Rouge tomato may avoid the fate of the Flavr Savr in the 1990s.

The Chinese pioneer of plant gene editing, Gao Caixia, has sensibly emphasised that there will be no single solution to the world's need for plants. She considers that the 'second Green revolution' she is striving to create will involve conventional plant breeding as well as gene editing.[35] However, base editing and prime editing – both of which Gao has successfully used in rice and wheat – present the possibility of overcoming the public perception that there is something unnatural about gene-edited crops, by emphasising the similarity between traditionally created plants and those produced by modern techniques.[36] The key to success, Gao argues, will be gaining public trust by transparency. That is important, but it will not be enough. Always, lurking in the background, is the need for funders to co-produce solutions with local communities – even if any of these clever ideas does eventually come good, unless the new crops are ones that fit the many, varied local ecologies and agricultural systems, it will all be for nought.[37]

Different considerations surround the use of transgenic micro-organisms, some of which are producing real benefits. Many consumers around the world (not me, regrettably) have been able to taste and enjoy the Impossible Burger, a plant-based burger that tastes meaty because it contains yeast which was engineered to produce leghaemoglobin, an iron-rich substance.[38] Given that much of humanity is apparently obsessed with eating grilled minced beef in a bun sandwich, this kind of ingenuity may lead to improvements in human health and animal welfare and reductions in the levels of CO_2 in the atmosphere.[39]

Other attractive prospects are the possibility of using engineered microbes to synthesise vanillin – the key compound in vanilla – from waste PET (the plastic found in bottles) or to produce ethanol from the dangerous indirect greenhouse gas carbon monoxide.[40] Similarly, yeast has been engineered to produce opioids, offering the possibility of safe routes for obtaining these essential drugs (all opioids are currently produced from opium poppies).[41] More fantastical still is the dream of altering fundamental biochemical pathways to improve photosynthesis so that plants can increase CO_2 fixation and biomass yield (this has already happened in evolution – not all plants use exactly the same process).[42] A similar hope revolves around using engineered microbes to produce sustainable biofuels that are carbon neutral or negative. Many approaches seem promising, but none have yet showed effectiveness at scale.

Turning these laboratory breakthroughs into viable industrial processes will be a massive challenge. And even if they succeed in the factory, not just in the laboratory, further problems may arise as shown by the successful industrial semi-synthesis of the plant-based malaria drug artemisinin via a recombinant yeast. Heralded as a great step forward for biopharmaceuticals, with a factory built in Italy to grow the yeast and produce the drug, the facility was eventually mothballed because naturally produced versions proved cheaper and more reliable.[43] Real-world problems, most of them social and not amenable to simple technofixes, have repeatedly brought a sharp dose of reality to the dreams of the genetic engineers.

*

Public responses to all this ingenuity highlight a fundamental issue with our global understanding of what this technology can do. Billions of people have been jabbed by needles containing COVID vaccines produced by recombinant bacteria or plasmids. Recombinant insulin is not only commonplace, it is virtually the only kind of insulin you can get. But compared with the widespread acceptance and even enthusiasm with regards to medicine, GM crops still provoke doubts. Some of this suspicion stems from the vague impression that 'they'* are doing something surreptitious which we would do well to avoid (the same floating anxieties may have been behind some COVID vaccine hesitancy). It is also probable that part of our acceptance of GM medication reflects ignorance (did you know that was how vaccines or insulin were produced?), lack of choice about the matter (animal insulin is hard to obtain and is more likely to provoke reactions) or a resigned assumption that physicians know best.

Other feelings may also be at work. Writing in 1997, pioneer molecular geneticist François Jacob argued that the then widespread popular unease about GM food flowed from the impression that genetically engineered organisms are mysterious and monstrous, almost supernatural.[44] That might be true of a spider-goat but it surely no longer applies to a Bt cotton plant. However, if we recast Jacob's analysis in terms of what is 'natural' it all starts to make a bit more sense.

For many people, 'natural' is good; not natural is weird and anxiety-inducing. No GMOs are 'natural', but then again very little in our modern world is. None of the 'normal' crops we consume are 'natural', any more than the animals we eat. They have all been shaped by thousands of years of artificial selection by humans who did not really know what they were doing. Their genes have been altered by us. On the other hand, many diseases are natural, but in these cases being 'natural' does not seem quite so advantageous. Antibiotics are not natural, yet they have saved millions of lives and we are rightly concerned that they may stop working. Perhaps we accept that most medical treatments are by definition not natural, so using genetic technology in medicine does not alter our attitudes much.

Philosophers, historians, sociologists and others have explored the complex ball of ideas and emotions that are wrapped up in the natural–unnatural conceptual dimension, revealing the contradictions and ambiguities of something that seems so simple.[45] Above

*Insert your preferred bogeyman.

all, the meanings and implications of 'natural', and its application to particular things, change over time and space. A few decades ago, IVF was seen as profoundly unnatural, leading to hostility and suspicion. Now it is an accepted form of assisted reproduction and while not 'natural', it is certainly normal and unremarkable.

The meanings of 'natural' are complex and our current focus on this concept surely relates primarily to the industrialisation and homogenisation of life over the last century or two, reinforced by the food safety fears such as those provoked by the 'mad cow' crisis at the end of the last century. The related concerns about the way that food is mass-produced and the lack of biodiversity in the countryside are all valid, but they apply to the whole of our food production system and are not simply due to the existence of GM crops. The GMOs have been designed to fit the production system, not the other way round.

Overcoming all these misconceptions and enabling people to make truly informed choices about the place of the products of genetic engineering in their lives will be a significant challenge in the coming decades. No matter how frustrating it might be to advocates of genetic engineering, it seems unlikely that simply explaining the many different and often contradictory meanings of 'natural' will persuade people – the roots of these attitudes lie beyond the reach of mere argument.

*

The application of genetic engineering to a range of increasingly commonplace uses and the decline in the expression of public hostility towards GMOs in the first decade or so of the twenty-first century may explain what looks like a contradiction: for all the excitement and alarm produced by the latest developments in genetic technology, there are clear signs that we are not taking any of this seriously. We may have become as gods, but we do not seem very interested in our new powers.

One of the main indicators that people consider an issue to be either threatening or exciting is its representation in culture. This particularly applies to technology, as seen by the significance of the railway in nineteenth-century novels, paintings and so on, and the omnipresence of the atomic bomb in post-war culture. But the way that genetic engineering has been represented in the twenty-first century is very different. It seems to be a non-issue for artists and creators of all kinds. That may mean that it is for the rest of us too.

The only serious cinematic attempt to engage with GM crops is *Little Joe* (2019), in which a scientist develops an antidepressant 'happy plant'; things inevitably go wrong as the plant turns out to alter accurate perceptions of threats. But even here, genetic engineering is not central to the plot – the plant could equally have been an alien device, or something natural that was discovered by accident. Literary fiction has occasionally flirted with cloning (the best exploration was Kazuo Ishiguro's 2005 *Never Let Me Go*, which was also turned into a film), but writers have generally been silent on genetic engineering. Even Margaret Atwood's dark post-apocalyptic 2003 novel *Oryx and Crake*, which refers to the spider-goats and features humanoids that have been genetically engineered to be docile, is probably more of a satire about consumerism.

Science fiction writers have continued to be alive to the possibilities of genetic engineering in distant futures – for example, *Light Alone* by Adam Roberts (2014) imagines what might happen if humans could photosynthesise (more inequality), while in *Austral* (2018), Paul McAuley describes the adventures of a genetically engineered woman living in a warming Antarctica. Two recent works of fiction focus on the dream, or rather nightmare, of resurrecting a Neanderthal – Tim Disney's 2019 film *William* and James Bradley's 2020 eco-thriller *Ghost Species*. Neither has been a major public success.

Popular culture has continued to use genetic engineering as a proxy for a kind of magical transformation rather than engaging with the possibilities and dangers of the techniques involved. This is what has happened in the endless film sequels to *Jurassic Park*, where increasingly absurd artificial dinosaurs are created by genetic magic and then inevitably escape and cause havoc. The problem has become the CGI dinosaurs, not the underlying genetic technology or scientific hubris. In the *Alien* prequel *Prometheus* (2012) alien Engineers (hardly an accidental name) are responsible for creating humanity, while in two Hollywood remakes of Marvel characters

– the 2002 version of *Spider-Man* and the 2003 film *The Hulk* – the power of the gene has replaced the power of the atom as the source of the hero's powers. In the *Planet of the Apes* reboot prequel *Rise of the Planet of the Apes* (2011), the chimps are altered by a virus-based treatment that is intended as a cure for Alzheimer's (in Pierre Boulle's original 1963 novel the apes took over when humanity degenerated, while in the 1968 film we inadvertently gave the apes space having destroyed civilisation through atomic warfare). The video game Bioshock features the paraphernalia of genetic engineering – plasmids and all – but these are simply the equivalents of potions in more magically focused games.

In none of these cases is the involvement of genetics significant. Similarly, in the long-delayed James Bond film *No Time to Die* (2021) the villain has developed an absurd viral weapon that enables him to control behaviour, but this could equally be about hypnosis or a mind-control ray (although there is a genetic twist). While the genetic modification of animals is at the heart of *Okja*, the 2017 Netflix film by Bong Joon-ho in which a young girl forms a close relationship with a 'super pig' and seeks to rescue it from mistreatment, the story would probably have been just as heart-rending if the pig had been an ordinary animal destined for the abattoir. And although the 2009 horror film *Splice* features an engineered human-animal hybrid, its body-horror themes and scenes of incest and rape obscure any attempt to engage with the deeper issues of genetic engineering.

About the only mention so far of CRISPR in Hollywood films is to be found in the 2018 film *Rampage*, which combined elements of *The Andromeda Strain* with gene editing, as a space station crashes to Earth releasing a CRISPR-based pathogen that transforms animals into hyper-intelligent beings. Based on the video game of the same name, the plot involves three CRISPRed animals – a wolf named Ralph, an alligator called Lizzie, and George, an albino lowland gorilla.

Occasionally there is a hint of the real issues at stake. In the light-hearted 2014 Marvel comics mini-series *Spider-Man and the X-Men*, supervillain Karl Lykos, who has transformed himself into a pterosaur through gene-editing, taunts Spider-Man with his evil plan of using his genetic technology to turn everyone into a dinosaur, spurning Spider-Man's suggestion that he should use his knowledge to cure cancer. Facetiously expressing the existential dilemma of genetic engineering in a comic-book word balloon, Lykos croaks his response to Spidey: 'But I don't *want* to cure cancer. I want to turn people into dinosaurs.'*

Artists have been even less inspired in their response to genetic technology. In 2000, the world was inexplicably excited by Alba, a female white GM rabbit that expressed green fluorescent protein. Created by French geneticist Louis-Marie Houdebine, Alba was an unremarkable recombinant animal, but in the hands of Chicago-based artist Eduardo Kac, she became a piece of transgenic art. Dramatic photos of a luminous green rabbit soon filled the media (and were sold by Kac), but they were clearly not real – only the animal's skin and eyes would have fluoresced, not its fur. Hyping his creation, Kac bombastically proclaimed: 'It is a new era, and we need a new kind of art. It makes no sense to paint as we painted in the caves.' In reality all he had done was to channel Marcel Duchamp's *Fountain* (1917) – take something commonplace, in Duchamp's case a urinal, show it to the public and state that it was art.[46] The global fascination with Alba led researchers in Singapore to sense a market opening and a few years later they created a fluorescing zebrafish; patented and marketed as GloFish, these animals eventually came in a variety of colours – red, orange, green, blue and purple.[47] Inevitably, the fish have now escaped into the wild and are breeding in Brazil.[48] In 2010, self-proclaimed artist and 'demented naturalist' Adam Zaretsky released some GloFish into the Gulf of Mexico, proclaiming 'transgenic life should have a chance to run wild for its own sake, not just for the sake of profit'.[49] The artistic impact was unclear; the impact on the animals will have been almost certain death through physiological stress in the brackish waters – they are freshwater fish.

To be frank, all this was tiresome crap.

Slightly more interesting was the work by Georg Tremmel, an Austrian artist at the University of Tokyo, who used CRISPR to genetically modify blue carnations – themselves GM plants, with the blue colour produced by insertion of a gene from a petunia – turning

*I'd like to be a stegosaur, please.

the flowers back to their original white. Working with artist Shiho Fukuhara, his intention was to make the public 'ponder whether these double modified carnations should be deemed any different from the unengineered plants with essentially the same genome'.[50]

The reason why artists have not been inspired by genetic engineering is not clear. Perhaps it feels like the main issues have been done to death by Michael Crichton in *The Andromeda Strain* and *Jurassic Park*, and writers and film-makers consider that there is nothing more to be said. However, these same creatives have clearly not felt constrained by the many works about, say, robots and artificial intelligence – the prospect of thinking machines has continually been explored in both serious fiction (for example, Kazuo Ishiguro's 2021 novel *Klara and the Sun* or *Ex Machina*, a film directed by Alex Garland in 2014) and endless pieces of popular culture. There is clearly an appetite for dramatic issues associated with science and the threats it may produce or reveal. But that does not appear to apply to genetic engineering.

It may be that half a century of safe application has led the public and artists of all kinds to conclude that genetic engineering does not represent any existential threat, to think that the stories it conjures do not contain enough jeopardy. Perhaps this makes genetic engineering and its applications in the laboratory, the field and the clinic seem dull and commonplace. And after all, neither railways nor atomic bombs are any longer the subject of major cultural attention. However, it is surely the case that the recent developments outlined in this book contain enough heft to alarm and inspire the artistic soul, or even to suggest a way of raking in the cash and producing a Crichtonesque blockbuster. Maybe a future hit film or novel or artwork about tragically mutated humans, an escaped gene drive or a gain-of-function twenty-first-century version of *The Andromeda Strain* is currently in the works.

This decline in interest appears to be more general, at least in the English-speaking world. Google Books' Ngram Viewer reveals changes in the frequency of 'gene' and 'genetic' in books in English over the last seventy years. If this is anything to go by, we seem to have reached peak interest in all things genetic in the two decades after 1990. The subsequent decline happened at exactly the same time as the three threatening areas of genetic engineering that impelled

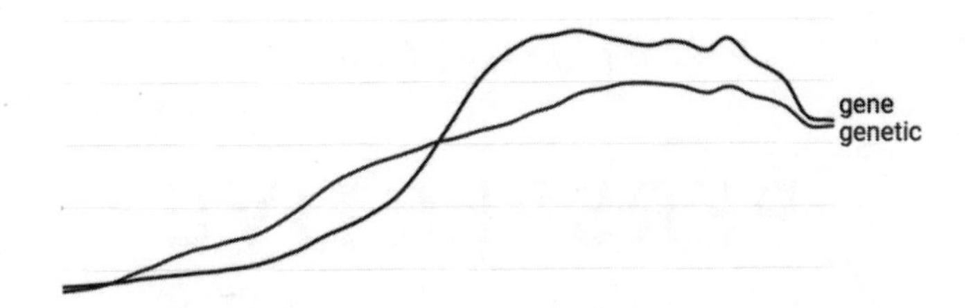

Frequency of the appearance of 'gene' and 'genetic' in Google Books 1950–2017, as reported by Ngram Viewer.

me to write this book – heritable human genetic engineering, gene drives and gain-of-function research on dangerous pathogens – were emerging. We seem to have dropped our guard.

– SEVENTEEN –

PERSPECTIVE

Our ability to precisely change genes has not only produced five decades of amazing and alarming discovery and application, it has also helped alter how we view life itself. As the 'engineering' part of genetic engineering suggests, this technology has reinforced a key interpretation of the natural world, which goes back centuries, according to which organisms can be considered as machines.

In the seventeenth century the French philosopher René Descartes argued that animals were 'beast machines', but this early parallel produced little insight apart from some basic mechanical descriptions – the heart is a pump, the musculoskeletal system is a series of levers, and so on. These were all true, but they did not change our understanding much.

As machines became more sophisticated, so too did our conception of life. In the nineteenth century, thinkers drew parallels between basic physiology and the way steam engines functioned, while the French scientist Claude Bernard argued that the fundamental difference between an organism and a machine lay not in the nature of their components, nor even their functioning, but rather in the contrast between growth and design.[1] Some began to dream of designing and controlling life, just as machines are designed and controlled. As early as 1891, Jacques Loeb, a German-born US biologist, had a vision of the future: 'the idea is now hovering before me that man himself can act as a creator, even in living Nature, forming it eventually according to his will'.[2]

Although it took many decades, that is where we are today. When the great theoretical physicist Richard Feynman died in 1988, he left a slogan written on the chalkboard in his office that has become a watchword for many genetic engineers, and in particular for those who identify as synthetic biologists: 'What I cannot create, I do not understand.' By enabling us to create new forms of life, genetic engineering has indeed allowed us to greatly increase our understanding and our mastery.

But although the term 'engineering' suggests a vision that Descartes might have recognised, a world of levers and gears, the machine-like understanding of life that has been reinforced by the past decades is utterly different and would bewilder the French philosopher. In 1953 Watson and Crick published a paper in *Nature* (not the double helix article but another one, which appeared six weeks later) outlining this new view. They claimed that in the DNA molecule 'the precise sequence of the bases is the code which carries the genetical information'.[3] As Crick put it later, life involves not only the flow of matter and energy, but also the flow of information.[4] Although there is a precise mathematical definition of information, established in the 1940s by Claude Shannon, this is not what Watson and Crick had in mind. Instead, their conception of information was metaphorical, expressed in the consistent, predictable relationship between the sequence of bases in a nucleic acid and the sequence of amino acids in a protein. In the 1960s, as the molecular detail of gene function became clear, scientists such as François Jacob and Ernst Mayr drew a more precise parallel, highlighting the similarities between how genes work and how a computer program functions. These two concepts – information and programme – have driven the successes of genetic engineering.

Remember how Arthur Riggs said he felt when a synthetic DNA sequence produced somatostatin in a bacterial cell:

> When it really worked, we had the final decisive proof that Watson and Crick were right, that the genetic code was right

> … you sit back and say, 'My God, that was all right! Science really works!'[5]

That is the quintessence of the informational view of life. Riggs and his team selected a series of bases from the genetic code that they knew was almost certainly not the sequence in the gene for somatostatin, but which corresponded to the known sequence of amino acids in the protein. The cell responded just as predicted and produced the desired molecule. The programme had been executed according to the information it contained.

Although the roots of this breakthrough went back to the late 1950s and Paul Zamecnick's in vitro studies of protein synthesis and the cracking of the genetic code by Marshall Nirenberg and Heinrich Matthaei in 1961, the somatostatin experiment marked a qualitatively new approach. The historian of science Hans-Jorg Rheinberger has called this 'soft technology':

> With gene technology, informational molecules are constructed according to an extracellular project and are subsequently translated into the intracellular environment. The organism itself transposes them, reproduces them, and 'tests' their characteristics. … The intact organism itself is turned into a laboratory.[6]

The successes of the last half-century have emphasised this particular vision of engineering, in which sequences of DNA act as programmes that produce cellular function. Adding a sequence to the genome of a cell – with the appropriate promoter sequences – will make it produce the corresponding protein. This technology is both incredibly sophisticated – it requires an exquisite degree of understanding and control of cell biology – and relatively crude. Although popular descriptions often present engineered microorganisms as cellular factories, they are unlike any real factory. The flow of information in such cases is extremely simple, and the outcome is more like turning on a tap, or, more accurately, producing alcohol through fermentation. Despite the computer inspiration of our modern view of life, we have far to go before our ability to manipulate organisms even begins to approach our ability to create inorganic information-processing devices.

In some cases, this lack of subtlety can undermine the effectiveness of the desired application, as for example in the Rothamsted wheat that was engineered to produce aphid alarm pheromone, but did so persistently, rather than in pulses. In the laboratory it is possible to acquire the kind of conditional control that might have made the transgenic wheat successful, by linking together different transgenic components and creating an organism in which only certain cells will perform a desired function under particular conditions. For example, in the fly *Drosophila*, which I have studied for over forty years, in the 1990s Andrea Brand and Norbert Perrimon developed a modular expression system involving two transgenes from yeast, *Gal4* and *UAS* (upstream activation sequence); the Gal4 protein is a transcription factor – it acts like a switch on the *UAS* sequence, activating it.[7] Researchers introduced *Gal4* into the *Drosophila* genome repeatedly, and on some occasions the transgene randomly inserted itself into the promoter region of a gene that was expressed only in certain cells, or in one sex, or at certain times. By combining *UAS* with a reporter gene, such as the gene encoding green fluorescent protein (from a jellyfish) and inserting that transgene into the fly's genome, the specific expression pattern of the fly gene to which *Gal4* is linked can be revealed. The Gal-4 protein activates the *UAS* sequence which activates the reporter gene, so only the cells in which the fly gene is active will glow green, making it possible to produce beautifully precise functional maps of the fly's anatomy. By using different *UAS* transgenes, scientists can equally alter the activity of the targeted cells in specific circumstances – for example, turning a gene on or off depending on the temperature, or even altering the sex of specific cells. This amazingly productive technique has led to profound discoveries about every aspect of the fly's biology and typifies a view of 'engineering' that is informational rather than mechanical.

Part of the idea that organisms are based on information is the widespread view that DNA is a text – the modern version of the book of life – with all the implications about what we can do with a text. This forms part of the underlying theme of this book. As David

Jackson, who published the first article on recombinant DNA with Paul Berg, argued in 1993:

> To be fluent in a language, one needs to be able to *read*, to *write*, to *copy*, and to *edit* in that language.[8]

That is what genetic engineering has enabled us to do, with increasing precision, over the last half-century. However, as seen in the recent discoveries about the substantial inadvertent genetic changes induced by 'editing' with CRISPR, such metaphors are deceptive. The reality of genetic engineering is far more complex and problematic than pressing a few keys on a computer keyboard. We are not yet fully fluent in the language of DNA, whether it be seen as computer program or text.

*

This helps explain why, despite repeated futuristic claims that genetic engineering will transform our world, the consequences of our growing fluency have been less dramatic than predicted. Andrew Jamison of Aalborg University has explained this provocatively, writing about all forms of information technology:

> As opposed to earlier generic technologies, or radical innovations – the steam engine, electricity, and atomic energy, for example, which were primarily attempts to find solutions to identified problems – these new types of technologies tend to be solutions in search of problems.[9]

That does not mean that these innovations are useless, nor that they cannot usher in major changes in how we produce, or cure, or understand. But it does lead us to question the point of genetic engineering. Its powerful role in scientific discovery is evident; its function – or functions – as a technology are less clear. This should prompt us to adopt a more nuanced view of the bold claims that are often made about our genetically engineered future, for it highlights the complex route from invention to application.

Social context can play a determinant role in any application, while in the long term some of our current solutions are only temporary. For example, genetically engineered human growth hormone became a useful and viable product only when the traditional source of the hormone – cadavers – risked infecting patients with Creutzfeldt–Jakob Disease, while virtually all GM crops are limited to one or two specific characters (Bt or herbicide tolerance) both of which will eventually fail because they will induce resistance in the pests they target. We need to temper the visions of utopia, as well the fears of future dystopia.

This in turn leads to the question of why genetic engineering solutions are often seen as a priority when the problems they seek to address may be extremely complex or the use of genetically engineered organisms to solve them may raise even more difficulties. Researchers, funders, governments and companies often seem to be in the grip of what the historian Lily Kay described as 'molecular utopianism'.[10] Many of the examples described here – the strong advocacy for GM crops in developing countries, the focus on heritable human gene editing, the emphasis on gene therapy and so on – seem to reveal an impatience with complex problems and an overwhelming desire to apply an apparently simple solution that our ingenuity has created. This produces frustration either when the complex situation bites back, or when people who are the subject of such proposals, or who live in the area where GMOs are to be released, raise difficulties.

Some humility may be in order – if, as is often argued, our fundamental concern is about saving human lives, then genetic engineering is surely not the most immediate nor the most cost-effective solution. Focusing funding efforts on, say, clean water and effective sewage systems around the world would surely be a better use of our resources. To use an example more directly related to genetic engineering, for all but a few dozen individuals around the world, pre-implantation embryo screening can respond to the desire for healthy biological offspring, a desire seen as so overwhelmingly important that it requires us to accept human heritable gene editing. We need to recognise that in some – many? – cases, genetic engineering may not be the best solution to a particular problem. Simpler, more sustainable, more equitable and more widely available answers may be wiser choices.

Despite this evident truth, the fascination with what the 2016 *Gene Drives on the Horizon* report described as an 'innovation thrill' – deploying what Oppenheimer called a 'technically sweet' solution and worrying about the consequences afterwards – clearly runs deeply among scientists, physicians and those involved in developing policy regarding the application of these techniques. I can appreciate the attraction of simple, neat solutions – using the *GAL4–UAS* system to investigate the sense of smell in *Drosophila* maggots has brought me the same kind of joy in an elegant technique that Jon Beckwith and his colleagues felt with their experimental manipulation of the *lac* gene back in 1969. But laboratory studies of maggots with only one smell cell in their nose are evidently very different from an application of genetic engineering to the real world. Furthermore, there is nothing inevitable about the use of any form of genetic engineering. We have a choice whether to employ it or not, whether to permit its deployment or not. Just because we *can* do something does not mean that we *should*.

These problems arise not from the basic techniques, but from their encounter with the larger, squishier and more complex issues that flow from technological application. They were nascent the moment that genetic engineering emerged from molecular biology and became a science of intervention and action.

*

To explore the worries that led me to write this book – my concerns about heritable human gene editing, gene drives and gain-of-function research – I talked to Sheila Jasanoff, one of the most thoughtful scholars of technology and its implications, and we discussed the dreams and nightmares that have peppered the last fifty years of genetic research. She began by highlighting the importance of optimism and its links to the power of science: 'There's a reason why the myth of Pandora's box sticks with us', she told me. 'At the bottom of the box, when all the horrible things came out, there was this thing called Hope.' But then she continued, exploring the other side, the need to be wary:

> I think that the fears, the nightmares, serve their own purpose. They are prods to make people think harder about the things that can go wrong. If the 'hope people' are the warriors, then the 'fear people' are the worriers. I think we might need both of them in society, to keep the one calibrating the other, and in some kind of balance.[11]

That seems about right. The dreams and nightmares must go hand in hand because no form of technology is simply good or bad. Not only must it be wielded by humans in particular social circumstances, but even in its intrinsic functions a given technology can have contradictory outcomes that simultaneously enrich and threaten us – the carbon economy is an obvious example. That implies that all of us need to have some say about how, where, when and if such techniques are used.

Throughout the history of genetic engineering, the role of scientists in sounding the alarm about potential dangers has been exemplary. No other branch of science has ever taken such action; geneticists have done so four times. Beckwith and his colleagues were not just posturing when they denounced their own manipulation of the *lac* gene back in 1969. They were also alerting the global public to what they saw as a substantial danger. That is to their immense credit. The same applies to Asilomar: whatever the limitations of the process, the 1974–75 moratorium showed that scientists took the threat seriously and the debates in front of the world's press highlighted the issue for the public.

In the new century, the alerts sounded by gene-drive researchers or the 2012 research pause decided by virologists performing gain-of-function studies are both reassuring and nowhere near enough. For example, leading gene-drive researchers have published a code of ethics for gene-drive research.[12] This is a step forward, but they are still not thinking hard enough about the implications of their work and above all about how and why the public should actively have a say in whether such research goes ahead. This is even more true of those working on gain-of-function manipulation of pathogens. All of the scientists involved are well-meaning – there are no Dr Strangeloves or Victor Frankensteins in their ranks – but they need to pay more attention to the social implications of what they do.

Although Asilomar rejected the idea of discussing the ethical or social aspects of genetic engineering, its very existence as an example of self-regulation, together with the myths that eventually grew around it, created the expectation that genetic engineering and its associated technologies would take social responsibility seriously. And in some ways, it did. When the Human Genome Project was developed in the 1990s, up to 5 per cent of the $3 billion budget from the NIH and the US Department of Energy was set aside for a framework known in the United States as ethical, legal and social issues (ELSI – in Europe the final noun was 'aspects', so 'ELSA'), including how genomic data might affect racial, ethnic and socioeconomic questions.[13]

The excitement around synthetic biology in the first decade of the new century led to a new framework for exploring the implications of science, known as responsible research and innovation (RRI), which has since become a subdiscipline of its own, complete with a journal, meetings and all the usual academic paraphernalia.[14] There is no accepted definition of RRI and there is the lurking suspicion in some quarters that it might be just another intellectual fad. Nevertheless, the focus on how to do research responsibly, in particular when it comes to the latest developments in genetic engineering, has led some scientists to explore the ethical aspects of their work even before the techniques are fully developed, which is, of course, exactly when they should be considered.[15]

The key issue in the coming decades is therefore not our ability to develop new forms of genetic technology – there will be many of these, some of which will undoubtedly make parts of this book look quaint – but rather how to control those innovations, ensuring that they are used appropriately for the benefit of the many, not the few. Like any other form of technology, genetic engineering is developed and applied in a particular social context, by particular individuals and social forces. And like other technologies, it also has its own, specific, social implications. As the political theorist Langdon Winner put it in 1980, 'artifacts have politics' – they embody specific forms of power and authority.[16] In the case of genetic engineering, it carries with it both an implicit assumption that such technically sweet solutions are best, and a view of life as something that can be endlessly manipulated to our ends. In that respect, it is the realisation of Sir Francis Bacon's dream from four centuries ago.

The past and present of genetic engineering suggest that the time has come to nuance that view, to ensure that our ingenuity is deployed not only to discover new potential applications, but also to find ways of rendering them cheap and accessible to all. And if that cannot be done safely, then the experiments and the applications should not take place.

So far, public decision-making with regards to genetic engineering has been limited. For some of the scientists doing the experiments, that is exactly how it should be. At Asilomar, policy was made by a couple of hundred people, without challenging what some of them were doing in secret – creating terrible bioweapons or standing to profit from the patenting of their discoveries. The imposition and lifting of a moratorium on gain-of-function experiments on dangerous pathogens similarly involved only a few dozen people, something that in retrospect seems crazy given the potential dangers were any of these organisms to escape. In contrast, the human gene editing 'community' cannot even agree on a moratorium on this pointless and dangerous procedure, never mind a ban, and while gene-drive researchers have repeatedly raised the need for a collective response to their technology, that has still not taken place.

Given the potential impact of each of these areas, it is clear to me that the model of self-regulation adopted in the 1970s cannot respond to the discoveries of the twenty-first century. Public involvement in decision-making, on the basis of open experimental data rather than secrecy and suspicion, needs to become widespread and routine. International agreements, backed up by inspection and sanction powers, need to become the rule – this remains a major weakness of the Biological Weapons Convention, where compliance is assessed by the parties, not by any external organisation.[17] This weakness must be overcome and the illicit development of bioweapons by the Soviet Union, South Africa and others must never be repeated. Facilities in countries that have a bioweapons capability, including where these are supposedly devoted to defence (Russia, the United States,

China, Israel) ... must come under international scrutiny and control. Applications that are designed to be released in nature, such as gene drives and engineered plants, need to be co-developed with local communities from the outset, meeting their needs and local ecologies, and every step of the project needs to be risk-assessed. It is only because public disquiet has prompted the introduction of regulatory control that genetic engineering has thus far been safely deployed. That control needs to be developed into even stronger forms in the future, to meet the prospects and challenges we face now and will face tomorrow.

At the beginning of this book, I said that in a way genetic engineering was on trial in these pages and that you were part of the jury. With the exception of human heritable gene editing, which seems pointless and foolish, I am enthusiastic about a technology that, as Bacon foresaw, has immensely enlarged the bounds of human empire through the tremendous insights it has provided into how organisms function and through its myriad applications. But I remain deeply concerned. If it were to be allowed, heritable gene editing would lead either to an increase in inequalities, or, more likely, to the creation of babies carrying genetic alterations with unknown effects, which could be catastrophic for them. On the other hand, we can see the real prospect of new, life-changing somatic therapies, which nonetheless need to overcome the massive health inequalities that exist around the world to truly fulfil their potential. We have the power to eradicate disease vectors and, if things go horribly wrong, to destroy ecosystems or to inadvertently unleash terrible diseases that we will have foolishly created in order to better predict future pandemics. Those possibilities are all excitingly and disturbingly real.

To return to Sheila Jasanoff's description of our responses to genetic engineering, I suppose that means that I am both a warrior and a worrier.

What about you?

ACKNOWLEDGEMENTS

This book was written during the dreadful COVID-19 pandemic. Locked down, working from home, anxious and preoccupied by doom-scrolling, I found myself unable to read fiction or watch films or television. But to my surprise, I was able to work on this book, most of which was written between April 2020 and August 2021.

I was also able to make a three-part BBC Radio programme – *Genetic Dreams, Genetic Nightmares* – which in 2021 was broadcast on both Radio 4 in the UK and on the World Service as well as being turned into a podcast. This was the third time I have worked with my brilliant producer, Andrew Luck-Baker, on programmes related to genetic engineering. The first was in 2016, when we made *Editing Life*, the BBC's first programme devoted to CRISPR; the second was in 2017, when we made a programme about the life and work of Sydney Brenner, *A Revolutionary Biologist*. Each time, as well as making the programme, Andrew arranged interviews with a number of people whose work is described in these pages. These conversations have been immensely useful to me in writing this book, providing additional insights, helping me to avoid some errors and fleshing out the story.

My thanks go to Andrew for his enthusiasm and professionalism, and to all the scientists and scholars who gave up their time to talk to me over the years: David Baltimore, Paul Berg, Mike Bonsall, Herb Boyer, the late Sydney Brenner, George Church, John Connolly, Andrea Crisanti, Jennifer Doudna, Kevin Esvelt, Robb Fraley, Benjamin Hurlbut, Sheila Jasanoff, Natalie Kofler, Jeantine Lunshof, Janet Mertz, Matthew Meselson, Tony Nolan, Bob Pollack, Rich Roberts, Michael Rogers, Matthew Schnurr, Ricarda Steinbrecher, Fydor Urnov, Marc van Montagu and Jim Watson. Evidently, they are in no way responsible for my interpretation of what, in many cases, were key moments of their lives.

Many other people, on Twitter, by email and – very occasionally, given the dreadful circumstances – in real life, helped me by reading the manuscript or providing me with information, insights and general support. My profound thanks go to: Antony Adamson, Philip Ball, Andrew Berry, Dominic Berry, Mike Bevan, Luis Campos, Nathaniel Comfort, Jerry Coyne, Kevin Davies, Andrew Doig, Tori Herridge, Brian Hilbush, Simon Hubbard (my boss while I was writing), Paul McAuley, Catherine McCrohan, Michel Morange, Nitika Mummidivarapu, Brigitte Nerlich, Adam Roberts, Adam Rutherford, Hallam Stevens, Carsten Timmerman, Jon Turney and Leslie Vosshall. I am also indebted to the thousands of first- and second-year students at the University of Manchester who since 2016 have been attending my lectures on the scientific and ethical aspects of the latest developments in genetic technology, helping me to clarify and develop my ideas.

My agent, Peter Tallack, expertly shaped the proposal and sharpened my ideas. My then UK editor, Ed Lake, helped me all along the way, giving insightful and useful advice and guidance, while my US editor, T. J. Kelleher, encouraged me to make the final version tauter and tarter. Nick Humphrey took over the reins from Ed for the final straight, whispering in the book's ears to get it over the finishing line. As ever, Penny Daniel expertly guided the manuscript through the production process.

Of course, any errors that remain are entirely the responsibility of our cats, Ollie, Pepper and Harry. And yet again, my greatest thanks go to my family for their forbearance: Tina, Lauren and Eve.

GLOSSARY AND ACRONYMS

ADA-SCID. Adenosine deaminase severe combined immunodeficiency. A genetic disease that has been the focus of gene therapy. In its most extreme form, SCID can require children to live in sterile environments ('bubbles') to avoid infection.

Allele. A version of a gene. Alleles differ by one or more base pairs. For a given gene there may be dozens of alleles, not all of which will produce any fitness differences.

Amino acid. A small molecule containing amine ($-NH_2$) and carboxylic acid ($-COOH$) groups. There are hundreds of different amino acids, but only twenty of them generally occur in organisms. They are strung together to make proteins.

Andromeda Strain, The. The title of a 1969 thriller by Michael Crichton and a 1971 film, this became a popular metaphor in the 1970s and 1980s, referring to an artificially created pandemic.

Bacteriophage. Often abbreviated to 'phage'; a virus that targets bacteria.

Base. A molecule – adenosine, cytosine, guanine, thymine or uracil – that forms part of a nucleotide in DNA or RNA. When coupled with other molecules it forms a nucleotide.

Base (or prime) editing. Recently developed forms of gene editing that alter a single base after inducing a single-strand break in the DNA molecule. Apparently less likely to induce undesired on- or off-target effects. There are subtle differences between the two forms of editing.

Biohazard. A threat to life – human or otherwise – produced by an organism, a virus or a biological molecule.

Biotechnology. The use of microorganisms in production, initially involving the unwitting use of naturally occurring yeasts to produce bread, beer, wine, etc., now more generally the industrial use of engineered microorganisms to produce a wide range of substances.

Bt. A naturally occurring soil bacterium, *Bacillus thuringiensis.* Generally used to describe an insecticide produced by this microbe, the gene for which is routinely used in GM crops.

BWC (Biological Weapons Convention). UN treaty intended to prevent the development of biological weapons, but which has very limited powers. Signed in 1972, came into effect in 1975. Did not stop either the Soviet Union or South Africa and perhaps other countries from developing biological weapons after signing the treaty.

Cas. CRISPR-associated genes found in bacteria, generally coding for enzymes that are either nucleases or helicases. There are a number of these genes; *Cas9* is the most widely known.

CBD (Convention on Biological Diversity). A United Nations (UN) multilateral treaty opened at the Rio de Janeiro Earth Summit in 1992. The United States is the only UN country not to have ratified the CBD.

CCR5. A human gene producing a protein expressed on the surface of white blood cells. One naturally occurring *CCR5* allele is associated with resistance to HIV. This was the gene targeted by He Jiankui in his catastrophic use of CRISPR on three human embryos.

cDNA (complementary DNA). A DNA sequence that is synthesised by scientists from a mature mRNA sequence using reverse transcriptase. In eukaryotes this does not necessarily correspond to the equivalent DNA in the genome because many eukaryotic genes have introns that are removed from the initial mRNA sequence by the eukaryotic cell, forming mature mRNA.

Chromosome. Cellular structures composed of DNA and proteins that contain genes. In prokaryotes they are generally circular; in eukaryotes they are normally linear.

Clone. A genetic copy. This has also been used as a verb, to describe two very different processes – the copying of bacterial DNA in plasmids or in cells, and the creation of an exact genetic copy of a multicellular organism, such as Dolly the Sheep.

Codon. A sequence of three bases in a DNA or RNA molecule that codes for an amino acid.

CRISPR. A method for editing genes in organisms. The acronym comes from a kind of bacterial DNA sequences and was coined before their function was apparent: Clustered Regularly Interspaced Short Palindromic Repeats. Used both as a noun and, increasingly, as a verb, and by extension as an adjective.

Cytoplasm. The contents of a eukaryotic cell, excluding the nucleus. Contains organelles involved in protein production (ribosomes, etc.) and energy production (mitochondria) as well as a wide range of chemicals.

Diploid. Adjective used to describe an organism that has two copies of each chromosome. Most animals are diploid.

Dominant. Adjective to describe an allele that is expressed if a diploid organism has different alleles of a particular gene on each chromosome. Dominance is always relative (a given allele will not be dominant against all others) and in many cases both alleles are expressed.

DNA. Deoxyribonucleic acid, a molecule formed of a double helix that is composed of a sugar/phosphate backbone and four bases: adenine, cytosine, guanine and thymine (A, C, G and T). The genetic material in all organisms and some viruses.

Drosophila. A tiny fly used by geneticists since the beginning of the twentieth century. I have studied this insect for over forty years.

E. coli. *Escherichia coli*, a bacterium found in the human gut and the most intensively studied microbe.

EcoR1. The first restriction enzyme to be isolated from *E. coli* by Herb Boyer, hence its number, R1.

Electrophoresis. The movement of molecules in a medium under a constant electric charge. Molecules of different sizes and different electrical charges will move at different rates, allowing them to be separated in a sample containing different sized molecules. Can be used with both nucleic acids and proteins.

Endonuclease. An enzyme that cleaves a nucleic acid at a point within the molecule. Restriction enzymes are endonucleases.

Enzyme. A protein or RNA molecule that catalyses (speeds up) a particular chemical reaction.

Eukaryote. An organism with a nucleus and mitochondria in its cells. A minority of eukaryotes are also multicellular. All multicellular organisms are eukaryotes.

Exon. A sequence in a eukaryotic gene that is expressed, i.e. turned into RNA.

FDA (Food and Drug Administration). Federal US body responsible for the control of all foods and drugs in the United States.

Gain-of-function. An experiment designed to increase the effect of a given gene, in particular those involved in pathogen virulence or transmissibility.

Gel. Semi-solid substance used in electrophoresis.

Gene. There are dozens of definitions of 'gene'. The one mainly used here is a sequence of DNA that enables the cell to produce a protein. Sometimes a gene can produce an RNA molecule that alters the activity of another gene.

Gene drive. A genetic construct in a diploid organism that copies itself onto the other chromosome in germline cells, thereby ensuring that all offspring carry the construct. This process repeats each generation, leading to an exponential growth in the frequency of the gene in the population.

Gene therapy. The medical application of genetic engineering, specifically the alteration of the activity of a particular gene in a somatic cell, usually by introducing a different version of the gene or, more recently, by changing the gene's sequence.

Genetic code. The relation between a DNA or RNA codon and an amino acid. Some amino acids are coded by several codons, others by only one or two.

Genome. The full heritable DNA sequence of an organism.

Germline. The DNA found in sex cells in animals or fungi, or the DNA of a single-cell embryo, which will therefore be passed down to the next generation. Generally contrasted with the soma or somatic cells, which do not pass their DNA on to any offspring. For humans, the phrase 'germline editing' is being replaced by heritable human gene editing (HHGE).

Globin. A protein that carries oxygen through a haem group. There are many different globins.

GMO (genetically modified organism). A transgenic plant or (more rarely) animal that has been produced by genetic engineering of one kind or another.

Guide RNA. In CRISPR, an RNA molecule that specifies the DNA target and directs a nuclease to that location.

H5N1. Highly pathogenic avian flu. Occurs frequently in birds but can spill over into humans where it can be highly contagious and virulent.

Haemoglobin. Protein that carries oxygen in blood; the iron (haem) group gives blood its red colour. Some animals have proteins that use other groups for this function and therefore have different colour blood.

Haploid. Adjective used to describe an organism that has a single copy of each chromosome. All bacteria and archaea are haploid.

HHGE (heritable human genome editing). Synonym for germline gene editing – the introduction of genetic changes that will be passed on to any descendants.

Homologous recombination. Also known as homology-directed repair. DNA repair mechanism in diploid organisms in response to a double strand break, in which the corresponding sequence on the other chromosome is used as a template to repair the missing bases. In multicellular organisms, this is the main form of repair in undifferentiated cells – different cells at different points in their cycle and in different organisms tend to preferentially use either homologous recombination or non-homologous end-joining. If an artificial DNA template guides the repair, the new sequence will be introduced into the cell's genome.

Intron. Part of a gene sequence in a eukaryotic organism that is not expressed. During gene expression these sequences are spliced out of the initial mRNA molecule, leaving a mature mRNA molecule.

IVF (in vitro fertilisation). A range of techniques involving the external fertilisation of an egg and the implantation of the resultant embryo into the mother. First successfully applied to humans by Robert Edwards, Patrick Steptoe and Jean Purdy in 1977, leading to the birth of Louise Brown in Oldham in July 1978.

***Jurassic Park*.** 1990 technothriller by Michael Crichton describing what might happen if dinosaurs were re-created using genetic engineering. Made into a film by Steven Spielberg in 1993, now a successful franchise. Fiction.

Landrace. A domesticated animal or plant that has become adapted to local conditions.

Meganuclease. An enzyme that cleaves a nucleic acid at a very precise location, specified by a dozen or more bases.

Moratorium. A temporary, voluntary ban on a particular process or action. Became widely used with regards to atmospheric tests of nuclear weapons in the 1950s and was subsequently applied to recombinant DNA research.

Mosaic. A multicellular organism in which not all cells have the same set of alleles.

mRNA. Messenger RNA. These molecules are formed as a complementary copy of the gene and move from the chromosome to the ribosome (in eukaryotes this involves the mRNA molecule moving from the nucleus to the cytoplasm), where each codon binds with a specific transfer RNA molecule which is attached to a corresponding amino acid.

MIT (Massachusetts Institute of Technology). Founded in 1861 and based on the banks of the Charles River in Cambridge, MA.

NIH (National Institutes of Health). The main US federal funder of genetic research.

Non-homologous end-joining. DNA repair mechanism in response to a double strand break, in which the cell fuses the two cut ends of the double helix. Different cells at different points in their cycle and in different organisms tend to preferentially use either homologous recombination or non-homologous end-joining. Non-homologous end-joining can lead to the deletion or addition of bases.

Nuclease. An enzyme that cleaves or cuts a nucleic acid.

Nucleic acid. RNA or DNA.

Nucleotide. A molecule that combines a base with a five-carbon sugar (ribose or deoxyribose) plus a phosphate group; forms the basis of the nucleic acid sequence.

Nucleus. A central structure within a eukaryotic cell, surrounded by a membrane, that contains the chromosomes.

Oncogene. A gene, generally involved in cell division, that causes cancer when it contains a mutation.

Operon. A set of functionally related genes, generally clustered together, that are involved in gene regulation in bacteria.

Phage. Short for bacteriophage. These are viruses that attack bacteria.

Plasmid. Small DNA molecule found in bacteria that is not part of the bacterial chromosome and is inherited separately. Generally circular, these molecules were essential in the development of recombinant DNA technology, allowing DNA to be transferred between cells.

Preprint. A scientific article that is made public, but which has not yet been subject to peer review.

Prime (or base) editing. Recently developed forms of gene editing that alter a single base after inducing a single-strand break in the DNA molecule. Apparently less likely to induce undesired on- or off-target effects. There are subtle differences between the two forms of editing.

Prokaryote. Single-celled organism without a nucleus, belonging to one of the two great branches of life, the bacteria and the archaea.

Protein. A large molecule consisting of chains of amino acids. Proteins come in a vast variety of forms and carry out many biological functions.

RAC (Recombinant DNA Advisory Committee). NIH committee set up in 1974, disbanded in 2019, responsible for guiding the regulation of genetic engineering funded by NIH. Despite this limited remit, it had influence both internationally and on US private-sector research.

Recessive. Adjective to describe an allele that is not expressed if a diploid organism has different alleles of a particular gene on each chromosome. Recessiveness is always relative (a given allele will not be recessive against all others) and in many cases, both alleles are expressed.

Recombinant. DNA from more than one source. Although technically all offspring of sexual reproduction are recombinant (including you), the term is generally used to apply to a novel form of DNA that has been deliberately created from different species.

rDNA. Recombinant DNA.

RNA. Ribonucleic acid. A helical molecule composed of a sugar/phosphate backbone and four bases: adenine, cytosine, guanine and uracil (A, C, G and U). The genetic material in some viruses; carries out a wide range of regulatory functions in all cells.

Restriction enzyme. A nuclease that cleaves DNA at a particular site by recognising a nucleotide sequence. In nature, restriction enzymes are used by bacteria as defence against bacteriophage infection, thereby 'restricting' the virus.

Reverse transcriptase. Enzyme that transcribes an RNA sequence into DNA, as part of the process of reverse transcription.

Science for the People. Radical US group primarily active in the 1970s and early 1980s. Also the title of the magazine produced by the group. Both have recently been revived.

Somatic. Cells in an animal that are not sex cells, so cannot pass their DNA onto the next generation.

Sticky end. Also 'cohesive end'. Often produced by the action of restriction enzymes, a protruding single strand of DNA that can be used to assemble a longer or circular DNA molecule by associating with the complementary, similarly protruding sequence. For example, bacteriophage DNA naturally has sticky ends which read GGGCGGCGACCT and AGGTCGCCGCCC. The GGG at the beginning of the first sequence will associate with the CCC at the end of the second sequence, and so on.

SV40 (simian virus 40). A primate virus that can induce tumours in mammalian cells, so once the focus of interest for cancer researchers. Subsequently used as a vector to introduce recombinant DNA into mammalian cells.

Synthetic biology. A form of genetic engineering that involves redesigning organisms (usually microbes) according to engineering principles, in order to meet human needs. Views organisms as devices or systems that can be engineered.

Synthetic DNA. DNA that is produced in the laboratory, now in a DNA synthesiser. Producing stretches of synthetic DNA over 200 base pairs in length is extremely difficult; synthetic DNA gene sequences have to be assembled from a series of component sequences.

TALEN (transcriptor-like effector nuclease). An efficient system of gene editing developed at the beginning of the twenty-first century that was rapidly overtaken by CRISPR.

Transfection. Alteration of the genome of a cell by the introduction of recombinant DNA.

Transgenic. One of many terms coined to describe an organism that has been engineered to contain DNA from different species.

Transposon. A mobile or 'jumping' gene that, with the aid of a transposase enzyme, can move from one place to another in the genome. Some transposons were originally viruses that have become trapped in the genome.

Venture capital. Speculative system of short-term private equity financing of small, innovative companies which became widespread in the United States in the late 1970s.

Virus. Submicroscopic infectious agent that replicates inside cells by hijacking the cell's biochemical processes. In many cases a molecule of DNA or RNA is contained within a protein coat. The nucleic acid contains instructions for making more virus.

ZFN (zinc finger nuclease). A nuclease enzyme that is guided to cleave a nucleic acid at a particular sequence by parts of the molecule containing zinc. By changing these zinc fingers, the enzyme can be programmed to target a desired sequence.

NOTES

Introduction

1 Jacob, F. (1998), *Of Flies, Mice, and Men* (Cambridge, MA, Harvard University Press), p. 69. The French edition appeared the year before.

2 Verulam, F. (1626), *Sylva Sylvarum: Or a Naturall Historie* (London, Lee). *The New Atlantis* ('A Worke unfinished') appears at the end of the 270-odd pages of *Sylva Sylvarum*, together with a brief note from Bacon's secretary, Rawley, who assembled the book for publication a few months after Bacon's death in 1626.

You probably won't believe it, but we weren't certain heredity existed until the nineteenth century. Cobb, M. (2006), *Nature Reviews Genetics* 7:953–8.

1 Prelude

1 Librado, P., et al. (2021), *Nature* 598:634–40. For an overview, see Shapiro, B. (2021), *Life as We Made It: How 50,000 Years of Human Innovation Refined – and Redefined – Nature* (London, Oneworld).

2 Bud, R. (1993), *The Uses of Life: A History of Biotechnology* (Cambridge, Cambridge University Press); Gaudillière, J.-P. (2009), *Studies in History and Philosophy of Biological and Biomedical Sciences* 40:20–8; Crowe, N. (2021), in M. Dietrich, et al. (eds), *Handbook of the Historiography of Biology* (Cham, Springer), pp. 217–41.

3 Cobb, M. (2006), *The Egg and Sperm Race: The Seventeenth-Century Scientists Who Unravelled the Secrets of Sex, Life and Growth* (London, Free Press).

4 For an exploration of the link between *Frankenstein* and genetic engineering, see Turney, J. (1998), *Frankenstein's Footsteps: Science, Genetics and Popular Culture* (Yale, Yale University Press).

5 Rutherford, A. (2022), *Control: The Dark History and Troubling Present of Eugenics* (London, Weidenfeld & Nicolson).

6 Williamson, J. (1952), *Dragon's Island* (New York, Popular Library), pp. 8, 12.

7 Ibid., p. 13.

8 Dobzhansky, T. (1941), *Genetics and the Origin of Species* (New York, Columbia University Press), pp. 49–50.

9 Cobb, M. (2015), *Life's Greatest Secret: The Race to Crack the Genetic Code* (London, Profile).
10 Tatum, E. (1959), *Science* 129:1711–5, p. 1714.
11 *Time*, 2 January 1961. This factoid is the opening sentence of Kelly Moore's excellent book *Disrupting Science* and was too good not to steal. Moore, K. (2008), *Disrupting Science: Social Movements, Americans Scientists, and the Politics of the Military, 1945–1975* (Princeton, Princeton University Press).
12 Ezrahi, Y., et al. (eds) (1994), *Technology, Pessimism and Postmodernism* (Dordrecht, Springer).
13 For details of Project Plowshare, see for example: O'Neill, D. (1994), *The Firecracker Boys: H-Bombs, Inupiat Eskimos and the Roots of the Environmental Movement* (New York, Basic Books) and Kaufmann, S. (2013), *Project Plowshare: The Peaceful Use of Nuclear Explosives in Cold War America* (London, Cornell University Press).
14 Frum, D. (2000), *How We Got Here. The 70s: The Decade That Brought You Modern Life (For Better or Worse)* (New York, Basic Books).
15 Meselson, M. (2017), in B. Friedrich, et al. (eds), *One Hundred Years of Chemical Warfare: Research, Deployment, Consequences* (Cham, Springer Open), pp. 335–48.
16 Ironically, Wyndham's central character imagines that the triffids were created using Lysenko's methods, which rejected genetics and were spurious. Link, M. (2015), *Irish Journal of Goth and Horror Studies* 14:63–80.
17 Mendelsohn, E. (1994), in Y. Ezrahit, et al. (eds), *Technology, Pessimism, and Postmodernism* (Amsterdam, Kluwer), pp. 151–73.
18 Rattray Taylor, G. (1968), *The Biological Time Bomb* (London, Thames & Hudson), p. 175. For an astute analysis of *The Biological Time Bomb*, including its debt to the 1959 book *Can Man be Modified?* by the French biologist Jean Rostand, see Turney, *Frankenstein's Footsteps*, pp. 155–9.
19 Rattray Taylor, *The Biological Time Bomb*, p. 237.
20 *Birmingham Daily Post*, 24 April 1968.
21 Crick's notes are in the Wellcome Library at PPCRI/E/1/16/13/1. The talk was not written up or published, but a brief summary appears in Anonymous (1968), *Nature* 220:429–30.
22 Nirenberg, M. (1967), *Science* 157:633.
23 Lederberg, J. (1967), *Science* 158:313.
24 Bud (1993), p. 171. I have not been able to consult this document, which has never been published.

2 Tools

1 Feenberg, A. (1998), *Questioning Technology* (London, Routledge).
2 See the excellent summary by Turney, J. (1998), *Frankenstein's Footsteps: Science, Genetics and Popular Culture* (Yale, Yale University Press) from which I have taken a number of examples. The opening two chapters owe much to both *Frankenstein's Footsteps* and to Agar, J. (2008), *British Journal for the History of Science* 41:567–600.
3 Maher, N. (2017), *Apollo in the Age of Aquarius* (London, Harvard University Press).
4 All details of Science for the People taken from Moore, K. (2008), *Disrupting Science: Social Movements, American Scientists, and the Politics of the Military, 1945–1975* (Princeton, Princeton University Press), pp. 158–89.
5 Salloch, R. (1969), *Bulletin of the Atomic Scientists*, May 1969, pp. 32–5.

6 DuBridge, L. (1969), *Bulletin of the Atomic Scientists*, May 1969, p. 26.
7 Roszak, T. (1970), *The Making of a Counter Culture: Reflections on the Technocratic Society and its Youthful Opposition* (London, Faber & Faber).
8 Goulian, M., et al. (1967), *Proceedings of the National Academy of Sciences USA* 58:2321–8.
9 Goulian, M. and Kornberg, A. (1967), *Proceedings of the National Academy of Sciences USA* 58:1723–30.
10 *Boston Globe*, 17 December 1967; *The Guardian*, 15 December 1967.
11 *Boston Globe*, 17 December 1967.
12 *Boston Globe*, 15 December 1967.
13 *Los Angeles Times*, 21 December 1967; Lederberg, J. (1969), in *Yearbook of Science and the Future* (Chicago, Encyclopaedia Britannica), p. 318. History – or rather, science – has not shared Lederberg's enthusiasm. The two papers have been cited only 82 and 112 times, respectively. In reality, the results were no more revolutionary than the description two years previously of the synthesis of a gene from an RNA virus – Spiegelman, S., et al. (1965), *Proceedings of the National Academy of Sciences USA* 54:919–27.
14 Shapiro, J., et al. (1969), *Nature* 224:768–74.
15 Beckwith, J. (1970), *Bacteriological Reviews* 34:222–7, pp. 224–5.
16 *Chicago Tribune*, 23 November 1969.
17 *New York Times*, 8 December 1969.
18 *Boston Globe*, 4 January 1970.
19 Shapiro et al. (1969); Beckwith (1970), p. 225.
20 Cited in Wick, G. (1970), *Science* 167:157–9, p. 157; *Los Angeles Times*, 23 November 1969.
21 *The Times*, 25 November 1969.
22 Beckwith, J. (2002), *Making Genes, Making Waves: A Social Activist in Science* (Harvard, Harvard University Press), p. 58. This interview was apparently with the *South Wales Echo*.
23 Glassman, J. (1970), *Science* 167:963–4. A few weeks later there were responses in the letters page of *Science*, with a group of young researchers accusing Shapiro of 'defection' – *Science* 167:1668.
24 Beckwith (1970), p. 224–5; Beckwith, *Making Genes, Making Waves*, pp. 59–64; *New York Times*, 29 April 1970. Half the money went to the Black Panther Free Health Movement, the other half to fund the defence of twenty-one New York Panthers who were being falsely tried. Beckwith knew one of the New York twenty-one, the biochemist Curtiss Powell – after a six-month trial and spending nearly two years in jail, Powell was acquitted, along with his comrades. It took the jury less than two hours to reach its verdict. For more on Powell, who went on to study sleeping sickness in Africa, see *New Scientist*, 15 February 1973, pp. 369–71.
25 Anonymous (1969), *Nature* 224:834–5.
26 Anonymous (1969), *Nature* 224:1241–2.
27 Anonymous (1969), *Nature* 223:985–8.
28 Maddox, J. (1972), *The Doomsday Syndrome* (London, Macmillan).
29 As Bob Dylan might have sung in his sneering voice on his 1965 composition 'Ballad of a Thin Man': 'And something is happening here, but you don't know what it is, do you, Mr Maddox?' I have stolen this joke from Jon Agar.
30 Alice Bell has written the first history of BSSRS; the details in this paragraph are from her account. Bell, A. (2017), *Radical History Review* 127:149–72.

31 Letter from Wright to Crick, 12 October 1962. Cold Spring Harbor Laboratory Archive, SB/11/1/180. http://libgallery.cshl.edu/items/show/52128 – Sydney Brenner confirmed to me what the 'confidential matter' referred to in the letter was.
32 Fuller, W. (ed.) (1971), *The Social Impact of Modern Biology* (London, Routledge & Kegan Paul). On 13 July 1970, Beckwith wrote to Wilkins explaining that Shapiro would be unable to attend as he would be in Cuba when the meeting took place (King's College London Archives, KPP178/11/1/15/1). This meeting forms the centrepiece of Agar, J. (2008), *British Journal for the History of Science* 41:567–600, but has not been fully studied. The papers and correspondence surrounding the meeting are available in the Wilkins papers at the King's College London Archives and have been digitised at the Wellcome Library. I have found no photos of the event. According to Wilkins's correspondence the meeting was recorded, but it is not clear whether the tapes survived.
33 Fuller, *The Social Impact of Modern Biology*, p. 5.
34 Ibid., p. 203.
35 Wilkins subsequently wrote to all participants smoothing any ruffled feathers with the hope that the 'abusive bits at the meeting may have a positive value as representing a significant awakening of concern' – letters of 2 and 3 December 1970 to speakers (King's College London Archives, KPP178/11/1/15/1); references to 'bollocks' and 'bullshit' being heard on the tape can be found in a manuscript addition to the letter of 2 December 1970.
36 Beckwith, J. (1971), *Science for the People*, May 1971, p. 7.
37 Bronowski, J. (1971), in W. Fuller (ed.), *The Social Impact of Modern Biology* (London, Routledge & Kegan Paul), pp. 233–46, p. 234.
38 Galston, A. (1971), in W. Fuller (ed.), *The Social Impact of Modern Biology* (London, Routledge & Kegan Paul), pp. 154–66; Leshem, Y. and Galston, A. (1979), *Phytochemistry* 10:2869–78; Hess, D. (1969), *Zeitschift für Pflanzenphysiologie* 60:348–53; Ledoux, L. and Huart, R. (1968), *Nature* 218:1256–9.
39 Galston, A. (1972), *Annals of the New York Academy of Sciences* 196:223–35.
40 Galston, in *The Social Impact of Modern Biology*, p. 163.
41 For example, they are absent from Morange, M. (2020), *The Black Box of Biology: A History of the Molecular Revolution* (Harvard, Harvard University Press).
42 Lurquin, P. (2001), *The Green Phoenix: A History of Genetically Modified Plants* (New York, Columbia University Press); see also Primrose, S. (1977), *Science Progress* 64:293–321. One of the decisive papers that debunked Ledoux's work was Kleinhofs, A., et al. (1975), *Proceedings of the National Academy of Sciences USA* 72:2748–52.
43 Ledoux's paper (Ledoux, L. and Huart, R. (1968), *Nature* 218:1256–9) has not been cited since 1991, with the exception of Lurquin, *The Green Phoenix*. There is a brief discussion of this period in Heimann, J. (2018), *Using Nature's Shuttle: The Making of the First Genetically Modified Plants and the People Who Did It* (Wageningen, Wageningen Academic Publishers).
44 Stent, G. (1968), *Science* 160:390–5.
45 Gros, F. (1986), *Les Secrets du gène* (Paris, Odile Jacob), p. 167.
46 Lewin, B. (1970), *Nature* 227:1009–13.
47 Ibid., p. 1012.
48 Baltimore, D. (1970), *Nature* 226:1209–11; Temin, H. and Mizutani, S. (1970),

Nature 226:1211–3. For a lively memoir of Temin's role, see Coffin, J. (2021), *Molecular Biology of the Cell* 32:91–7. The discovery of reverse transcriptase led some people (in particular an excited *Nature* editorialist, who according to Coffin, ibid., was John Tooze) to argue that it contradicted one of the foundations of molecular biology, known as the central dogma. This was enunciated by Francis Crick in 1957 and states that genetic information passes only from DNA to RNA to proteins, and not the other way. In reality there was no contradiction and *Nature*'s initial crowing attracted Crick's well-directed ire – Cobb, M. (2017), *PLoS Biology* 15:e2003243.

49 Meselson, M. and Yuan, R. (1968), *Nature* 217:1110–4; Smith, H. and Wilcox, K. (1970), *Journal of Molecular Biology* 51:379–91.

50 Abelson, P. (1971), *Science* 173:285; Boffey, P. (1971), *Science* 171:874–6, p. 875.

3 Biohazards

1 Cold Spring Harbor Laboratory (1971), *Annual Report 1971* (Cold Spring Harbor, NY: Cold Spring Harbor Laboratory Press), p. 21.

2 Mertz's CV can be found at the Cold Spring Harbor Laboratory Archive, SB/1/1/414/4. http://libgallery.cshl.edu/items/show/63403

3 Friedberg, E. (2014), *A Biography of Paul Berg: The Recombinant DNA Controversy Revisited* (Singapore, World Scientific Publishing), p. 127; Berg, P. (2000), *A Stanford Professor's Career in Biochemistry, Science, Politics, and the Biotechnology Industry. An Oral History Conducted in 1997 by Sally Smith Hughes* (Berkeley, Regional Oral History Office, The Bancroft Library, University of California), p. 98.

4 SV40 had been discovered in 1961 following the observation that the widely used polio vaccine caused cancer if injected into hamsters. The vaccine was contaminated with SV40, raising fears of an epidemic of cancer in humans – thankfully, this did not occur. A potential carcinogenic role for SV40 seems extremely unlikely as no spike in cancers has been observed from the period when millions of US children were vaccinated with SV40-contaminated polio vaccine. Modern polio vaccine is free of SV40.

5 Many versions of these discussions have been given by the participants over the years. None of them will be precisely accurate. This is taken from Friedberg, *A Biography of Paul Berg*, p. 206.

6 They both recalled these events in *Genetic Dreams, Genetic Nightmares* (BBC Radio, 2021), episode 1: https://www.bbc.co.uk/programmes/m000xzdp. According to Lear, J. (1978), *Recombinant DNA – the Untold Story* (New York, Crown), Pollack merely asked 'Do you really mean to put SV40 into *E. coli*?' (p. 26). The rest of this description of the phone call is from Berg, *A Stanford Professor's Career*.

7 Friedberg, *A Biography of Paul Berg*, p. 210; Berg, *A Stanford Professor's Career*, p. 91.

8 Berg, *A Stanford Professor's Career*, p. 73.

9 Friedberg, *A Biography of Paul Berg*, p. 135.

10 The film director was Robert Alan Weiss of the University of California San Diego; the Stanford student newspaper reviewed the film under the title 'Molecular Movie Made for Masses'. *Stanford Daily*, 11 October 1971.

11 https://www.youtube.com/watch?v=u9dhOoiCLww

12 Lear, *Recombinant DNA*, p. 27.

13 Ibid., p. 28.

14 This document was kindly shared with me by Bob Pollack. A PDF can be found at Cold Spring Harbor Laboratory Archive. Bob and I published a brief letter referring to it, Cobb, M. and Pollack, R. (2021), *Nature* 594:496, as well as a longer article: Pollack, R. and Cobb, M. (2022), *PLoS Biology* 20:e3001539.
15 Friedberg, *A Biography of Paul Berg*, p. 208. This quote is taken from Mertz's interview for the MIT Oral History of Recombinant DNA project in 1977, which I was not able to consult directly because of the pandemic.
16 Lear, *Recombinant DNA*, pp. 37–8.
17 Ibid., p. 38.
18 Friedberg, *A Biography of Paul Berg*, p. 213.
19 Baldwin, C. and Runkle, R. (1967), *Science* 158:264–5.
20 The culprit was a young molecular virologist, James Robb, of the University of California San Diego. See Hellman, A., et al. (eds) (1973), *Biohazards in Biological Research* (Cold Spring Harbor, NY, Cold Spring Harbor Laboratory), p. 137.
21 Lear, *Recombinant DNA*, p. 44.
22 Jensen, R., et al. (1971), *Biochemical and Biophysical Research Communications* 43:384–92. Berg was presented with this paper in an oral history interview in 1997 and averred that he had never heard of it (Berg, *A Stanford Professor's Career*, pp. 10–11). The paper has been cited only nine times; it was cited four times in the 1970s, once by Lobban. Berg provides a discussion of the paper, and the journal to which it was submitted (which he once edited), in *A Stanford Professor's Career*, pp. 129–31. The purity of the ligase used in the Stanford biochemistry department was far superior; it had been synthesised by PhD student Paul Modrich, who later shared the 2015 Nobel Prize in Chemistry for his work on DNA repair.
23 Lobban's work was referred to at several points by Jackson, D., et al. (1972), *Proceedings of the National Academy of Sciences USA* 69:2904–9. Referring to other methods for joining DNA molecules, Jackson et al. also cite an article by Lobban and Sgaramella, which was allegedly in press in *Nature*. No such article exists, nor have I found any trace of it. It is presumably a drafting error that got overlooked during publication.
24 Ibid.
25 Ibid., p. 2904.
26 Morange, M. (2020), *The Black Box of Biology: A History of the Molecular Revolution* (Cambridge, MA, Harvard University Press), p. 183.
27 Berg, *A Stanford Professor's Career*, p. 69.
28 Sgaramella, V. (1972), *Proceedings of the National Academy of Sciences USA* 69:3389–93.
29 Mertz, J. and Davis, R. (1972), *Proceedings of the National Academy of Sciences USA* 69:3370–4.
30 You can hear this part of my interview with Mertz here: *Genetic Dreams, Genetic Nightmares* (BBC Radio, 2021), episode 1: https://www.bbc.co.uk/programmes/m000xzdp
31 Hedgpeth, J., et al. (1972), *Proceedings of the National Academy of Sciences USA* 69:3448–52.
32 Mertz and Davis (1972), p. 3374.
33 Anonymous (1972), *Nature* 240:73–4, p. 74. Not only was this extremely unlikely, it turned out to be impossible – according to Berg, *A Stanford Professor's Career*, the procedure had inadvertently disrupted the ability of the phage DNA sequence to infect *E. coli*. However, some of the recombinants

developed by Janet Mertz could have been successfully cloned in E. coli; indeed, that was her aim in creating them. She explained to me in 2022 that she did not carry out these experiments in 1972 because of the Berg laboratory's self-imposed moratorium on creating recombinant DNA.

34 Cohen, S. (2013), *Proceedings of the National Academy of Sciences USA* 110:15521–9. The image is credited to Dick Adair, *Honolulu Advertiser*. Stanley Falkow had a slightly different recollection, highlighting the significance of Boyer's description of the recent work by Mertz and Davis and by Peter Lobban, which led to the penny dropping. Falkow, S. (2001), *ASM News* 67:555–9.

35 Cohen, S., et al. (1973), *Proceedings of the National Academy of Sciences USA* 70:3240–4. The ethidium bromide technique was developed at Cold Spring Harbor by Phil Sharp and Joe Sambrook – Sharp, P., et al. (1973), *Biochemistry* 12:3055–63.

36 Hogan, A. (2016), *Life Histories of Genetic Disease: Patterns and Prevention of Postwar Medical Genetics* (Baltimore, Johns Hopkins University Press), pp. 120–46.

37 Cohen et al. (1973), p. 3244.

38 Berg, *A Stanford Professor's Career*, p. 78.

39 Krimsky, S. (1982), *Genetic Alchemy: The Social History of the Recombinant DNA Controversy* (London, MIT Press), pp. 70–80 provides a forensic analysis of the various drafts of the letter. Rich Roberts, who in 1993 would share the Nobel Prize in Physiology or Medicine for the discovery of introns, was at the Gordon Conference and shared his memories with me in 2022.

40 Singer, M. and Söll, D. (1973), *Science* 181:1114.

41 Ziff, E. (1973), *New Scientist*, 25 October 1973, pp. 274–5.

42 Dixon, B. (1973), *New Scientist*, 25 October 1973, p. 236.

43 Wade, N. (1973), *Science* 182:566–7, p. 567.

44 Morrow, J., et al. (1974), *Proceedings of the National Academy of Sciences USA* 71:1743–7, p. 1747.

45 Hughes, S. (2011), *Genentech: The Beginnings of Biotech* (Chicago, Chicago University Press), p. 20.

46 Berg, *A Stanford Professor's Career*, p. 115.

47 Brenner, S. (1974), *Nature* 248:785–7. The next article in the issue described, in great detail, Rosalind Franklin's contribution to the discovery of the double helix: Klug, A. (1974), *Nature* 248:785–8.

48 Morange, *The Black Box of Biology*, p. 197. The same was true of Brenner's friend and close colleague, Francis Crick, who in 1969 was invited by *Nature* to celebrate the journal's centenary by predicting the state of molecular biology in 2000 – Crick, F. (1970), *Nature* 228:613–5. Although he did accurately predict many developments in our understanding, it is striking that Crick did not interest himself much in the application of those discoveries even though he had highlighted the possibility in a lecture the previous year (see chapter 2). Some academics have argued that there was no such change in the nature of molecular biology, and that the interventionist, entrepreneurial tendencies that marked genetic engineering were there all along (e.g. Kay, L. (1998), in Thackray, A. (ed.) *Private Science: Biotechnology and the Rise of the Molecular Sciences* (Philadelphia, University of Pennsylvania Press), pp. 20–38). I think that is mistaken; while there were clearly continuities, the mid-1970s represent a qualitatively new development.

49 *New York Times*, 27 August 1973. The title of the article, which was about the

International Congress of Genetics, held at Berkeley in August 1973, was 'Challenge of Genetics Fades for Scientists'.

4 Asilomar

1 Berg, P., et al. (1974), *Science* 185:303.
2 Bosley, K., et al. (2015), *Nature Biotechnology* 33:478–86, p. 483.
3 Watson, J. (1979), in J. Morgan and W. Whelan (eds), *Recombinant DNA and Genetic Experimentation* (Oxford, Pergamon Press), pp. 187–92, p. 189.
4 Material in this paragraph from Watson, ibid., p. 190.
5 This was recalled by Berg, and is quoted in Lear, J. (1978), *Recombinant DNA – The Untold Story* (New York, Crown), p. 82.
6 Lear, ibid., provides a detailed description of the genesis of the letter.
7 Berg et al. (1974), p. 303.
8 There had been contradictory manoeuvring by various university press offices to give their favoured journalists an exclusive; this led to the story breaking early in the *New York Times* and the *Washington Post*. Lear, *Recombinant DNA*, pp. 89–94 has the grisly details.
9 Quoted in Lear, ibid., p. 91.
10 *New York Times*, 18 July 1974.
11 *Washington Post*, 18 July 1974.
12 *The New York Times*, 30 May 1974; *Newsweek*, 17 June 1974.
13 Krimsky, S. (1982), *Genetic Alchemy: The Social History of the Recombinant DNA Controversy* (London, MIT Press), pp. 90–5.
14 Moore, K. (2008), *Disrupting Science: Social Movements, American Scientists, and the Politics of the Military, 1945–1975* (Princeton, Princeton University Press), pp. 170–5.
15 For summaries of the activity of the working groups, see Krimsky, *Genetic Alchemy*, pp. 126–34.
16 Letter from Donald Comb to Paul Berg, 29 July 1974, reproduced in J. Watson and J. Tooze (eds) (1981), *The DNA Story: A Documentary History of Gene Cloning* (San Francisco: W. H. Freeman), p. 14.
17 Rogers, M. (1977), *Biohazard* (New York, Alfred Knopf), p. 48.
18 Letter from Berg to the signatories of the 'Berg letter' and Maxine Singer, 2 August 1974, reproduced in Watson and Tooze, *The DNA Story*, p. 13.
19 Kirby, D. (2007), *Literature and Medicine* 26:83–108.
20 *New York Times*, 26 September 1974.
21 *Genetic Engineering: Evolution of a Technological Issue, Supplemental Report 1*, prepared for the Committee on Science and Astronautics by the Science Policy Research Division, Congressional Research Service, Library of Congress, December 1974 (Washington, DC, US Government Printing Office), p. v.
22 Ibid., p. 23.
23 Wright, S. (1994), *Molecular Politics: Developing American and British Regulatory Policy for Genetic Engineering, 1972–1982* (London, University of Chicago Press), p. 140.
24 Ibid., p. 142. Ashby chaired the Royal Commission on Environmental Pollution from 1970 to 1973. The President of the Royal Society, Alan Hodgkin, was not consulted on the composition of the Ashby working party, which caused a few ruffled feathers. Letter from Eric Ashby to Paul Berg, 5 November 1974, Cold Spring Harbor Laboratory Archive, SB/1/2/43/73. http://libgallery.cshl.edu/items/show/71394

25 Ashby, E. (1972), *Sociological Review* 18, Supplement 1: 209–26.
26 *Le Monde*, 18 September 1974.
27 All quotes from *The Times*, 3 September 1974.
28 *Daily Mail*, 3 September 1974.
29 *Radio Times*, 12 September 1974. The programme was broadcast on Monday 16 September 1974.
30 Bud, R. (1994), in A. Thackray (ed.), *Private Science: Biotechnology and the Rise of the Molecular Sciences* (Philadelphia, University of Pennsylvania Press), pp. 3–19, p. 13.
31 *New Scientist*, 19 September 1974, p. 755.
32 Both letters are quoted in Wright, *Molecular Politics*, p. 143.
33 Brenner evidence to Ashby Working Party, 26 September 1974. Cold Spring Harbor Laboratory Archive, SB/1/2/40/15. http://libgallery.cshl.edu/show/71340
34 Ibid., p. 7.
35 Ibid.
36 *Report of the Working Party on the Experimental Manipulation of the Composition of Micro-Organisms* (Command Paper 5880) (London, HMSO, 1975).
37 Lear, *Recombinant DNA*, p. 112.
38 Dixon, B. (1975), *New Scientist*, 23 January 1975, p. 86.
39 Anonymous (1975), *Nature* 253:295.
40 *The Times*, 22 January 1975.
41 Lewin, R. (1974), *New Scientist*, 17 October 1974, p. 163.
42 Ibid.
43 Gottweis, H. (1998), *Governing Molecules: The Discursive Politics of Genetic Engineering in Europe and the United States* (London, MIT Press), p. 86.
44 *Le Monde*, 24 July 1975.
45 Gottweis, *Governing Molecules*, p. 86.
46 Brenner, S. (1974), *Nature* 248:785–7, p. 786.
47 Letter from Berg to Ashby, 1 November 1974. Cold Spring Harbor Laboratory Archive, SB/1/2/43/74. http://libgallery.cshl.edu/items/show/71395 – a full list of attendees and their affiliations can be found in Fredrickson, D. (2000), *The Recombinant DNA Controversy: A Memoir* (Washington, DC, ASM), Appendix 1.1.
48 The chaotic approval of press credentials at Asilomar, which included the threat of lawsuits, is described in Lear, *Recombinant DNA*, pp. 116–21.
49 Rogers, M. (1975), *Rolling Stone*, 19 June 1975, pp. 37–82. This article was reproduced in Watson and Tooze, *The DNA Story*, pp. 28–40, but judicious googling will lead you to a PDF.
50 Pollack's offering of his place is described in Lear, *Recombinant DNA*. For more on the attempts to involve Science for the People in Asilomar, see Krimsky, *Genetic Alchemy*, pp. 110–1. Surprisingly, there is no mention of Asilomar in Beckwith's memoirs – Beckwith, J. (2002), *Making Genes, Making Waves: A Social Activist in Science* (London, Harvard University Press). The Science for the People statement is reproduced in Watson and Tooze, *The DNA Story*, p. 49.
51 Rogers, M. (1977), *Biohazard* (New York, Alfred Knopf), p. 68.
52 Rogers (1975), p. 40.
53 Rogers, *Biohazard*, p. 56.
54 Wright, *Molecular Politics*, pp. 148–9. The quote is from the tape of the opening session.

55 Hughes, S. (2001), *Isis* 92:541–75.
56 *Washington Post*, 9 March 1975. According to Rogers, *Biohazard*, p. 52, it was Baltimore who said this; when the tapes are released in 2025, this minor riddle will be resolved.
57 Interview with Brenner as part of the MIT Oral History Project, cited in Friedberg, E. (2014), *A Biography of Paul Berg: The Recombinant DNA Controversy Revisited* (Singapore, World Scientific Publishing), p. 280.
58 *Washington Post*, 9 March 1975.
59 Berg, P. (2007), *Reflections on California's Stem Cell Research Initiative. An Oral History Conducted in 1997 by Sally Smith Hughes* (Berkeley, Regional Oral History Office, The Bancroft Library, University of California), p. 10.
60 Lear, *Recombinant DNA*, p. 131.
61 Wade, N. (1977), *The Ultimate Experiment: Man-Made Evolution* (New York, Walker and Company), p. 46.
62 *Washington Post*, 9 March 1975.
63 Rogers, *Biohazard*, p. 49.
64 Different versions of this story can be found in the *Washington Post*, 9 March 1975 and Rogers, *Biohazard*, p. 101.
65 Goodfield, J. (1977), *Playing God: Genetic Engineering and the Manipulation of Life* (London, Hutchinson), pp. 110–11.
66 Letter from Berg to Brenner, 15 March 1975, Cold Spring Harbor Laboratory Archive, SB/4/1/25. http://libgallery.cshl.edu/items/show/74250
67 Rogers, *Biohazard*, pp. 70–1; Wade, *The Ultimate Experiment*, pp. 31–2.
68 Rogers (1975); Lear, *Recombinant DNA*, pp. 140–1. A decade later, Robert Sinsheimer had a rather different recollection: 'Occasional monitory comments by the lawyers present received little attention' – Sinsheimer, R. (1984), *BioEssays* 1:83–4. Sinsheimer's view is very much an outlier; none of the contemporary accounts I have read suggest this was the case, nor does it accord with the recollections of Paul Berg, David Baltimore and Michael Rogers when I interviewed them in 2021.
69 *Washington Post*, 9 March 1975.
70 In 2021 I asked Paul Berg if he recalled who typed up the summary. He could not remember, but, sharp as a tack at ninety-five years old, in his mind's eye he could see the Asilomar conference centre duplicator churning off the copies.
71 Letter from Berg to Brenner, 15 March 1975, Cold Spring Harbor Laboratory Archive, SB/4/1/25. http://libgallery.cshl.edu/items/show/74250
72 The opposition of Lederberg and Watson was a foregone conclusion; Cohen said he voted against because he had not had time to read the final document. The reasons why Kourilsky voted against the summary are not so clear – his report on Asilomar, which appeared in the French journal *Biochimie*, criticised the meeting for not having philosophers present to guide the conference towards a deeper understanding of science and its role in society, complained of 'remarkable hysteria' around the potential dangers (he accused the organisers of indulging in 'terror-fiction'), and snootily suggested that matters would have been helped had the organisers first talked to suitably knowledgeable colleagues. So far, so French. Kourilsky, F. (1975), *Biochimie* 57(3):vii–xii. Kourilsky later said that he had been profoundly mistaken in his initial critique of Asilomar, and that the key issues were not philosophical but scientific. Had experts in microbial ecology been present, he claimed, things might have been different, as they would have realised that recombinant

microbes posed no threat. Kourilsky, P. (1987), *Les Artisans de l'hérédité* (Paris, Odile Jacob), p. 152.

73 Berg, P., et al. (1975), *Nature* 255:442–4; *Science* 188:991–4; *Proceedings of the National Academy of Sciences USA* 71:1981–4. The final report, together with various pieces of correspondence, can be seen at the Cold Spring Harbor Laboratory Archive: http://libgallery.cshl.edu/items/show/74249

74 Weiner, C. (2001), *Perspectives in Biology and Medicine* 44:208–20.

75 *Genetic Roulette* (BBC TV, 1977); *Genetic Dreams, Genetic Nightmares* (BBC Radio, 2021), episode 1: https://www.bbc.co.uk/programmes/m000xzdp

5 Politics

1 Wright, S. (1986), *Social Studies of Science* 16:593–620, p. 593. See also the definitive account: Wright, S. (1994), *Molecular Politics: Developing America and British Regulatory Policy for Genetic Engineering, 1972–1982* (London, University of Chicago Press). Much of the politico-legal jiggery-pokery had little consequence and is of limited interest to the non-specialist. Many relevant contemporary documents are reproduced in Watson, J. and Tooze, J. (1981), *The DNA Story: A Documentary History of Gene Cloning* (San Francisco, W. H. Freeman). As soon as the controversy began, researchers at MIT started to collect oral testimonies, which ballooned into interviews with over 120 participants and recordings of key events. The contents of this Recombinant DNA History Collection are not available online, and because of the COVID-19 pandemic I was unable to consult them in situ.

2 Rogers, M. (1977), *Biohazard* (New York, Alfred Knopf), p. 42.

3 Krimsky, S. (1982), *Genetic Alchemy: The Social History of the Recombinant DNA Controversy* (London, MIT Press), p. 157.

4 Ibid., p. 183.

5 Zilinskas, R. and Zimmerman, B. (eds) (1986), *The Gene-Splicing Wars: Reflections on the Recombinant DNA Controversy* (London, Collier Macmillan).

6 Wright (1986), p. 235.

7 Culliton, B. (1975), *Science* 188:1187–9.

8 Chargaff, E. (1976), *Science* 192:938–9.

9 Krimsky, *Genetic Alchemy*, p. 266.

10 Cohen, C. (1978), *Southern California Law Review* 51:1081–114.

11 Campos, L. (2021), in L. Campos, et al. (eds), *Nature Remade: Engineering Life, Envisioning Worlds* (Chicago, Chicago University Press), pp. 151–72.

12 Norman, C. (1976), *Nature* 262:2–4.

13 See, for example, Mendelsohn, E. (1984), in E. Mendelsohn (ed.), *Transformation and Tradition in the Sciences: Essays in Honor of I. Bernard Cohen* (Cambridge, Cambridge University Press), pp. 317–36.

14 Hall, S. (1988), *Invisible Frontiers: The Race to Synthesize a Human Gene* (London, Sidgwick & Jackson), p. 26.

15 Kaiser, D. (2010), in D. Kaiser (ed.), *Becoming MIT* (Cambridge, MA, MIT Press), pp. 145–63, p. 151. Kaiser reproduces some of the exchanges between Vellucci and the scientists, giving an impression of the knockabout nature of some of the discussions.

16 Hall, *Invisible Frontiers*, p. 44.

17 Krimsky, *Genetic Alchemy*, p. 301.

18 Lear, J., *Recombinant DNA: The Untold Story* (New York, Crown, 1978), p. 156.

19 See, for example: Waddell, C. (1989), *Science, Technology, & Human Values*

14:7–25; Feldman, M. and Lowe, N. (2008), *European Planning Studies* 16:395–410; Kaiser, in *Becoming MIT*.

20 Mendelsohn, E. (1984), in E. Mendelsohn (ed.), *Transformation and Tradition in the Sciences: Essays in Honor of I. Bernard Cohen* (Cambridge, Cambridge University Press), pp. 317–36.

21 Wilson, E. (1994), *Naturalist* (Washington, DC, Island), p. 283.

22 Cavalieri, L. (1976), *New York Times Magazine*, 22 August 1976.

23 Malzberg, B. (1973), *Phase IV* (New York, Pocket).

24 Waddell, C. (1989), *Science, Technology, & Human Values* 14:7–25, p. 12.

25 *Time*, 18 April 1977.

26 Wald, G. (1976), *The Sciences*, September 1976; Hubbard, R., *Science* 193:834; Krimsky, *Genetic Alchemy*, p. 280.

27 See photo in Watson and Tooze, *The DNA Story*, p. 134.

28 Hall, S. (1988), *Invisible Frontiers: The Race to Synthesize a Human Gene* (London, Sidgwick & Jackson), p. 127.

29 *New Scientist*, 1 July 1976, p. 15.

30 *The Real Paper*, 15 January 1977. Reproduced in Watson and Tooze, *The DNA Story*, p. 103.

31 Park, B. and Thacher, S. (1977), *Science for the People*, September–October 1977, pp. 28–35. In 1979 Rae Goodell published a critical account of the Cambridge controversy, highlighting the role of Harvard lobbyists and describing the uneven power dynamics within the CERB, but even she admitted that the final report was 'strikingly well-reasoned and cogent'. Goodell, R. (1979), *Science, Technology, & Human Values* 4:36–43, p. 40.

32 *The Times*, 9 February 1977.

33 Botelho, A. (2021), *Science as Culture* 30:74–104, p. 97, n. 7; Hall, *Invisible Frontiers*, p. 317.

34 Owen-Smith, J. and Powell, W. (2005), *Organization Science* 15:5–21.

35 Letter from Vellucci to Handler, 16 May 1977. Reproduced in Watson and Tooze, *The DNA Story*, p. 206.

36 For an excellent brief summary, see Gibson, K. (1986), in R. Zilinskas and B. Zimmerman (eds), *The Gene-Splicing Wars: Reflections on the Recombinant DNA Controversy* (London, Collier Macmillan), pp. 55–71. For more extensive accounts, see Wright, *Molecular Politics*; Gottweis, H. (1998), *Governing Molecules: The Discursive Politics of Genetic Engineering in Europe and the United States* (London, MIT Press); McKechnie, S. (1978), *Nature* 276:7; Denselow, J. (1982), *New Scientist*, 26 August 1982, pp. 558–61, p. 561.

37 See the advert for the meeting in *New Scientist*, 28 September 1978, p. 990.

38 Anonymous (1978), *Nature* 276:2.

39 Brenner, S. (1978), *Nature* 276:2–4; King, J. (1978), *Nature* 276:4–7.

40 Fredrickson, D. (2001), *The Recombinant DNA Controversy: A Memoir. Science, Politics, and the Public Interest 1974–1981* (Washington, DC, ASM Press).

41 Ibid., pp. 246–9.

42 Wright, *Molecular Politics*. See also Powledge, T. (1977), *Hastings Center Report*, December 1977, pp. 8–10.

43 US Senate, Committee on Commerce, Science and Transportation, Subcommittee on Science, Technology and Space (1978), *Regulation of Recombinant DNA Research (2, 8 & 10 November 1977)* (Washington, DC, US Government Printing Office).

44 *Le Monde*, 12 June 1975; Anonymous (1975), *Nature* 256:5; Robert, B. (2014),

A History of the Molecular Biology Department, https://hal-pasteur.archives-ouvertes.fr/pasteur-01719506/document; see also Könninger, S. (2016), *Genealogie der Ethikpolitik: Nationale Ethikkomitees als neue Regierungstechnologie. Das Beispiel Frankreichs* (Bielefeld, Transcript).

45 *Le Monde*, 17 June 1975; Gottweiss, *Governing Molecules*.

46 *Le Monde*, 24 July 1975.

47 The intricate politics of this period are explored in all their gory detail by Wright, *Molecular Politics*.

48 Abelson, P. (1977), *Science* 197:721; Halvorson, H. (1984), in R. Zilinskas and B. Zimmerman (eds), *The Gene-Splicing Wars: Reflections on the Recombinant DNA Controversy* (London, Collier Macmillan, 1984), pp. 73–91.

49 Wright, *Molecular Politics*, p. 334; Statement from EMBO, 30 November 1977, reproduced in Watson and Tooze, *The DNA Story*, p. 300.

50 Royal Society (1979), *Nature* 277:509–10. This view was particularly evident at a meeting held at Wye College in the UK, jointly sponsored by COGENE and the Royal Society – see Morgan, J. and Whelan, W. (eds) (1979), *Recombinant DNA and Genetic Experimentation* (Oxford, Pergamon Press). For an analysis of the Wye College meeting, see Wright, *Molecular Politics*, pp. 341–51.

51 Stoker, M. (1979), in Morgan and Whelan, *Recombinant DNA and Genetic Experimentation*, pp. xix–xx, p. xx.

52 US House of Representatives, Committee on Science and Technology, Subcommittee on Science, Research and Technology (1978), *Report on Science Policy Implications of DNA Recombinant Molecule Research (March 1978)* (Washington, DC, US Government Printing Office), p. ix.

53 Chang, S. and Cohen, S. (1977), *Proceedings of the National Academy of Sciences USA* 74:4811–5.

54 There was a lot of dispute about both the nature of this evidence and the reality of the 'consensus'. The position that was widely quoted was '*E. coli* K12 is inherently enfeebled and not capable of pathogenic transformation by DNA inserts'. See Krimsky, *Genetic Alchemy*, p. 216.

55 Wade, N. (1977), *Science* 197:348–9.

56 Israel, M., et al. (1979), *Science* 203:883–7; Chan, H., et al. (1979), *Science* 203:887–92. Although injection of phage DNA or *E. coli* DNA containing the viral sequence did lead to some cancerous growth, the effect was several orders of magnitude weaker than that for injecting the actual virus or naked viral DNA.

57 For a critique of this conclusion, see Rosenberg, B. and Simon, L. (1979), *Nature* 282:773–4. They argued that the results of the experiment provided 'no basis for reassurance on the risks of recombinant DNA'.

58 Wright, S. (1986), *Osiris* 2:303–60, p. 320.

59 Fisher, E. (1985), *Journal of Hazardous Materials* 10:241–61; Lewin, M. (1982), *Recombinant DNA Technical Journal* 5:177–80.

60 Singer, M. (1979), in Morgan and Whelan, *Recombinant DNA and Genetic Experimentation*, pp. 185–6, p. 185.

61 Brenner (1978), p. 2.

62 Morgan and Whelan, *Recombinant DNA and Genetic Experimentation*, p. 236.

63 Grobstein, C. (1986), in R. Zilinskas and B. Zimmerman (eds), *The Gene-Splicing Wars: Reflections on the Recombinant DNA Controversy* (London, Collier Macmillan), pp. 3–10, p. 7.

64 Berg, P. (2000), *A Stanford Professor's Career in Biochemistry, Science Politics,*

and the Biotechnology Industry. An Oral History Conducted in 1997 by Sally Smith Hughes (Berkeley, Regional Oral History Office, The Bancroft Library, University of California), p. 80.

6 Business

1 I have particularly relied upon Hall, S. (1988), *Invisible Frontiers: The Race to Synthesise a Human Gene* (London, Sidgwick & Jackson); Hughes, S. (2011), *Genentech: The Beginnings of Biotech* (Chicago, Chicago University Press); Rasmussen, N. (2016), *Gene Jockeys: Life Science and the Rise of Biotech Enterprise* (Baltimore, Johns Hopkins University Press); Yi, D. (2015), *The Recombinant University: Genetic Engineering and the Emergence of Stanford Biotechnology* (London, University of Chicago Press).
2 Disappointed at the lack of initiative shown by Cetus, Kleiner and Perkins eventually sold their stake in the company. Hughes, *Genentech*, p. 32.
3 Ibid., p. 37. The San Francisco Genentech campus now hosts a life-size bronze statue depicting the meeting, showing an enthusiastic Swanson and a pensive Boyer; see ibid., p. 37.
4 Hall, *Invisible Frontiers*, p. 66.
5 Hughes, *Genentech*, p. 20.
6 For a full account of the history of insulin, which covers much of the material in this chapter, see Hall, K. (2022), *Insulin – The Crooked Timber: A History from Thick Brown Muck to Wall Street Gold* (Oxford, Oxford University Press).
7 Hall, *Invisible Frontiers*, p. 67.
8 Rasmussen, *Gene Jockeys*, p. 118.
9 Owen, G. and Hopkins, M. (2016), *Science, the State, and the City: Britain's Struggle to Succeed in Biotechnology* (Oxford, Oxford University Press), p. 28.
10 Isaacson, W. (2011), *Steve Jobs* (London, Little, Brown), p. 65. In fact, three signatories set up Apple – the third was Ronald Wayne, who quickly got cold feet and withdrew his signature a couple of weeks later. Had he kept with his initial decision he would eventually have been a billionaire.
11 Robertson, M. (1974), *Nature* 251:564–5.
12 Hughes, *Genentech*, p. 41. With his ungainly suggestion, Swanson may have been trying to riff on the name of a famous *Washington Post* cartoonist, Herb Block, who signed as 'Herblock' (see image page 133).
13 Ibid., p. 26.
14 The choice of Cohen and Boyer as the sole applicants, which was to cause a great deal of conflict, is explained in Cohen, S. (2009), *Science, Biotechnology, and Recombinant DNA: A Personal History. An Oral History Conducted by Sally Smith Hughes in 1995* (Berkeley, Regional Oral History Office, The Bancroft Library, University of California), pp. 150–1. Cohen insists that he had no part in that decision.
15 For the grisly details, see Yi, *The Recombinant University*.
16 Reimers, N. (1998), *Stanford's Office of Technology Licensing and the Cohen/Boyer Cloning Patents. An Oral History Conducted in 1997 by Sally Smith Hughes* (Berkeley, Regional Oral History Office, The Bancroft Library, University of California), pp. 13–4. This fact was squirrelled out by the eagle-eyed Doogab Yi: Yi, *The Recombinant University*, p. 136.
17 Berg, P. (2000), *A Stanford Professor's Career in Biochemistry, Science Politics, and the Biotechnology Industry. An Oral History Conducted in 1997 by Sally*

Smith Hughes (Berkeley, Regional Oral History Office, The Bancroft Library, University of California), p. 116.

18 Hall, *Invisible Frontiers*, p. 22.

19 Efstratiadis, A., et al. (1975), *Cell* 4:367–78. Efstratiadis, A., et al. (1976), *Cell* 7:279–88.

20 Hall, *Invisible Frontiers*, p. 19.

21 The first paper in the series was Khorana, H., et al. (1976), *Journal of Biological Chemistry* 251:565–70. The other articles follow in the same issue.

22 Heyneker, H., et al. (1976), *Nature* 263:748–52.

23 Ibid., p. 752.

24 They had submitted a grant to NIH to carry out this groundbreaking work. The grant was not funded – the reviewers said it was unfeasible in the proposed three-year time span and that it had no practical implications. Hughes, *Genentech*, p. 54. For Riggs's memoir of how they were able to synthesise human insulin, see Riggs, A. (2021), *Endocrine Reviews* 42:374–80.

25 Itakura, K., et al. (1977), *Science* 198:1056–63.

26 Hughes, *Genentech*, p. 63.

27 Hall, *Invisible Frontiers*, p. 176.

28 Quotes in this and subsequent two paragraphs from US Senate, Committee on Commerce, Science and Transportation, Subcommittee on Science, Technology and Space (1978), *Regulation of Recombinant DNA Research (2, 8 & 10 November 1977)* (Washington, DC, US Government Printing Office), p. 36.

29 Guillemin, R. and Lemke, G. (2013), *Annual Review of Physiology* 75:1–22.

30 Hall, *Invisible Frontiers*, p. 233.

31 Hopson, J. (1977), *Smithsonian* 8 (March 1977), pp. 55–62, p. 58.

32 Hughes, *Genentech*, p. 53.

33 Hall, *Invisible Frontiers*, p. 200.

34 Ullrich, A., et al. (1977), *Science* 196:1313–9.

35 Hall, *Invisible Frontiers*, pp. 195–212.

36 Ibid., p. 195.

37 Villa-Komaroff, L., et al. (1978), *Proceedings of the National Academy of Sciences USA* 75:3727–31.

38 Hall, *Invisible Frontiers*, p. 245.

39 A few years later the *New England Journal of Medicine* – deeply protective of the commercial interests of journals, then as now – was still complaining about 'gene cloning by press conference', tut-tutting about this apparent circumvention of scientific convention and journal exclusivity. Andreopolous, S. (1980), *New England Journal of Medicine* 302:743–6.

40 The lack of peer review might seem outrageous, but it was routine in this publication at the time.

41 Goeddel, D., et al. (1979), *Proceedings of the National Academy of Sciences USA* 76:106–10.

42 Hughes, *Genentech*, p. 94.

43 There is a graphic account of this experiment from hell in Hall, *Invisible Frontiers*, pp. 249–65. Gilbert's assessment is from p. 265.

44 Keen, H., et al. (1980), *Lancet* 2(8191):398–401.

45 Both factories are still going strong, and there are now many others around the world as the demand for insulin grows with the diabetes epidemic.

46 Hall, *Invisible Frontiers*, p. 302.

47 Hopson (1977), pp. 57, 58. The forensic senatorial quizzing can be found in US

Senate, Committee on Commerce, Science and Transportation, Subcommittee on Science, Technology and Space (1978), pp. 176–243.

48 Wade, N. (1977), *Science* 197:1342–5.
49 Ullrich et al. (1977).
50 Hughes, *Genentech*, p. 127.
51 Marshall, E. (1997), *Science* 277:1028–30; Cook-Deegan, R. (1997), *Science* 278:557–61.
52 *New York Times Magazine*, 17 February 1980.
53 Seeburg, P., et al. (1977), *Nature* 270:486–94.
54 This was Seeburg's testimony in a 1999 patent infringement trial. Marshal, E. (1999), *Science* 284:883–6.
55 *New York Times Magazine*, 17 February 1980.
56 Martial, J., et al. (1979), *Science* 205:602–7; Goeddel, D., et al. (1979), *Nature* 281:544–8.
57 Hughes, *Genentech*, p. 127.
58 Henner, D., et al. (1999), *Science* 284:1465; Barinaga, M. (1999a), *Science* 284:1752–3; Barinaga, M. (1999b), *Science* 286:1655.
59 Dalton, R. and Schiermeier, Q. (1999), *Nature* 402:335.
60 Wilsden, W. (2016), *Frontiers in Molecular Neuroscience* 9:133.
61 Timmermann, C. (2019), *Moonshots at Cancer: The Roche Story* (Basel, Editiones Roche).
62 Teitelman, R. (1989), *Gene Dreams: Wall Street, Academia, and the Rise of Biotechnology* (New York, Basic), pp. 25–6.
63 Hughes, *Genentech*, p. 158.
64 Ibid., p. 159.
65 *New York Times*, 1 January 1981.
66 Isaacson, *Steve Jobs*, p. 23.
67 Cohen, *Science, Biotechnology, and Recombinant DNA*, p. 133.
68 Gitschier, J. (2009), *PloS Genetics* 5:e1000653, p. 4. In an interview with me in 2021, Boyer said that 'total blind luck seems to have defined my life'.

7 Bio-riches

1 Kevles, D. (1998), in A. Thackray (ed.), *Private Science: Biotechnology and the Rise of the Molecular Sciences* (Philadelphia, University of Pennsylvania Press), pp. 65–79, p. 65.
2 Sherkow, J. and Greely, H. (2015), *Annual Review of Genetics* 49:161–82, p. 164.
3 Heidelberger, C. and Duschlasky, R. (1957), US Patent 2,802,005; Spiegelman S. and Haruna, I. (1972), US Patent 3,661,893.
4 Kevles, D. (1994), *History and Studies of the Physical and Biological Sciences* 25:111–35.
5 Ibid., p. 118.
6 Krimsky, S. (1999), *Chicago-Kent Law Review* 75:15–39.
7 Lyon & Lyon, Thomas D. Kiley (1980), Brief on Behalf of Genentech, Inc., Amicus Curiae.
8 https://ipmall.info/content/diamond-v-chakrabarty-peoples-business-commission
9 Krimsky (1999).
10 US Supreme Court, *Diamond v. Chakrabarty* (1980) 447 US 303.
11 *Chicago Tribune*, 18 June 1982.

12 Cohen, S. and Boyer, H. (1980), US Patent 4,237,224.
13 Feldman, M., et al. (2007), in A. Krattiger, et al. (eds), *Intellectual Property Management in Health and Agricultural Innovation: A Handbook of Best Practices* (Oxford, MIHR), pp. 1797–807.
14 Creager, A. (1998), in A. Thackray (ed.), *Private Science: Biotechnology and the Rise of the Molecular Sciences* (Philadelphia, University of Pennsylvania Press, 1998), pp. 39–63; Rasmussen, N. (2016), *Gene Jockeys: Life Science and the Rise of Biotech Enterprise* (Baltimore, Johns Hopkins University Press).
15 Creager, in *Private Science*; Pickstone, J. (2001), *Ways of Knowing: A New History of Science, Technology, and Medicine* (Chicago, University of Chicago Press).
16 Kevles, in *Private Science*.
17 Jon Agar has argued that there was a deep entrepreneurial streak in the supposedly alternative world of the 1960s, and that this contributed to the booms in personal computing and genetic engineering. Agar, J. (2008), *British Journal for the History of Science* 41:567–600.
18 Culliton, B. (1977), *Science* 195:759–63.
19 Yi, D. (2008), *Journal of the History of Biology* 41:589–636; Yi, D. (2015), *The Recombinant University: Genetic Engineering and the Emergence of Stanford Biotechnology* (London, University of Chicago Press).
20 Culliton, B. (1982a), *Science* 216:960–2.
21 Culliton, B. (1982b), *Science* 216:1295–6.
22 Hughes, S. (2011), *Genentech: The Beginnings of Biotech* (Chicago, Chicago University Press); Kornberg, A. (1995), *The Golden Helix: Inside Biotech Ventures* (Sausalito: Science Books).
23 Culliton (1982a), p. 961.
24 Ibid.
25 Kenney, M. (1986), *Biotechnology: The University-Industrial Complex* (London, Yale University Press), p. 104.
26 Wade, N. (1980), *Science* 208:688–92, p. 689.
27 Wright, S. (1986), *Osiris* 2:303–60, p. 320.
28 Kenney, M. (1998), in A. Thackray (ed.), *Private Science: Biotechnology and the Rise of the Molecular Sciences* (Philadelphia, University of Pennsylvania Press), pp. 131–43.
29 For example, the rise and fall of Genetic Systems, a start-up focused on diagnostics, is told in Teitelman, R. (1989), *Gene Dreams: Wall Street, Academia, and the Rise of Biotechnology* (New York, Basic Books). Kornberg, *The Golden Helix*, provides a largely hagiographic account of the industry, based on his experience with DNAX.
30 Yoxen, E. (1983), *The Gene Business: Who Should Control Biotechnology?* (London, Pan), p. 69. Within four years, Genex had lost its contract to produce aspartame and was laying off 40 per cent of its workforce (*Washington Post*, 14 June 1985; *New York Times*, 1 November 1985).
31 Owen, G. and Hopkins, M. (2016), *Science, the State, and the City: Britain's Struggle to Succeed in Biotechnology* (Oxford, Oxford University Press). From 1981 onwards the Stock Exchange created alternative markets to meet this need.
32 Bud, R. (1998), in A. Thackray (ed.), *Private Science: Biotechnology and the Rise of the Molecular Sciences* (Philadelphia, University of Pennsylvania Press), pp. 3–19, pp. 4, 15, 8.
33 Dickson, D. (1980), *Nature* 283:128–9.

34 Anonymous (1982), *Nature* 298:599.
35 Anonymous (1984), *Nature* 307:201.
36 Kenney, *Biotechnology*, p. 157.
37 Gekko, played by Michael Douglas, actually said: 'Greed, for lack of a better word, is good.' I have not seen this film.
38 *The Economist*, 13 May 1989.
39 Kenney, in *Private Science*.
40 Brock, M. (1989), *Biotechnology in Japan* (London, Routledge).
41 *Biotechnology: Report of a Joint Working Party* (London, HMSO, 1980), p. 3. See Owen and Hopkins, *Science, the State, and the City*.
42 Gottweis, H. (1998), in A. Thackray (ed.), *Private Science: Biotechnology and the Rise of the Molecular Sciences* (Philadelphia, University of Pennsylvania Press), pp. 105–30, p. 111. For a survey of the development of British biotechnology, in particular the role of the Cambridge Laboratory of Molecular Biology, see de Chadarevian, S. (2011), *Isis* 102:601–33.
43 Wright, S. (1998), in A. Thackray (ed.), *Private Science: Biotechnology and the Rise of the Molecular Sciences* (Philadelphia, University of Pennsylvania Press), pp. 80–104, p. 99; Gottweis, in *Private Science*, p. 124.
44 Gottweiss, in *Private Science*, p. 117.
45 Ibid., p. 107.
46 Wade, N. (1977), *Science* 197:1342–5, p. 1342.
47 Kenney, *Biotechnology*, pp. 122–3.
48 Ibid., p. 124.
49 Krimsky, S., et al. (1996), *Science and Engineering Ethics* 2:395–410.
50 Anonymous (1997), *Nature* 385:469.
51 Yoxen, E. (1981), in L. Levidow and B. Young (eds), *Science Technology and the Labour Process: Marxist Studies*, Volume 1 (London, CSE Books), pp. 66–122, p. 112.
52 Wigler, M., et al. (1979), *Cell* 16:777–85, p. 77.
53 Colaianni, A. and Cook-Deegan, R. (2009), *Millbank Quarterly* 87:683–715.
54 Doetschman, T., et al. (1987), *Nature* 330:576–8, p. 578.
55 Smithies, O. (2001), *Nature Medicine* 7:1083–6.
56 Palmiter, R. and Brinster, R. (1986), *Annual Review of Genetics* 20:465–99; Hanahan, D., et al. (2007), *Genes & Development* 21:2258–70. For a historian's perspective, see Myelnikov, D. (2019), *History and Technology* 35:425–52.
57 Gordon, J. and Ruddle, F. (1981), *Science* 214:1244–6.
58 Brinster, R., et al. (1984), *Cell* 37:367–79.
59 Leder, P. and Stewart, T. (1988), US Patent 4,736,866.
60 Kevles, in *Private Science*, p. 175.
61 Abbot, A. (1992), *Nature* 360:286.
62 Parthasarathy, S. (2017), *Patent Politics: Life Forms, Markets & the Public Interest in the United States and Europe* (London, University of Chicago Press); Sherkow and Greely (2015); Jasanoff, S. (2016), *The Ethics of Invention: Technology and the Human Future* (London, Norton), pp. 194–6.
63 Sherkow and Greeley (2015), p. 166.
64 Anonymous (1986), *Biotechnology Law Report*, May 1986, pp. 136–76.
65 Adams, M., et al. (1991), *Science* 252:1651–6.
66 Venter, C.(2007), *A Life Decoded. My Genome: My Life* (London, Allen Lane).
67 Roberts, L. (1992a), *Science* 254:184–6.
68 Roberts, L. (1992b), *Science* 256:301–2; Sherkow and Greely (2015).

69 Marshall, E. (1997), *Science* 278:2046–8.
70 Krimsky (1999), p. 26.
71 Marshall, E. (2013), *Science* 340:1387–8.
72 https://twitter.com/NIHDirector/status/345194268840849411
73 Calvert, J. and Joly, P.-B. (2011), *Social Science Information* 50:157–77.
74 Heller, M. and Eisenberg, R. (1998), *Science* 280:698–701, p. 698.
75 Ibid.
76 Saiki, R., et al. (1985), *Science* 230:1350–4. For the background to this discovery and the history of Cetus, see Rabinow, P. (1996), *Making PCR: A Story of Biotechnology* (Chicago, University of Chicago Press).
77 Anonymous (1988), *Biotechnology Newswatch*, 5 September 1988, p. 7.
78 Heller and Eisenberg (1998).
79 Hanahan et al. (2007), p. 2268.
80 Williams, H. (2013), *Journal of Political Economy* 121:1–27.
81 Nelsen, L. (2004), *Nature Reviews: Molecular Cell Biology* 5:1–5; Sampat, B. (2010), *Nature* 468:755–6.
82 Contreras, J. (2018), *Science* 361:335–7.

8 Frankenfood

1 Charles, D. (2001), *Lords of the Harvest: Biotech, Big Money, and the Future of Food* (Cambridge, MA, Perseus); Heimann, J. (2018), *Using Nature's Shuttle: The Making of the First Genetically Modified Plants and the People Who Did It* (Wageningen, Wageningen Academic Publishers); Lurquin, P. (2001), *The Green Phoenix: A History of Genetically Modified Plants* (New York, Columbia University Press); Somssich, M. (2019a), *PeerJ Preprints*, https://doi.org/10.7287/peerj.preprints.27096v3; Somssich, M. (2019b), *PeerJ Preprints*, https://doi.org/10.7287/peerj.preprints.27556v2.
2 Chilton, M.-D., et al. (1977), *Cell* 11:263–71.
3 Van Montagu, M. (2011), *Annual Review of Plant Biology* 62:1–23; Chilton, M.-D. (2018), *Annual Review of Plant Biology* 69:1–20, p. 18.
4 Kranakis, E. (2019), *Isis* 110:701–25, p. 705; Chilton (2018), p. 18.
5 *Genetic Dreams, Genetic Nightmares* (BBC Radio, 2021), episode 2: https://www.bbc.co.uk/sounds/play/m000y6jb
6 Barton, K., et al. (1983), *Cell* 32:1033–43; Fraley, R., et al. (1983), *Proceedings of the National Academy of Sciences USA* 80:4803–7; Herrera-Estrella, L., et al. (1983), *Nature* 303:209–13.
7 *The Times*, 27 January 1983; *Wall Street Journal*, 20 January 1983. While *The Times* mangled the story by suggesting that all three groups had worked together, the *WSJ* explained that the Chilton laboratory and the Ghent team had made the breakthrough 'about the same time'. For researchers, Chilton's group gained the greatest immediate accolade – the front cover of the journal *Cell* showed a photo of their transgenic tobacco plant.
8 Somssich, M. (2019b).
9 Horsch, R., et al. (1985), *Science* 227:1229–31.
10 Paszkowski, J., et al. (1984), *EMBO Journal* 3:2717–22.
11 *The Guardian*, 20 March 1987.
12 This point was made by Lurquin, *The Green Phoenix*, p. 98.
13 Klein, T., et al. (1987), *Nature* 327:70–3.
14 All quotes from Sanford, J. (2000), *In Vitro Cellular & Developmental Biology – Plant* 36:303–8, p. 305.

15 Charles, *Lords of the Harvest*, p. 84.
16 Klein, T., et al. (1988), *Proceedings of the National Academy of Sciences USA* 85:4305–9. For technical reasons, the two articles on organelle transformation were both done on single-cell organisms, not plants: Boynton, J., et al. (1988), *Science* 240:1534–8; Johnston, S., et al. (1988), *Science* 240:1538–41.
17 These sequences were patented by Monsanto (US patents 5,034,322 and 5,352,605) following the realisation of their significance by British scientist Michael Bevan. Bevan worked in Mary-Dell Chilton's laboratory in the early 1980s and was also a Monsanto consultant, carrying out the work in parallel with Steve Rogers and others in the Monsanto laboratory. Although they did discuss their findings, there was not much communication. In an email to me in 2020 Bevan explained that he did not apply for a patent himself because 'there was no interest or support for patenting' from his institution. According to Charles, *Lords of the Harvest*, pp. 18–9, he later recalled: 'I was terribly naive in those days.' See Somssich (2019a), although the role of Bevan is not mentioned.
18 Bevan, M., et al. (1983), *Nature* 304:184–7.
19 Charles, *Lords of the Harvest*, p. 25.
20 *Wall Street Journal*, 13 May 1998.
21 Charles, *Lords of the Harvest*, p. 192.
22 Vaeck, M., et al. (1987), *Nature* 328:33–7.
23 Pechlaner, G. (2012), *Corporate Crops: Biotechnology, Agriculture, and the Struggle for Control* (Austin, University of Texas Press), pp. 178–9.
24 Charles, *Lords of the Harvest*, pp. 181–5.
25 Zangerl, A., et al. (2001), *Proceedings of the National Academy of Sciences USA* 98:11908–12.
26 Charles, *Lords of the Harvest*, p. 179.
27 Cao, C. (2018), *GMO China: How Global Debates Transformed China's Agricultural Biotechnology Policies* (New York, Columbia University Press), p. 71
28 Material in this paragraph from Charles, *Lords of the Harvest*, pp. 60–1.
29 Kranakis (2019); Shah, D., et al. (1986), *Science* 233:478–81.
30 Charles, *Lords of the Harvest*, pp. 109–25.
31 Ibid., p. 110.
32 Ibid., p. 187.
33 Kranakis (2019), p. 722. Kranakis's analysis of the case revealed fundamental flaws in Monsanto's case, unnoticed by the courts or by Schmeiser's inexperienced lawyers, in particular the company's reliance on the original 1986 patent for overproducing the EPSPS enzyme, which is not how Roundup Ready plants work.
34 See the typically caustic account by Miller, H. (1997), *Policy Controversy in Biotechnology: An Insider's View* (Austin, Landes Bioscience); Jukes, T. (1988), *Journal of Chemical Technology and Biotechnology* 43:245–55.
35 Love, J. and Lesser, W. (1989), *Northeastern Journal of Agricultural and Resource Economics* 18:1–9.
36 Skirvin, R., et al. (2000), *Scientia Horticulturae* 84:179–89.
37 Ibid., p. 183.
38 Martineau, B. (2001), *First Fruit: The Creation of the Flavr Savr™ Tomato and the Birth of Genetically Engineered Food* (New York, McGraw-Hill), p. 194.
39 Smith, C., et al. (1988), *Nature* 334:724–6; Sheehy, R., et al. (1988), *Proceedings of the National Academy of Sciences USA* 85:8805–9.

40 Martineau, *First Fruit*.
41 Ibid., pp. 104–12.
42 Ibid., p. 170.
43 Charles, *Lords of the Harvest*, p. 95.
44 *Sunday Times*, 3 December 1989.
45 *New York Times*, 16 June 1992.
46 *New York Times*, 28 June 1981. Charles, *Lords of the Harvest*, p. 12 claims that when the quote appeared, 'Apple was mortified', and that what he had meant was that the nutrients found in a pork chop would be found in a plant.
47 According to songwriter Andy McCluskey, 'I was very positive about the subject! I didn't expect someone like Monsanto to come along and say, "Fuck it, we can make money out of cross-pollination".' *The Guardian*, 7 March 2008.
48 Penney, D., et al. (2013), *PLoS ONE* 8:e73150.
49 The version of this speech in the novel is much less punchy: 'Scientists are actually preoccupied with accomplishment. So they are focused on whether they can do something. They never stop to ask if they should do something. They conveniently define such considerations as pointless.' Six years before *Jurassic Park* appeared, a very similar plot formed the centrepiece of *Carnosaur*, by Harry Adam Knight (the pseudonym of John Brosnan). In an example of the magic of publishing, *Jurassic Park* became a global phenomenon, whereas *Carnosaur* languished as a cult classic.
50 Itakura, K. and Riggs, A. (1980), *Science* 209:1401–5.
51 Hall, S. (1988), *Invisible Frontiers: The Race to Synthesize a Human Gene* (London, Sidgwick and Jackson), p. 55–6.
52 Kritikos, M. (2018), *EU Policy-making on GMOs: The False Promise of Proceduralism* (London, Palgrave Macmillan).

9 Suspicion

1 The most thorough account of this period can be found in Schurman, R. and Munro, W. (2010), *Fighting for the Future of Food: Activists versus Agribusiness in the Struggle over Biotechnology* (London, University of Minnesota Press).
2 Jasanoff, S. (2005), *Designs on Nature: Science and Democracy in Europe and the United States* (Oxford, Princeton University Press).
3 Will, R., et al. (1996), *Lancet* 347:921–5.
4 Colling, J., et al. (1996), *Nature* 383:685–90.
5 Wilmut, I., et al. (1997), *Nature* 385:810–3.
6 Willadsen, S. (1986), *Nature* 320:63–5.
7 McLaren, A. (2000), *Science* 288:1775–80.
8 Pennisi, E. (1997), *Science* 278:2038–9; Shapiro, H. (1997), *Science* 277:195–6. For an exploration of visions of human cloning in fiction and films, see Nerlich, B., et al. (2001), *Journal of Literary Semantics* 30:37–52; Cormick, C. (2006), *História, Ciências, Saúde – Manguinhos* 13 (Supplement):181–212.
9 Hurlbut, J. (2017), *Experiments in Democracy: Human Embryo Research and the Politics of Bioethics* (New York, Columbia University Press).
10 Lynas, M. (2018), *Seeds of Science: Why We Got It So Wrong on GMOs* (London, Bloomsbury Sigma), p. 26.
11 Sinclair, K., et al. (2016), *Nature Communications* 7:12359. The article includes a photo of the four animals – 2260, 2261, 2262 and 2263 or, more personally, Daisy, Diana, Debbie and Denise. They all look exactly the same, but then, they are sheep.

12 Pennisi (1997).
13 DEFRA Farm Animal Genetic Resources Committee (2016), Statement on Cloning of Farm Animals.
14 Tachibana, M., et al. (2013), *Cell* 153:1228–38; Ma, H., et al. (2014), *Nature* 511:177–83. For the Korean affair, see Kennedy, D. (2006), *Science* 311:335.
15 National Research Council (2002), *Scientific and Medical Aspects of Human Reproductive Cloning* (Washington, DC, The National Academies Press), p. 1.
16 Oliver, M., et al. (1998), US Patent 5,723,765.
17 Hybrid maize was introduced in the United States in the 1930s and transformed agriculture, increasing yields massively, soon dominating US maize production. Every year, the farmers bought seeds, leading to a burgeoning new sector for US capitalism in the form of the seed industry. Kloppenburg, J. (1988), *First the Seed: The Political Economy of Plant Biotechnology 1492–2000* (Cambridge, Cambridge University Press), p. 11.
18 *Daily Telegraph*, 8 June 1998.
19 Enserink, M. (1998), *Science* 281:1124–5.
20 Ewen, S. and Pusztai, A. (1999), *Lancet* 354:1353–4; Horton, R. (1999), *Lancet* 354:1729.
21 Schurman and Monro, *Fighting for the Future of Food*, p. 108.
22 https://www.pewresearch.org/global/2003/06/20/broad-opposition-to-genetically-modified-foods – this link is archived at the Wayback Machine at archive.org.
23 Schurman and Monro, *Fighting for the Future of Food*; the role of transcendental meditation followers is highlighted by Grohman, G. (2021), *Annals of Iowa* 80:1–34.
24 *Le Monde*, 18 January 1998.
25 *Le Monde*, 14 August 1999.
26 *Le Monde*, 1 September 1999.
27 *Le Monde*, 9 September 1999.
28 Lynas, *Seeds of Science*, pp. 29–31. A photo of the demonstration appears in Charles, D. (2001), *Lords of the Harvest: Biotech, Big Money, and the Future of Food* (Cambridge, MA, Perseus).
29 For a depressing survey of how the UK press treated GM food in the first half of 1999, see Parliamentary Office of Science and Technology, *The 'Great GM Food Debate'*, Report 138, May 2000.
30 *Daily Mirror*, 16 February 1999.
31 Charles, *Lords of the Harvest*, p. 272.
32 Ibid., p. 259.
33 *Horizon: Is GM Safe?* (BBC2, 9 March 2000). http://www.bbc.co.uk/science/horizon/1999/gmfood_script.shtml
34 Schurman and Munro, *Fighting for the Future of Food*, pp. 140–6.
35 Jasanoff, S. (2016), *The Ethics of Invention: Technology and the Human Future* (London, Norton), pp. 100–3.
36 Clancy, K. (2017), *The Politics of Genetically Modified Organisms in the United States and Europe* (London, Palgrave Macmillan).
37 Siefert. F. (2009), *Sociologia Ruralis* 49:20–40.
38 Clancy, *The Politics of Genetically Modified Organisms*, examines the imagery used by anti-GM campaigners in various European countries and the United States, which often involves syringes, weird hybrids and skulls.
39 Regis, E. (2019), *Golden Rice: The Imperiled Birth of a GMO Superfood* (Baltimore,

Johns Hopkins University), pp. 52–72, has an extensive discussion of the protocol and its problems. For a comparison of regulation in the European Union and the United States, see Gronvall, G. (2015), *Health Security* 13:378–89.

40 *Los Angeles Times*, 23 September 2000.

41 *The Guardian*, 13 February 2001.

42 Cao, C. (2018), *GMO China: How Global Debates Transformed China's Agricultural Biotechnology Policies* (New York, Columbia University Press), p. 10.

43 Quist, D. and Chapela, I. (2001), *Nature* 414:541–3. For an analysis of the history of maize, see Curry, H. (2022), *Endangered Maize: Industrial Agriculture and the Crisis of Extinction* (Oakland, University of California Press).

44 Metz, M. and Fütterer, J. (2002), *Nature* 416:600–1; Kaplinsky, N., et al. (2002), *Nature* 416:601–2; Quist, D. and Chapela, I. (2002), *Nature* 416:602.

45 Bonneuil, C., et al. (2014), *Social Studies of Science* 44:901–29, p. 911; Ortiz-García, S., et al. (2005), *Proceedings of the National Academy of Sciences USA* 102:12338–43; Cleveland, D., et al. (2005), *Environmental and Biosafety Research* 4:197–208; Ortiz-García, S., et al. (2005), *Environmental and Biosafety Research* 4:209–15.

46 Bonneuil et al. (2014), p. 924.

47 Agapito-Tenfen, S., et al. (2017), *Ecology and Evolution* 7:9461–72; Garcia Ruiz, M., et al. (2018), *GM Crops & Food* 9:152–68.

48 Losey, J., et al. (1999), *Nature* 399:214.

49 The six articles were published in pp. 11908–42 of volume 98 of the *Proceedings of the National Academy of Sciences USA*, along with a summary article: Scriber, J. (2001), *Proceedings of the National Academy of Sciences USA* 98:12328–30.

50 Romeis, J., et al. (2008), *Nature Biotechnology* 26:203–8; Malcolm, S. (2018), *Annual Review of Entomology* 63:277–302.

51 Brower, L., et al. (2012), *Insect Conservation and Diversity* 5:95–100.

52 Schnurr, M. (2019), *Africa's Gene Revolution: Genetically Modified Crops and the Future of African Agriculture* (London, McGill-Queen's University Press), p. 21.

53 Ibid., p. 90.

54 Ibid., pp. 71–8, gives a close reading of the situation.

55 Ibid., pp. 109–33.

56 Ibid., pp. 54–6.

57 Ibid., p. 203.

58 Schurman and Munro, *Fighting for the Future of Food*, p. 177.

59 Schnurr, *Africa's Gene Revolution*, p. 194.

60 For example, biofortification – increased nutritional content – is advocated by donors but comes very low on farmers' list of priorities. See Schnurr, M., et al. (2020), *The Journal of Peasant Studies* 47:326–45.

61 Ba, M., et al. (2018), *Journal of Pest Science* 91:1165–79; Addae, P., et al. (2020), *Journal of Economic Entomology* 113:974–9.

62 Schurman, R. (2017), *Journal of Agrarian Change* 17:441–58.

63 Juma, C. (2014), https://geneticliteracyproject.org/2014/12/09/global-risks-of-rejecting-agricultural-biotechnology/. This article caused a minor kerfuffle because the subject was suggested to Juma by Monsanto as part of a pro-GM campaign they were running – *Boston Globe*, 1 October 2015. Juma's vision for international development was enthusiastically based on innovation and a controlled relaxation of regulation. See, for example, Juma, C. (2016), *Innovation and Its Enemies: Why People Resist New Technologies* (Oxford, Oxford University Press).

64 Schnurr, *Africa's Gene Revolution*, p. 209.
65 Aga, A. (2021), *Genetically Modified Democracy: Transgenic Crops in Contemporary India* (London, Yale University Press).
66 Ibid., p. 252.
67 Cao, *GMO China*, p. 66.
68 Ibid., p. 59.
69 Shen, X. (2010), *Journal of Development Studies* 46:1026–46.
70 Cao, *GMO China*, p. 81.
71 Ibid., p. 99.
72 Huang, J., et al. (2002), *Australian Journal of Agricultural and Resource Economics* 46:367–87.
73 *China Daily*, 23 January 2009.
74 Chen, N. (2015), in S. Jasanoff and S.-H. Kim (eds), *Dreamscapes of Modernity: Sociotechnical Imaginaries and the Fabrication of Power* (London, University of Chicago Press), pp. 219–32.
75 Cao, *GMO China*, pp. 110–12.
76 Ibid., p. 185, p. 123.
77 For opposing one-sided examples, see: Miller, H. and Conko, G. (2004), *The Frankenfood Myth: How Protest and Politics Threaten the Biotech Revolution* (London, Praeger) and Druker, S. (2015), *Altered Genes, Twisted Truth: How the Venture to Genetically Engineer Our Food Has Subverted Science, Corrupted Government, and Systematically Deceived the Public* (Salt Lake City, Clear River Press). A balanced account can be found in Krimsky, S. (2019), *GMOs Decoded: A Skeptic's View of Genetically Modified Foods* (London, MIT Press). For an interesting blend of genetics and organic farming, see Ronald, P. and Adamchak, R. (2018), *Tomorrow's Table: Organic Farming, Genetics, and the Future of Food* (Oxford, Oxford University Press). However, Maywa Montenegro de Wit argues strongly that whatever the apparent attraction of a complementarity between organic farming and biotechnology, the underlying problem is 'the politics of technology and the worldmaking of coloniality'. Montenegro de Wit, M. (2021), *Agriculture and Human Values*, https://link.springer.com/article/10.1007/s10460-021-10284-0
78 All quotes from Grove-White, R., et al. (1997), *Uncertain World: Genetically Modified Organisms, Food and Public Attitudes* (Lancaster, Centre for the Study of Environmental Change, Lancaster University).
79 Blancke, S., et al. (2015), *Trends in Plant Science* 20:414–8.
80 Regis, *Golden Rice*, p. 20.
81 Datta, S., et al. (1990), *Bio/Technology* 8:736–40.
82 Ye, X., et al. (2000), *Science* 287:303–5.
83 Bollinedi, H., et al. (2017), *PLoS ONE* 12:e0169600.
84 Cotter, J. (2013), *Golden Illusion: The Broken Promises of 'Golden' Rice* (Amsterdam, Greenpeace International).
85 Regis, *Golden Rice*, pp. 172–8; *New York Times*, 30 June 2016. The text of the letter can be found at https://www.supportprecisionagriculture.org/nobel-laureate-gmo-letter_rjr.html – the current number of signatories stands at 156. The letter has had no discernible effect.
86 Regis, *Golden Rice*, pp. 187–96 describes this process and untangles the US FDA's lacklustre description of the benefits of Golden Rice, which would be minimal for US consumers.
87 Stone, G. and Glover, D. (2017), *Agriculture and Human Values* 34:87–102.

88 Aga, A. and Montenegro de Wit, M. (2021), *Scientific American*, https://www.scientificamerican.com/article/how-biotech-crops-can-crash-and-still-never-fail/
89 Dong, O., et al. (2020), *Nature Communications* 11:1178.
90 Fagerström, T., et al. (2012), *EMBO Reports* 13:493–7.
91 Beale, M., et al. (2006), *Proceedings of the National Academy of Sciences USA* 103:10509–13.
92 Panagiotou, A. (2017), *Structure, Agency and Biotechnology: The Case of the Rothamsted GM Wheat Trials* (London, Anthem Press).
93 Bruce, T., et al. (2015), *Scientific Reports* 5:11183.
94 Kunert, G., et al. (2010), *BMC Ecology* 10:23. This paper was not cited by the Rothamsted researchers (Bruce et al., 2015). Steinbrecher, R., Submission to DEFRA, August 2011, https://www.econexus.info/publication/ge-wheat-trial-rothamsted-research.
95 Fernandez-Cornejo, J., et al. (2014), *Genetically Engineered Crops in the United States, ERR-162* (Washington, DC, USDA Economic Research Service), p. 12. In the early years, GM crops sometimes even produced lower yields than non-GM crops, because the genes were not introduced into the highest-yielding varieties.
96 Pixley, K., et al. (2019), *Annual Review of Phytopathology* 57:165–88.
97 Aga, *Genetically Modified Democracy*, p. 7.
98 Anonymous (2020), *Science* 370:747.
99 Ledford, H. (2013), *Nature* 497:17–8.
100 Lazaris, A., et al. (2002), *Science* 295:472–6.
101 For a lively account of the spider-goats, see Rutherford, A. (2013), *Creation: The Future of Life* (London, Viking), pp. 13–17.
102 Martyn-Hemphill, R. (2019), https://agfundernews.com/what-happened-to-those-gm-spider-goats-with-the-silky-milk.html
103 Zhang, X., et al. (2019), *Biomacromolecules* 20:2252–64.
104 Brookes, G. and Barfoot, P. (2020), *GM Crops & Food* 11:215–41.
105 Fernandez-Cornejo et al., *Genetically Engineered Crops in the United States*, p. 24.
106 Tabashnik, B., et al. (2021), *Proceedings of the National Academy of Sciences USA* 118:e2019115118.
107 Perry, E., et al. (2016), *Science Advances* 2:e1600850.
108 Bhardwaj, A. (2010), in D. Taylor (ed.), *Environment and Social Justice: An International Perspective* (Bingley, Emerald), pp. 241–59; Plewis, I. (2014), *Significance* February 2014:14–8.
109 Kennedy, J. and King, L. (2014), *Globalization and Health* 10:16.
110 See for example: Kranthi, K. and Stone, G. (2020), *Nature Plants* 6:188–96; Plewis, I. (2020), *Nature Plants* 6:1320.
111 United States Environmental Protection Agency (2020), *Glyphosate: Interim Registration Review Decision Case Number 0178, Docket Number EPA-HQ-OPP-2009-0361*. For ecological effects, see for example: Relyea, R. (2012), *Ecological Applications* 22:634–47; Annett, R., et al. (2014), *Journal of Applied Toxicology* 34:458–79.
112 Gilbert, N. (2013), *Nature* 497:24–6.
113 Séralini, G.-E., et al. (2014), *Environmental Sciences Europe* 26:14; Coumoul, X., et al. (2019), *Toxicological Sciences* 168:315–38.
114 *Financial Times*, 27 May 2021.
115 Anonymous (2013), *Nature* 497:5–6.

10 Therapy

1 *New York Times*, 28 November 1999.
2 *New York Times*, 29 September 1999.
3 *New York Times*, 9 December 1999.
4 Cousin, J. and Kaiser, J. (2005), *Science* 307:1028.
5 Davies, K. (2020), *Human Gene Therapy* 31:135–9.
6 Szybalska, E. and Szybalski, W. (1962), *Proceedings of the National Academy of Sciences USA* 48:2026–34.
7 Szybalski, W. (2013), *Gene* 525:151–4; Friedmann, T. (1992), *Nature Genetics* 2:93–6.
8 Davis, B. (1970), *Science* 170:1279–83.
9 Rogers, S. and Pfuderer, P. (1968), *Nature* 219:749–51; Friedmann, T. (2001), *Molecular Therapy* 4:285–8.
10 *New York Times*, 21 September 1970; Friedmann, T. and Roblin, R. (1972), *Science* 175:949–55; *New York Times*, 1 March 1975.
11 Terheggen, H., et al. (1975), *Zeitschrift für Kinderheilkunde* 119:1–3.
12 Friedmann (1992), p. 94.
13 Giri, I., et al. (1985), *Proceedings of the National Academy of Sciences USA* 82:1580–4.
14 Rogers, S. (1972), *Science* 178:648–9.
15 Fox, M. and Littlefield, J. (1971), *Science* 173:195; Aposhian, H. (1970), *Perspectives in Biology and Medicine* 14:98–108; Friedmann and Roblin (1972).
16 Cline, M. (1985), *Pharmacology & Therapeutics* 29:69–82. The fullest account of the Cline affair can be found in Thompson, L. (1994), *Correcting the Code: Inventing the Genetic Cure for the Human Body* (London, Simon & Schuster).
17 Wade, N. (1980), *Science* 210:509–11; *Le Monde*, 15 October 1980.
18 Beutler, E. (2001), *Molecular Therapy* 4:396–7.
19 Thompson, *Correcting the Code*, p. 202.
20 Sun, M. (1981), *Science* 214:1220; Dickson, D. (1981), *Nature* 291:369.
21 President's Commission for the Study of Ethical Problems in Medicine and Biomedical and Behavioral Research (1982), *Splicing Life: A Report on the Social and Ethical Issues of Genetic Engineering with Human Beings* (Washington, DC, US Government), the quotes are from pp. 2, 4.
22 Ibid., p. 85.
23 Wivel, N. (2014), *Human Gene Therapy* 25:19–24.
24 Walters, L., et al. (2021), *CRISPR Journal* 4:469–76.
25 Anderson, W. and Fletcher, J. (1984), *New England Journal of Medicine* 303:1293–7, p. 1296.
26 Miller, A., et al. (1983), *Proceedings of the National Academy of Sciences USA* 80:4709–13.
27 Hammer, R., et al. (1984), *Nature* 311:65–7.
28 Anderson, W. (1992), *Human Gene Therapy* 3:251–2.
29 *Le Monde*, 3 February 1988.
30 Anderson, W. (1993), *Human Gene Therapy* 4:401–2.
31 Rosenberg, S. (1992), *The Transformed Cell: Unlocking the Mysteries of Cancer* (London, Chapmans), p. 279.
32 Anderson, W. (1993), *Human Gene Therapy* 4:555–6.
33 All information on the experiment on Kuntz can be found in Lyon, J. and Gorner, P. (1995), *Altered Fates: Gene Therapy and the Retooling of Human Life* (London, Norton), p. 162–73. Rosenberg, in *The Transformed Cell*, gave his

patients pseudonyms out of respect for their privacy; Kuntz was 'Lester Franks'.

34 Rosenberg, S., et al. (1990), *New England Journal of Medicine* 323:570–8.
35 Lyon and Gorner, *Altered Fates*, p. 187.
36 Rosenberg, *The Transformed Cell*.
37 Hershfield, M., et al. (1987), *New England Journal of Medicine* 316:589–96.
38 Marshall, R. (1995), *Science* 269:1050–5, p. 1051. The article described this scandalous situation as a 'misfortune'.
39 Lyon and Gorner, *Altered Fates*, p. 273.
40 Marshall (1995), p. 1050.
41 Philippidis, A. (2016), *Genetic Engineering & Biotechnology News*, 1 April 2016.
42 Blaese, R., et al. (1995), *Science* 270:475–80.
43 Anderson, W. (2000), *Science* 288:627–8; *New York Times*, 3 July 2005; Wilson, J. (2016), *Human Gene Therapy Clinical Development* 27:53–6.
44 Rosenberg, S., et al. (1993), *Annals of Surgery* 218:455–64. Rosenberg, *The Transformed Cell*, contains moving accounts of his relationships with his patients.
45 Anderson, C. (1992), *Nature* 360:399–400; Gershon, D. (1994), *Nature* 369:598.
46 Miller, A. (1992), *Nature* 357:455–60.
47 *Le Monde*, 3 February 1988; Anderson, W. (1993), *Human Gene Therapy* 4:125–6.
48 Lyon and Gorner, *Altered Fates*, quotes in this paragraph are from pp. 49, 24, 62.
49 Ibid., p. 281.
50 Ibid., p. 292.
51 Bauer, G. and Anderson, J. (2014), *Gene Therapy for HIV: From Inception to a Possible Cure* (Berlin, Springer).
52 Wivel, N. and Walters, L. (1993), *Science* 262:533–8.
53 *The Economist*, 25 April 1992.
54 *New Scientist*, 25 January 1992.
55 Lyon and Gorner, *Altered Fates*, pp. 284–5.
56 *New York Times*, 28 November 1999.
57 Orkin, S. and Motulsky, A. (1995), *Report and Recommendations of the Panel to Assess the NIH Investment in Research on Gene Therapy* (Bethesda, MD, National Institutes of Health).
58 Marshall (1995), p. 1050.
59 Jenks, S. (1996), *Journal of the National Cancer Institute* 88:9–10.
60 Touchette, N. (1996), *Nature Medicine* 2:7–8, p. 7.
61 Fox, J. (1996), *Bio/Technology* 14:14–5. The roles of the RAC and the FDA were also clarified at this time – not only was there a long-running turf war between these two institutions, AIDS activists and the biotech industry were hostile to the RAC because they felt it was delaying the development of novel treatments. The RAC remained, but the FDA was given sole authority to approve gene therapy protocols.
62 Anonymous (1997), *Nature Biotechnology* 15:815.
63 Fox, J. (1998), *Nature Biotechnology* 16:407.
64 *New York Times*, 28 November 1999.
65 Thompson, L. (2000), *FDA Consumer*, September–October 2000, pp. 19–24.
66 Friedmann, T. and Steele, F. (2001), *Molecular Therapy* 4:284.
67 *New York Times*, 28 November 1999.
68 Wilson, J. (2009), *Science* 324:727–8, p. 728.
69 Cavazzana-Calvo, M., et al. (2000), *Science* 288:669–72.

70 Morange, M. (2020), *The Black Box of Biology: A History of the Molecular Revolution* (Cambridge, MA, Harvard University Press), p. 303.
71 *Le Monde*, 9 April 2002.
72 *Le Monde*, 13 September 2002.
73 *Le Monde*, 16 January 2003.
74 Nathan, D. and Orkin, S. (2009), *Genome Medicine* 1:38, p. 3.
75 Sheridan, C. (2011), *Nature Biotechnology* 29:121–8.
76 Wirth, T., et al. (2013), *Gene* 525:162–9; Zhang, W.-W., et al. (2018), *Human Gene Therapy* 29:160–79.
77 Regalado, A. (2016), *MIT Technology Review*, 4 May 2016.
78 Wang, D., et al. (2019), *Nature Reviews Drug Discovery* 18:358–78.
79 Collins, F. and Gottlieb, S. (2018), *New England Journal of Medicine* 379:1393–5.
80 Orkin, S. and Reilly, P. (2016), *Science* 352:1059–61.
81 *Le Monde*, 24 November 1994.
82 Johnson, M. and Gallagher, K. (2016), *One in a Billion: The Story of Nic Volker and the Dawn of Genomic Medicine* (New York, Simon & Schuster).
83 Owen, M., et al. (2021), *New England Journal of Medicine* 384:2159–61.
84 Wirth et al. (2013), p. 167.
85 Maldonado, R., et al. (2021), *Journal of Community Genetics* 12:267–76.
86 Nogrady, B. (2019), *Nature Index 2019: Biomedical Sciences* S23–5.
87 Kohn, D., et al. (2021), *New England Journal of Medicine* 384:2002–13.
88 Melenhorst, J., et al. (2022), *Nature* 602:503–9.
89 Anonymous (2006), *Nature* 442:341; Begley, S. (2018), *STAT*, 23 July 2018; Lyon and Gorner, *Altered Fates*, p. 277.
90 Sheridan (2011).
91 Dunbar, C. (2018), *Science* 359:eaan4672.
92 Kirby, D. (2000), *Science Fiction Studies* 27: 193–215; Kirby, D. (2004), *Literature and Medicine* 23:184–200; Kirby, D. (2007), *Literature and Medicine* 26:83–108.
93 Simcoe, M., et al. (2021), *Science Advances* 7:eabd1239.
94 Hsu, S. (2014), *Nautilus* 18.
95 Karavani, E., et al. (2019), *Cell* 179:1424–35.
96 Kaiser, J. (2021), *Science* 372:776.
97 Khamel, R. (2020), *Nature* 583:S12–4.
98 Sheridan (2011), p. 128.

11 Editing

1 Capecchi, M. (1989), *Science* 244:1288–92.
2 Adli, M. (2018), *Nature Communications* 9:1911.
3 Rouet, P., et al. (1994), *Proceedings of the National Academy of Sciences USA* 91:6064–8. A key reagent used in this experiment was provided by Bernard Dujon in Paris who had been thinking along similar lines for some time; for a lively account of how it felt for the French group to be scooped, see Choulika, A. (2016), *Réécrire la vie: la fin du destin génétique* (Paris, Hugo Doc).
4 Morange, M. (2017), *Journal of Biosciences* 42:527–30.
5 Kim, Y.-G., et al. (1996), *Proceedings of the National Academy of Sciences USA* 93:1156–60.
6 Chandrasegaran, S. and Smith, J. (1999), *Biological Chemistry* 380:841–8, p. 847.
7 Chandrasegaran, S. and Carroll, D. (2016), *Journal of Molecular Biology* 428:963–89, p. 972.

8 Bibikova, M., et al. (2003), *Science* 300:764; Porteus, M. and Baltimore, D. (2003), *Science* 300:763.
9 Mani, M., et al. (2005), *Biochemical and Biophysical Research Communications* 335:447–57; Urnov, F., et al. (2005), *Nature* 435:646–51.
10 Carroll, D. (2008), *Gene Therapy* 15:1463–8.
11 Tebas, P., et al. (2014), *New England Journal of Medicine* 370:901–10.
12 Moscou, M. and Bogdanove, A. (2009), *Science* 326:1501; Boch, J., et al. (2009), *Science* 326:1509–12.
13 Boch et al. (2009), p. 1512.
14 Balbas, P. and Gosset, G. (2001), *Molecular Biotechnology* 19:1–12; Stark, W. and Akoplan, A. (2003), *Discovery Medicine* 3:34–5; Gruenert, D., et al., *Journal of Clinical Investigation* 112:637–41. For an overview of the shift to 'editing' see Morange, M. (2016), *Journal of Biosciences* 41:9–11.
15 Urnov et al. (2005); Urnov, F., et al. (2010), *Nature Reviews Genetics* 11:636–46. Kevin Davies pointed out the use of 'gene editing' on the cover of *Nature* in 2005 – Davies, K. (2020), *Editing Humanity: The CRISPR Revolution and the New Era of Genome Editing* (New York, Pegasus), p. 116.
16 Ding, Q., et al. (2013), *Cell Stem Cell* 13:238–51. The article was submitted on 23 August 2012.
17 Jinek, M., et al. (2012), *Science* 337:816–21, p. 820. The article was published online on 17 August 2012.
18 Ishino, Y., et al. (1987), *Journal of Bacteriology* 169:5429–33. For all information on Mojica, see Davies, K. and Mojica, F. (2018), *CRISPR Journal* 1:1–5.
19 Mojica, F. and Rodriguez-Valera, F. (2016), *FEBS Journal* 283:3162–9.
20 Jansen, R., et al. (2002), *Molecular Microbiology* 43:1565–75; Makarova, K., et al. (2002), *Nucleic Acids Research* 30:482–96.
21 Mojica, F., et al. (2005), *Journal of Molecular Evolution* 60:174–82.
22 Bolotin, A., et al. (2005), *Microbiology* 151:2551–61; Pourcel, C., et al. (2005), *Microbiology* 151:653–63.
23 Barrangou, R., et al. (2007), *Science* 315:1709–12.
24 Ibid., p. 1712.
25 Brouns, S., et al. (2008), *Science* 321:960–4.
26 Davies and Mojica (2018), p. 4.
27 Marraffini, L. and Sontheimer, E. (2008), *Science* 322:1843–5.
28 Ibid., p. 1845. For an explanation of why some still doubted that DNA was the target, see Horvath, P. and Barrangou, R. (2010), *Science* 327:167–70.
29 US Provisional Patent Application 61/009,317, filed 23 September 2008; published on 25 March 2010 as US2010/0076057.
30 Makarova, K., et al. (2011), *Nature Reviews Microbiology* 9:467–77; Deltcheva, E., et al. (2011), *Nature* 471:602–7; Sapranauskas, R., et al. (2011), *Nucleic Acids Research* 39:9275–82; Wiedenheft, B., et al. (2011), *Nature* 477:486–9. For an excellent overview of this period, see Morange, M. (2015), *Journal of Biosciences* 40:829–32. The Deltcheva et al. and Wiedenheft et al. articles were among the first CRISPR articles to be published by *Nature*. By accident or design, having rejected Mojica's article in 2003, *Nature* had been out of the CRISPR loop for seven years.
31 Doudna, J. and Sternberg, S. (2017), *A Crack in Creation: The New Power to Control Evolution* (London, Bodley Head), p. 72.
32 Ibid., pp. 75–6.
33 *Le Monde*, 1 August 2016.

34 Jinek, M., et al. (2012), *Science* 337:816–21.
35 Gasiunas, G., et al. (2012), *Proceedings of the National Academy of Sciences USA* 109:E2579–86.
36 Ibid., p. E2585; Jinek et al. (2012), p. 820.
37 Jinek et al. (2012), p. 820.
38 Cong, L., et al. (2013), *Science* 339:819–23; Mali, P., et al. (2013), *Science* 339:823–6.
39 *Le Monde*, 8 August 2016.
40 The clearest account of this rather murky series of events can be found in Isaacson, W. (2021), *The Code Breaker: Jennifer Doudna, Gene Editing, and the Future of the Human Race* (London, Simon & Schuster).
41 Jinek, M., et al. (2013), *eLife* 2:e00471; Cho, S., et al. (2013), *Nature Biotechnology* 31:230–2.
42 Doudna and Sternberg, *A Crack in Creation*, p. 95.
43 Jao, L.-E., et al. (2013), *Proceedings of the National Academy of Sciences USA* 110:13904–9; Ren, X., et al. (2013), *Proceedings of the National Academy of Sciences USA* 110:19012–7; Tan, W., et al. (2013), *Proceedings of the National Academy of Sciences USA* 110:16526–31; Hou, Z., et al. (2013), *Proceedings of the National Academy of Sciences USA* 110:15644–9; Schwank, G., et al. (2013), *Cell Stem Cell* 13:653–8.
44 Choulika, *Réécrire la vie*, p. 66.
45 Shalem, O., et al. (2014), *Science* 343:84–7.
46 Ding, Q., et al. (2013), *Cell Stem Cell* 12:393–4.
47 Musunuru, K. (2019), *The CRISPR Generation: The Story of the World's First Gene-Edited Babies* (n.p.), pp. 82–3.
48 Pennisi, E. (2013), *Science* 341:833–6.
49 https://twitter.com/UoM_GEU/status/1263057453320671235?s=20
50 Leslie has subsequently deleted this tweet, along with much of her Twitter feed, and changed her Twitter handle, but she has given me permission to reproduce the tweet here.
51 Trible, W., et al. (2017), *Cell* 170:727–35; Yan, H., et al. (2017), *Cell* 170:736–47.
52 Isaacson, *The Code Breaker*, pp. 113–18.
53 Cohen, J. (2017), *Science* 355:681–4.
54 Hopkins, M., et al. (2013), *Industrial and Corporate Change* 22:903–52.
55 For a clear account of the situation in the Broad/MIT/Harvard versus Berkeley US patent row up until 2020, see Isaacson, *The Code Breaker*, pp. 231–41.
56 Martin-Laffon, J., et al. (2019), *Nature Biotechnology* 37:613–20.
57 Egelie, K., et al. (2016), *Nature Biotechnology* 14:1025–31.
58 Anonymous (2021), *Nature* 597:152.
59 Egelie et al. (2016) explore the possibilities of a more restrictive patent licensing scheme.
60 *Le Monde*, 8 August 2016.
61 Charlesworth, C., et al. (2019), *Nature Medicine* 25:249–54.
62 Moreno, A., et al. (2019), *Nature Biomedical Engineering* 3:806–16.
63 *The Independent*, 6 November 2013; *Boston Globe*, 25 November 2013.
64 *Le Monde*, 16 December 2013.
65 Anonymous (2013), *Science* 342:1434–5.
66 *New York Times*, 4 March 2014.
67 *Washington Post*, 11 November 2014.
68 *San Francisco Chronicle*, 7 September 2014.

69 *Editing Life* (BBC Radio, 2016), https://www.bbc.co.uk/programmes/b06zr3zj
70 *Le Monde*, 1 August 2016.
71 Ibid.
72 Doudna and Sternberg, *A Crack in Creation*; Isaacson, *The Code Breaker*; Davies, *Editing Humanity*.
73 Lander, E. (2016), *Cell* 164:18–28.
74 https://www.michaeleisen.org/blog/?p=1825 and https://genotopia.scienceblog.com/573/a-whig-history-of-crispr – these pages have been saved to the Wayback Machine at archive.org.
75 Vence. T. (2016), *The Scientist*, 19 January 2016.

12 #CRISPRbabies

1 Doudna, J. and Sternberg, S. (2017), *A Crack in Creation: The New Power to Control Evolution* (London, Bodley Head), p. 199.
2 Niu, Y., et al. (2014), *Cell* 156:836–43.
3 *Le Monde*, 15 August 2016.
4 Manheimer, K., et al. (2018), *Human Genetics* 137:183–93.
5 Doudna, J. (2015a), *Nature* 528:469–71, p. 470.
6 Ibid., p. 471.
7 Regalado, A. (2015), *MIT Technology Review*, 5 March 2015.
8 Greely, H. (2021), *CRISPR People: The Science and Ethics of Editing Humans* (London, MIT Press), pp. 60–5.
9 Lanphier, E., et al. (2015), *Nature* 519:410–11.
10 Baltimore, D., et al. (2015), *Science* 348:36–8.
11 Zhang, X. (2015), *Protein & Cell* 6:313.
12 Liang, P., et al. (2015), *Protein & Cell* 6:363–72.
13 Regalado, A. (2015), *MIT Technology Review*, 22 April 2015.
14 Bosley, K., et al. (2015), *Nature Biotechnology* 33:478–86.
15 Ibid., p. 479.
16 Regalado, A. (2018), *MIT Technology Review*, 11 December 2018.
17 *Boston Globe*, 1 August 2015.
18 Miller, H. (2015), *Science* 348:1325.
19 Pollack, R. (2015), *Science* 348:871.
20 *Le Monde*, 22 August 2016.
21 Society for Developmental Biology (2015), Position statement from the Society for Developmental Biology on Genomic Editing in Human Embryos, 24 April 2015. https://www.sdbonline.org/uploads/files/SDBgenomeeditposstmt.pdf
22 Membres Comité d'Éthique de l'INSERM (2016), Saisine concernant les questions liées au développement de la technologie CRISPR (clustered regularly interspaced short palindromic repeat)-Cas9. https://www.hal.inserm.fr/inserm-02110670/document
23 Quotes from Bosley et al. (2015), p. 481.
24 *Editing Life* (BBC Radio, 2016), https://www.bbc.co.uk/sounds/play/b06zr3zj
25 An excellent summary of the different legal and ethical frameworks around the world can be found in Greely, *CRISPR People*.
26 Ledford, H. (2015), *Nature* 526:310–11.
27 National Academies of Sciences, Engineering, and Medicine (2015), *International Summit on Human Gene Editing: A Global Discussion* (Washington, DC, The National Academies Press).
28 Ibid.

29 Reardon, S. (2015), *Nature* 528:173.
30 https://www.statnews.com/2015/12/02/gene-editing-summit-embryos/
31 Doudna, J. (2015b), *Nature* 528:S6.
32 For example, Hogan, A. (2016), *Endeavour* 40:218–22.
33 Parthasarathy, S. (2015), *Ethics in Biology, Engineering & Medicine* 6:305–12, pp. 306–7, 308, 309, 310–11.
34 Hurlbut, J., et al. (2015), *Issues in Science and Technology* 32(1).
35 Ceccarelli, L. (2018), *Life Sciences, Society and Policy* 14:24.
36 Ibid., p. 9.
37 Kang, X., et al. (2016), *Journal of Assisted Reproduction and Genetics* 33:581–8.
38 Callaway, E. (2016), *Nature* 532:289–90.
39 Doudna (2015b).
40 Ma, H., et al. (2017), *Nature* 548:413–9.
41 Baylis, F. (2019), *Altered Inheritance: CRISPR and the Ethics of Human Genome Editing* (London, Harvard University Press), p. 109. Baylis showed that these eggs were obtained from paid volunteers and – without payment – from women undergoing fertility treatment. The relevant phrase stated that 'Study staff will try to sync the timing of your egg donation with that of another study participant.' The implication drawn by Baylis was that the egg might be implanted into another woman at a similar point in her hormonal cycle. Even if the inclusion of the phrase was an error, it raises questions about whether the participants in the project fully understood what they were signing up to.
42 National Academies of Sciences, Engineering, and Medicine (2017), *Human Genome Editing: Science, Ethics, and Governance* (Washington, DC, The National Academies Press.
43 Ibid., p. 188.
44 Ibid., p. 189.
45 Kaiser, J. (2017), *Science* 355:675.
46 Ibid.
47 Nuffield Council on Bioethics (2018), *Genome Editing and Human Reproduction: Social and Ethical Issues* (London, Nuffield Council on Bioethics).
48 Anonymous (2018), *Nature Medicine* 24:1081.
49 This and subsequent quotes from Gregorowius, D., et al. (2017), *EMBO Reports* 18:355–8.
50 Baylis, *Altered Inheritance*, pp. 135–6.
51 *New York Times*, 23 January 2019.
52 Cohen, J. (2019), *Science* 365:430–7, p. 433.
53 Regalado, A. (2018), *MIT Technology Review*, 25 November 2018; *Washington Post*, 26 November 2018; *The He Lab* (2018), https://www.youtube.com/watch?v=thovnOmFltc
54 Davies, K. (2020), *Editing Humanity: The CRISPR Revolution and the New Era of Genome Editing* (New York, Pegasus), p. 237.
55 Greely, *CRISPR People*, p. 100.
56 Yong, E. (2018), *The Atlantic*, 15 December 2018. https://www.theatlantic.com/science/archive/2018/12/15-worrying-things-about-crispr-babies-scandal/577234/; Ryder, S. (2018), *CRISPR Journal* 1:355–7.
57 Davies, *Editing Humanity*; Greely, H. (2019), *Journal of Law and the Biosciences* 13:111–183; Greely, *CRISPR People*.
58 Greely (2019), p. 113.

59 *Genetic Dreams, Genetic Nightmares* (BBC Radio, 2021), episode 3: https://www.bbc.co.uk/sounds/play/m000ycvv
60 Wang, C., et al. (2019), *Lancet* 393:25–6.
61 Yi, L. (2018), https://www.yicai.com/news/100067069.html (in Chinese).
62 Wang et al. (2019); Zhang, B., et al. (2019), *Lancet* 393:25; Zhang, L., et al. (2019), *Lancet* 393:26–7.
63 Cohen, J. (2019), *Science* 8 May 2019, http://doi.org/10.1126/science.aax9733
64 Cyranoski, D. (2020), *Nature* 577:154–5.

13 Aftermath

1 Nie, J.-B. (2018), *The Hastings Center Forum*, https://www.thehastingscenter.org/jiankuis-genetic-misadventure-china; Lei, R., et al. (2019), *Nature* 569:184–6.
2 Ben Ouagrham-Gormley, S. and Vogel, K. (2020), *Bulletin of the Atomic Scientists* 76:192–9.
3 Wang, D., (2020), *Gene Therapy* 27:338–48; Ben Ouagrham-Gormley and Vogel (2020).
4 Gao, F., et al. (2016), *Nature Biotechnology* 34:768–72; Cyranoski, D. (2018), *Nature*, https://doi.org/10.1038/d41586-018-06163-0
5 Davies, K. (2020), *Editing Humanity: The CRISPR Revolution and the New Era of Genome Editing* (London, Pegasus), p. 268.
6 Hurlbut, J. (2020), *Perspectives in Biology and Medicine* 63:177–94, pp. 182–3.
7 *New York Times*, 29 June 2015.
8 Lei et al. (2019), p. 186.
9 Cohen, J. (2019a), *Science* 365:430–7, p. 432.
10 Ibid., pp. 432–3.
11 Ibid., pp. 434–5.
12 Ibid., p. 434.
13 Hurlbut (2020), p. 191.
14 Baylis, F. (2019), *Altered Inheritance: CRISPR and the Ethics of Human Genome Editing* (London, Harvard University Press), p. 140.
15 O'Keefe, M., et al. (2015), *American Journal of Bioethics* 15:3–10.
16 Cohen (2019a), p. 436.
17 Yong, E. (2018), *The Atlantic*, 15 December 2018. https://www.theatlantic.com/science/archive/2018/12/15-worrying-things-about-crispr-babies-scandal/577234/
18 Davies, *Editing Humanity*, p. 257.
19 Friedmann, T. (1983), *Gene Therapy: Fact and Fiction in Biology's New Approaches to Disease* (Plainview, NY, Cold Spring Harbor Laboratory Press). Quotes from Baltimore can be found on pp. 58–60.
20 Hurlbut (2020), p. 185.
21 Lander, E., et al. (2019), *Nature* 567:165–8.
22 *Washington Post*, 13 March 2019. For a more developed presentation of this view, see Knoppers, B. and Kleiderman, E. (2019), *CRISPR Journal* 2:285–92.
23 Cohen, J. (2019b), *Science* 363:1130–1, p. 1131.
24 Davies, *Editing Humanity*, p. 274.
25 Ibid., p. 271.
26 Hough, S. and Ajetunmobi, A. (2019), *CRISPR Journal* 2:343–5.
27 Normile, D. (2018), *Science*, 29 November 2018. http://doi.org/10.1126/science.aaw2223

28 Dzau, V., et al. (2018), *Science* 362:1215.
29 Hess, M. (2020), *Notre Dame Law Review* 95:1369–97, p. 1395; Dryzek, J., et al. (2020), *Science* 369:1435–7.
30 Dryzek et al. (2020), p. 1437.
31 Cyranoski, D. (2019a), *Nature* 570:145–6, p. 146.
32 Cyranoski, D. (2019b), *Nature* 574:465–6, p. 466.
33 I am not going to give these characters the satisfaction of identifying them any more precisely.
34 Regalado, A. (2019), *MIT Technology Review*, 1 February 2019.
35 Egli, D., et al. (2018), *Nature* 560:E5–7 (2018); Adikusama, F., et al. (2018), *Nature* 560:E8–9 (2018); Callaway, E., *Nature* 8 August 2018, https://doi.org/10.1038/d41586-018-05915-2.
36 Ma, H., et al. (2018), *Nature* 560:E10–16; Kosicki, M., et al. (2018), *Nature Biotechnology* 36:765–71; Cullot, G., et al. (2019), *Nature Communications* 10:1136; Owens, D., et al. (2019), *Nucleic Acids Research* 47:7402–17; Przewrocka, J., et al. (2020), *Annals of Oncology* 31:1270–1273; Alanis-Lobatoa, G., et al. (2020), *Proceedings of the National Academy of Sciences USA* 118:e2004832117; Zuccaro, M., et al. (2020), *Cell* 183:1650–64; Papathansiou, S., et al. (2021), *Nature Communications* 12:5855.
37 Boutin, J., et al. (2022), *CRISPR Journal* 5:19–30.
38 Turocy, J., et al. (2021), *Cell* 184:1561–74.
39 Zhang, M., et al. (2019), *Genome Biology* 20:101.
40 *South China Morning Post*, 1 June 2019.
41 Costas, J. (2018), *American Journal of Medical Genetics B: Neuropsychiatric Genetics* 177:274–83.
42 National Academy of Sciences (2020), *Heritable Human Genome Editing* (Washington, DC, The National Academies Press), pp. ix–x.
43 *Le Monde*, 21 March 2016.
44 Isaacson, W. (2019), *Air Mail*, 27 July 2019.
45 Angrist, M., et al. (2020), *CRISPR Journal* 3:333–49, p. 342.
46 National Academy of Sciences, *Heritable Human Genome Editing*, p. x.
47 Angrist et al. (2020), p. 336.
48 Greely, H. (2021), *CRISPR People: The Science and Ethics of Editing Humans* (London, MIT Press), p. 227.
49 Cohen, J. (2020), *Science*, 3 September 2020. http://doi.org/10.1126/science.abe6341; Baylis, *Altered Inheritance*, p. 31. In 2018, genetic counsellors working with the Nuffield Council stated that in their collective experience, which added up to hundreds of years of discussions with patients, they had never encountered an example of a couple in which both people were homozygous for the same disease gene – Hurlbut (2020), p. 190.
50 WHO (2021), *Human Genome Editing: A Framework for Governance* (Geneva, WHO), p. 22.
51 WHO (2021), *Human Genome Editing: Position Paper* (Geneva, WHO), p. 3.
52 Baylis, *Altered Inheritance*, p. 165.
53 Hess (2020).
54 Andorno, R., et al. (2020), *Trends in Biotechnology* 38:351–4.
55 Wang, D., et al. (2020), *Gene Therapy* 27:338–48, p. 345.
56 Saha, K., et al. (2018), *Trends in Biotechnology* 36:741–3.
57 Suzuki, D. and Knudtson, P. (1989), *Genethics: The Ethics of Engineering Life* (London, Harvard University Press), p. 355.

58 Baylis, *Altered Inheritance*, p. 33.
59 Cohen (2020).
60 If you still want to read such a book, try Metzl, J. (2019), *Hacking Darwin: Genetic Engineering and the Future of Humanity* (Naperville, IL, Sourcebooks), or Mason, C. (2021), *The Next 500 Years: Engineering Life to Reach New Worlds* (London, MIT Press), or turn to some science fiction.

14 Ecocide

1 Serebrovskii, A. (1940), *Zoologicheskiĭ zhurnal* 19:618–30 (in Russian). English translation (1969) in: Serebrovskii, A. (1940), *Panel Proceedings Series No. STI/PUB/224* (Vienna, IAEA), pp. 123–37; Vanderplank, F. (1944), *Nature* 154:607–8; Adkisson, P. and Tumlinson, J. (2003), *National Academy of Sciences Biographical Memoirs* 83:1–15.
2 For example, to prevent the medfly pest from becoming established in Mexico in the 1980s, the government built a factory capable of producing 500 million sterile medflies per week. Curtis, C. (1985), *Biological Journal of the Linnean Society* 26:359–74; Hamilton, W. (1967), *Science* 156:477–88.
3 Craig, G., et al. (1960), *Science* 132:1887–9; Foster, G., et al. (1972), *Science* 176:875–80.
4 Curtis (1985), p. 372.
5 WHO Special Programme for Research and Training in Tropical Diseases (1991), Report of the meeting 'Prospects for malaria control by genetic manipulation of its vectors' – TDR/BCV/MAL-ENT/91.3; Morel, C., et al. (2002), *Science* 298:79.
6 Alphey, L., et al. (2002), *Science* 298:119–21; Scott, T., et al. (2002), *Science* 298:117–9. The best bet at the time was the use of transposons or jumping genes. One kind of transposon (the P element) had recently been found to be invading wild populations of *Drosophila*, causing some strains to produce no viable offspring if they were crossed. Transposons were also used by molecular geneticists to delete or add DNA sequences in laboratory studies of gene function, and were soon found in many species of insect, including malaria mosquitoes. All this raised the prospect of engineering transposons to drive genes, such as those for parasite resistance, into pest populations. Attempts to employ this system revealed it was too fragile and the idea was abandoned.
7 Laven, H. (1967), *Nature* 216:383–4, p. 384.
8 Anonymous (1975), *Nature* 256:355–7, p. 357.
9 Powell, K. and Jayaraman, K. (2002), *Nature* 419:867.
10 Beeman, R., et al. (1992), *Science* 256:89–92. Surprisingly, given the obvious potential application, the researchers merely described their discovery as 'an unusual strategy for the self-propagation of selfish DNA'.
11 Chen, C., et al. (2007), *Science* 316:597–600.
12 Gould, F. (2007), *Evolution* 62:500–10, p. 504.
13 Burt, A. (2003), *Proceedings of the Royal Society B* 270:921–8.
14 Deredec, A., et al. (2008), *Genetics* 179:2013–26.
15 Hammond, A. and Galizi, R. (2018), *Pathogens and Global Health* 111:412–23.
16 Burt (2003), p. 927.
17 Hurlbut, J., (2018) in I. Braverman (ed.), *Gene Editing, Law and the Environment: Life Beyond the Human* (Abingdon, Routledge), pp. 77–94.
18 Windbichler, N., et al. (2011), *Nature* 473:212–5, p. 212.
19 Galizi, R., et al. (2014), *Nature Communications* 5:3977.

20 Fang, J. (2010), *Nature* 466:432–4.
21 Esvelt, K., et al. (2014), *eLife* 3:e03401.
22 Valderrama, J., et al. (2019), *Nature Communications* 10:5726.
23 Webber, B., et al. (2015), *Proceedings of the National Academy of Sciences USA* 112:10565–7.
24 Esvelt, K. (2018), in I. Braverman (ed.), *Gene Editing, Law and the Environment: Life Beyond the Human* (Abingdon, Routledge), pp. 21–38, p. 27.
25 *New York Times Magazine*, 8 January 2020.
26 Gantz, V. and Bier, E. (2015), *Science* 348:442–4, pp. 443–4.
27 Gantz, V., et al. (2015), *Proceedings of the National Academy of Sciences USA* 112:E6736–43.
28 Hammond, A., et al. (2016), *Nature Biotechnology* 34:78–83.
29 Hammond, A., et al. (2017), *PLoS Genetics* 13:e1007039.
30 Champer, J., et al. (2017), *PLoS Genetics* 13:e1006796.
31 Unckless, R., et al. (2017), *Genetics* 205:827–41.
32 Drury, D., et al. (2017), *Science Advances* 3:e1601910, p. 5; The Anopheles gambiae 1000 Genomes Consortium (2017), *Nature* 552:96–100; Buchman, A., et al. (2018), *Proceedings of the National Academy of Sciences USA* 115:4725–30.
33 Kyrou, K., et al. (2018), *Nature Biotechnology* 36:1062–6.
34 Hammond, A., et al. (2021), *Nature Communications* 12:4589.
35 Bassett, L., et al. (2019), *Proceedings of the Royal Society B* 286:20191515.
36 Grunwald, H., et al. (2019), *Nature* 566:105–9; Conklin, B. (2019), *Nature* 566:43–5.
37 Pfitzner, C., et al. (2020), *CRISPR Journal* 3:388–97.
38 Scudellari, M. (2019), *Nature* 571:160–2, p. 161.
39 Kolbert, E. (2021), *New Yorker*, 18 January 2021.
40 National Academies of Science, Engineering and Medicine (2016), *Gene Drives on the Horizon: Advancing Science, Navigating Uncertainty, and Aligning Research with Public Values* (Washington, DC, National Academies Press), p. 15.
41 Evans, S. and Palmer, M. (2018), *Journal of Responsible Innovation* 5:S223–42, p. S224.
42 National Academies of Science, Engineering and Medicine, *Gene Drives on the Horizon*, p. 9.
43 Kaebnick, G., et al. (2016), *Science* 354:710–1, p. 711.
44 Akbari, O., et al. (2015), *Science* 349:927–9.
45 DiCarlo, J., et al. (2015), *Nature Biotechnology* 33:1250–5.
46 Esvelt, K. and Gemmell, N. (2017), *PLoS Biology* 15:e2003850; Noble, C., et al. (2019), *Proceedings of the National Academy of Sciences USA* 116:8275–82.
47 Taxiarchi, C., et al. (2021), *Nature Communications* 12:3977.
48 Garthwaite, J. (2016), *Scientific American*, 18 November 2016.
49 Noble, C., et al. (2018), *eLife* 7:e33423, p. 7.
50 Esvelt and Gemmell (2017), p. 4.
51 *New York Times*, 16 November 2017.
52 Palme, M., et al. (2015), *Science* 350:1471–3.
53 Kofler, N., et al. (2018), *Science* 362:527–9.
54 Hokanson, K. (2019), *Frontiers in Bioengineering and Biotechnology* 7:82.
55 Walters, L., et al. (2021), *CRISPR Journal* 4:469–76.
56 Guerrini C., et al. (2017), *Nature Biotechnology* 35:22–4.
57 Callaway, E. (2016), *Nature*, 21 December 2016. https://doi.org/10.1038/nature.2016.21216.

58 ETC Group (2017), *The Gene Drive Files*, 4 December 2017; Cohen, J. (2017), *Science*, 11 December 2017. https://www.science.org/content/article/there-really-covert-manipulation-un-discussions-about-regulating-gene-drives
59 *The Guardian*, 4 December 2017. DARPA's budget allocations are opaque; a year earlier the $100 million figure was quoted as covering DARPA's whole synthetic biology budget – see Garthwaite (2016).
60 *Common Call for a Global Moratorium on Genetically-Engineered Gene Drives*. https://www.synbiowatch.org/gene-drives/gene-drives-moratorium/
61 ETC Group and Heinrich Böll Stiftung (2018), *Forcing the Farm: How Gene Drive Organisms Could Entrench Industrial Agriculture and Threaten Food Sovereignty*. https://www.etcgroup.org/content/forcing-farm
62 *New York Times Magazine*, 8 January 2020.
63 World Health Organization (2021), *Guidance Framework for Testing Genetically Modified Mosquitoes* (Geneva, WHO).
64 Meghani, Z. and Kuzma, J. (2017), *Journal of Responsible Innovation* 5:S203–22; Waltz, E. (2022), *Nature* 604:608–9.
65 Connolly, J., et al. (2021), *Malaria Journal* 20:170.
66 *Le Monde*, 4 July 2019.
67 *Le Monde*, 29 June 2018.
68 *Libération*, 18 November 2018.
69 *Daily Telegraph*, 8 October 2019.
70 *New York Times Magazine*, 8 January 2020.
71 *The Guardian*, 18 November 2019.
72 Barry, N., et al. (2020), *Malaria Journal* 19:199, p. 5.
73 *Daily Telegraph*, 8 October 2019; *Le Monde*, 29 June 2018.
74 *Le Monde*, 4 July 2019.
75 Moloo, Z. (2018), *Project Syndicate*, December 2018. https://www.project-syndicate.org/commentary/target-malaria-gene-drive-experiments-lack-of-consent-by-zahra-moloo-2018-12. See also the video by Moloo, *A Question of Consent: Exterminator Mosquitoes in Burkina Faso*. https://www.youtube.com/watch?v=nD_1noCf2x8
76 *Le Monde*, 4 July 2019.
77 Beisel U. and Ganle J. (2019), *African Studies Review* 62:164–73.
78 George, D., et al. (2019), *Proceedings of the Royal Society B* 286:20191484, p. 5.
79 Meghani, Z. and Boëte, C. (2018), *PLoS Neglected Tropical Diseases* 12:e0006501.
80 *Le Monde*, 29 June 2018.
81 Thizy, D., et al. (2021), *Gates Open Research* 5:19.
82 Collins, C., et al. (2019), *Medical and Veterinary Entomology* 33:1–15.
83 Courtier-Orgogozo, V., et al. (2020), *Evolutionary Applications* 13:1888–905.
84 Datoo, M., et al. (2021), *The Lancet* 397:1809–18; Mwakingwe-Omari, A., et al. (2021), *Nature* 595:289–94; Moreira, L., et al. (2009), *Cell* 139:1268–78.
85 Utarini, A., et al. (2021), *New England Journal of Medicine* 384:2177–86.
86 Montenegro de Wit, M. (2019), *Agroecology and Sustainable Food Systems* 43:1054–74.
87 *New Scientist*, 16 June 2016.
88 Montenegro de Wit (2019), p. 1071.

15 Weapons

1 Michael Rogers has given two slightly different accounts of this event: Rogers, M. (1975), *Rolling Stone*, 19 June 1975, p. 40; Rogers, M. (1977), *Biohazard*

(New York, Alfred Knopf), pp. 71–2. Goldfarb was able to emigrate from the Soviet Union later in 1975 and continued his scientific career, eventually moving to the United States.

2 Information here from: van Courtland Moon, J. (2002), in M. Wheelis, et al. (eds), *Deadly Cultures: Biological Weapons Since 1945* (London, Harvard University Press), pp. 9–46; Meselson, M. (2001), *New York Review of Books*, 20 December 2001. The reasons behind Nixon's decision are still debated; one factor may have been the friendship between Henry Kissinger, who was Nixon's National Security Advisor and Matthew Meselson, molecular biologist and long-time opponent of biological warfare – Dyson, F. (2003), *New York Review of Books*, 13 February 2003. Surprisingly, Nixon did not refer to the abandonment of bioweapons in his memoirs (van Courtland Moon, ibid., p. 36).

3 Wright, S. (1994), *Molecular Politics: Developing America and British Regulatory Policy for Genetic Engineering, 1972–1982* (London, University of Chicago Press), p. 149.

4 A full list of Asilomar attendees and their affiliations can be found in Fredrickson, D. S. (2000), *The Recombinant DNA Controversy: A Memoir* (Washington, DC, ASM), Appendix 1.1.

5 Zilinskas, R. (2018), *The Soviet Biological Weapons Programme and its Legacy in Today's Russia* (Washington, DC, National Defense University Press); Rimmington, A. (2021a), *The Soviet Union's Biowarfare Programme: Ploughshares to Swords* (Cham, Palgrave Macmillan).

6 Rimmington, A. (2021b), *The Soviet Union's Invisible Weapons of Mass Destruction: Biopreparat's Covert Biological Warfare Programme* (Cham, Palgrave Macmillan).

7 Leitenberg, M. and Zilinskas, R. (2012), *The Soviet Biological Weapons Program: A History* (London, Harvard University Press), pp. xi, 71. Remaining information in this paragraph from pp. 154 and 164.

8 Rogers (1975), p. 40.

9 Ovchinnikov is still renowned for his 1978 study of the amino acid sequence of a particularly important class of cellular receptors and is fêted in Russia for his civilian work. For a memoir of Ovchinnikov which makes no mention of bioweapons, see Ivanov, V., et al. (1989), *Journal of Membrane Biology* 110:97–101.

10 Leitenberg and Zilinskas, *The Soviet Biological Weapons Program*, p. 61.

11 Ibid., p. 65.

12 *Los Angeles Times*, 20 February 1977.

13 Leitenberg and Zilinskas, *The Soviet Biological Weapons Program*, p. 73.

14 Material in this paragraph from ibid., pp. 156, 157, 703. A terrifying list of objectives on pp. 77–8 is reproduced from a 1985 document stamped Top Secret.

15 Ben Ouagrham-Gormley, S. (2014), *Barriers to Bioweapons: The Challenges of Expertise and Organization for Weapons Development* (London, Cornell University Press), pp. 91–121, details the organisational limitations on the Soviet bioweapons programme.

16 For a summary, see Leitenberg and Zilinskas, *The Soviet Biological Weapons Program*, pp. 157–9.

17 Alibek, K. and Handelman, S. (1999), *BioHazard: The Chilling True Story of the Largest Covert Biological Weapons Programme in the World – Told from the Inside By the Man Who Ran It* (London, Hutchinson), p. 164; Leitenberg and Zilinskas, *The*

Soviet Biological Weapons Program, pp. 194–6. The researcher was Serguei Popov, who defected in 1992 – Pontin, M. (2006), *MIT Technology Review*, March 2006. The existence of this work was confirmed in 1992 – see Adams, J. (1994), *The New Spies: Exploring the Frontiers of Espionage* (London, Hutchinson), pp. 278–9.

18 Rogers, *Biohazard*, p. 40.

19 The document is heavily redacted; Leitenberg and Zilinskas, *The Soviet Biological Weapons Program*, p. 361.

20 Ibid., p. 365; Guillemin, J. (2002), *Proceedings of the American Philosophical Society* 146:18–36.

21 Leitenberg and Zilinskas, *The Soviet Biological Weapons Program*, p. 365.

22 See, for example, the article by S. Solarz in the *Wall Street Journal*, 22 June 1983. Solarz was chair of the House Foreign Affairs Subcommittee on Asian and Pacific Affairs.

23 Maddox, J. (1984), *Nature* 309:207; Nowicke, J. and Meselson, M. (1984), *Nature* 309:205–6.

24 The series began in the *Wall Street Journal* on 23 April 1984 with the running title 'Beyond "Yellow Rain"' and ended on 8 May 1984. See Cole, L. (1984), *Bulletin of the Atomic Scientists* 40:36–38.

25 *Wall Street Journal*, 1 May 1984.

26 Tucker, J. (1984–5), *Foreign Policy* 57:58–79.

27 *Los Angeles Times*, 20 February 1977.

28 Miller, J., et al. (2001), *Germs: The Ultimate Weapon* (London, Simon & Schuster), p. 96.

29 All material regarding Pasechnik from Adams, *The New Spies*, pp. 274–6. There is no way of knowing if any of this is true. Adams claims that Pasechnik did not even know of the existence of the Biological Weapons Convention (p. 273).

30 *New Yorker*, 9 March 1998, pp. 52–65, p. 59.

31 Ibid.

32 Ibid., pp. 63–4.

33 Hart, J. (2006), in M. Wheelis, et al. (eds), *Deadly Cultures: Biological Weapons Since 1945* (London, Harvard University Press), pp. 132–56.

34 Leitenberg and Zilinskas, *The Soviet Biological Weapons Program*, pp. 165–9.

35 Pomerantsev, A., et al. (1997), *Vaccine* 15:1846–50.

36 Leitenberg and Zilinskas, *The Soviet Biological Weapons Program*, p. 699; Ackerman, G. (2018), *Nature* 555:162–3.

37 Balmer, B. (2006), in M. Wheelis, et al. (eds), *Deadly Cultures: Biological Weapons Since 1945* (London, Harvard University Press), pp. 47–83; Agar, J and Balmer, B. (2016), in D. Leggett and C. Sleigh (eds), *Scientific Governance in Britain, 1914–79* (Manchester, Manchester University Press), pp. 122–43; Lepick, O. (2006), in M. Wheelis, et al. (eds), *Deadly Cultures: Biological Weapons Since 1945* (London, Harvard University Press), pp. 108–31.

38 Croddy, E. (2002), *Nonproliferation Review* 9:16–47; Alibek and Handelman (2000), p. 273.

39 Pearson, G. and Chevrier, M. (1999), in J. Lederberg (ed.), *Biological Weapons: Limiting the Threat* (London, MIT Press), pp. 113–32.

40 All information from Gould, C. and Hay, A. (2003), *Project Coast: Apartheid's Chemical and Biological Warfare Programme* (Geneva, United Nations Institute for Disarmament Research) and Gould, C. and Hay, A. (2006), in M. Wheelis, et al. (eds), *Deadly Cultures: Biological Weapons Since 1945* (London, Harvard University Press), pp. 191–212.

41 Alves, G., et al. (2014), *Anaerobe* 30:102–7.
42 Cohen, A. (2001), *Nonproliferation Review* 8:27–53.
43 van Aken, J. and Hammond, E. (2003), *EMBO Reports* 4:S57–60.
44 Pearson, G. and Chevrier, M. (1999), in J. Lederberg (ed.), *Biological Weapons: Limiting the Threat* (London, MIT Press), pp. 113–32. Material in the rest of this section on Iraq from Vogel, K. (2013), *Phantom Menace or Looming Danger? A New Framework for Assessing Bioweapons Threats* (Baltimore, Johns Hopkins University Press), pp. 131–47; Ben Ouagrham-Gormley, *Barriers to Bioweapons*, pp. 123–31.
45 United Nations Monitoring, Verification and Inspection Commission (2006), *Summary of the Compendium of Iraq's Proscribed Weapons Programmes in the Chemical, Biological and Missile Areas.* United Nations Security Council S/2006/420.
46 Rose, S. (1989), *Naval War College Review* 42:6–29, quotes from pp. 15, 23.
47 Wheelis, M. and Sugishima, M. (2006), in M. Wheelis, et al. (eds), *Deadly Cultures: Biological Weapons Since 1945* (London, Harvard University Press), pp. 284–303; Török, T., et al. (1997), *Journal of the American Medical Association* 278:389–95.
48 Cohen, W. (1999), in J. Lederberg (ed.), *Biological Weapons: Limiting the Threat* (London, MIT Press), pp. xi–xvi, p. xi.
49 Vogel, *Phantom Menace or Looming Danger?*, p. 3.
50 Jackson, R., et al. (2001), *Journal of Virology* 75:1205–10.
51 MacKenzie, D. (2003), *New Scientist*, 29 October 2003. The study was led by Mark Buller; I could find no trace of it being published.
52 Finkel, E. (2001), *Science* 291:585; Ball, P. (2001), *Nature* 411:232–5.
53 Pontin (2006).
54 Stanford, M. and McFadden, G. (2005), *Trends in Immunology* 26:339–45, p. 339.
55 Lane, H., et al. (2001), *Nature Medicine* 7:1271–3.
56 Dance, A. (2021), *Nature* 598:554–7.
57 Enserink, M. (2003), *Science*, 8 October 2003; Tucker, J. (2006), *International Security* 31:116–50.
58 Beck, V. (2003), *EMBO Reports* 4:S53–6.
59 Ledford, H. (2012), *Nature* 481:9–10.
60 Pontin (2006).
61 Rosengard, A., et al. (2002), *Proceedings of the National Academy of Sciences USA* 99:8808–13.
62 Cello, J., et al. (2002), *Science* 297:1016–8.
63 Shimono, N., et al. (2003), *Proceedings of the National Academy of Sciences USA* 100:15918–23; Smith, H., et al. (2003), *Proceedings of the National Academy of Sciences USA* 100:15440–5.
64 Wang, N., et al. (2018), *Virologica Sinica* 33:104–7.
65 Tumpey, T., et al. (2005), *Science* 310:77–80.
66 van Aken, J. (2007), *Heredity* 98:1–2; von Bubnoff, A. (2005), *Nature* 437:794–5.
67 Atlas, R. (2002), *Science* 298:753–4.
68 Rappert, B. (2015), *Frontiers in Public Health* 2:74.
69 Wein, L. and Liu, Y. (2005), *Proceedings of the National Academy of Sciences USA* 102:9984–9.
70 Davies, J. (2018), *Synthetic Biology: A Very Short Introduction* (Oxford, Oxford University Press).
71 Drexler, K. (1981), *Proceedings of the National Academy of Sciences USA* 78:5275–8.

72 Gardner, T., et al. (2000), *Nature* 403:339–42; Elowitz, M. and Leibler, S. (2000), *Nature* 403:335–8.
73 Carlson, R. (2010), *Biology is Technology: The Promise, Peril, and New Business of Engineering Life* (London, Harvard University Press).
74 Davies, *Synthetic Biology*, p. 118.
75 Carlson, R. (2003), *Biosecurity and Bioterrorism: Biodefence Strategy, Practice, and Science* 1:203–14.
76 Tucker, J. (2010), *Issues in Science and Technology* 26(3).
77 *The Guardian*, 14 June 2006.
78 Wolinetz, C. (2012), *Science* 336:1525–7.
79 Andrianantoandrou, E., et al. (2006), *Molecular Systems Biology* 2006:0028.
80 Marris, C., et al. (2014), *BioSocieties* 9:393–420. For recent revivals of enthusiasm for synthetic biology, see for example: *New York Times*, 23 November 2021; *New Yorker*, 7 March 2022; Webb, A. and Hessel, A. (2022), *The Genesis Machine: Our Quest to Rewrite Life in the Age of Synthetic Biology* (New York, Public Affairs).
81 For measured explorations of synthetic biology's promises and significance (or not), see these two excellent collections: Boldt, J. (ed.) (2016), *Synthetic Biology: Metaphors, Worldviews, Ethics, and Law* (Wiesbaden, Springer); Schmidt, M., et al. (eds) (2010), *Synthetic Biology: The Technoscience and its Societal Consequences* (London, Springer).
82 Acevedo-Rocha, C. (2016), in K. Hagen, et al. (eds) *Ambivalences of Creating Life: Societal and Philosophical Dimensions of Synthetic Biology* (Cham, Springer), pp. 9–44, p. 24.
83 Gibson, D., et al. (2010), *Science* 329:52–6.
84 Hutchison, C., et al. (2016), *Science* 351:aad6253.
85 Pelletier, J., et al. (2021), *Cell* 184:2430–40.
86 Piccirilli, J., et al. (1990), *Nature* 343:33–7.
87 Eremeeva, E. and Herdewijn, P. (2019), *Current Opinion in Biotechnology* 57:25–33; Krueger, A., et al. (2011), *Journal of the American Chemical Society* 133:18447–51; Lajoie, M., et al. (2013), *Science* 342:357–60; Malyshev, D., et al. (2014), *Nature* 509:385–8.
88 Robertson, W., et al. (2021), *Science* 372:1057–62.
89 Duffy, K., et al. (2020), *BMC Biology* 18:112.
90 Jefferson, C., et al. (2014), *Frontiers in Public Health* 2:115. At the time of writing, the diybio.org site, home to what was one of the most flourishing DNA hacker communities, had not been updated since 2018.
91 See, for example, National Academies of Sciences, Engineering, and Medicine (2018), *Biodefense in the Age of Synthetic Biology* (Washington, DC, The National Academies Press).
92 Garrett, L. (2013), *Foreign Affairs* 92:28–46.
93 Imai, M., et al. (2012), *Nature* 486:420–8; Herfst, S., et al. (2012), *Science* 336:1534–41.
94 Enemark, C. (2017), *Medical Law Review* 25:293–313. For a detailed account of this period see Rappert (2015) and Murdock, K. and Koepsell, D. (2014), *Frontiers in Public Health* 2:109.
95 Fouchier, R., et al. (2012a), *Nature* 481:443.
96 Butler, D. (2012), *Nature* 482:447–8.
97 Anonymous (2013), *Nature* 493:451–2; Fouchier, R., et al. (2012b), *Nature* 493:609. The end of the pause was announced by the signatories to the original letter, who now used the word 'moratorium'.

98 Fouchier, R., et al. (2013a), *Nature* 500:150–1; Fouchier, R., et al. (2013b), *Science* 341:612–3.
99 Reardon, S. (2014), *Nature* 11 July 2014, doi.org/10.1038/nature.2014.15544
100 Henkel, R., et al. (2012), *Applied Biosafety* 17:171–80; Duprex, W., et al. (2015), *Nature Reviews Microbiology* 13:58–64.
101 Watanabe, T. (2014), *Cell Host Microbe* 15:692–705.
102 *The Guardian*, 11 June 2014.
103 Kaiser, J. (2020), *Science*, 24 January 2020. https://www.sciencemag.org/news/2020/01/after-criticism-federal-officials-revisit-policy-reviewing-risky-virus-experiments
104 Koblentz, G. (2020), *Washington Quarterly* 43:177–96; Palmer, M. (2020), *Science* 367:1057.
105 Noyce, R., et al. (2018), *PLoS ONE* 13:e0188453.
106 Kupferschmidt, K. (2017), *Science*, 6 July 2017. https://www.sciencemag.org/news/2017/07/how-canadian-researchers-reconstituted-extinct-poxvirus-100000-using-mail-order-dna; Koblentz (2020).
107 Noyce, R. and Evans, D. (2018), *PLoS Pathogens* 14:e1007025, p. 2.
108 Maddalo, D., et al. (2014), *Nature* 516:423–7.
109 Ledford, H. (2015), *Nature* 522:20–4, p. 21.
110 Yang, D. (2021), *Asian Perspective* 45:7–31.
111 Thorp, H. (2021), *Science* 374:793.
112 Holmes, E., et al. (2021), *Cell* 184:4847–56; Lytras, S., et al. (2021), *Science* 373:968–70.
113 For example: Andersen, K., et al. (2020), *Nature Medicine* 26:450–2.
114 Lewis, G., et al. (2020), *Nature Communications* 11:6294; Appelt, S., et al. (2021), *Nature Communications* 12:3078.
115 Ben Ouagrham-Gormley, *Barriers to Bioweapons*, p. 17.
116 *The Guardian*, 6 September 2021.
117 Ben Ouagrham-Gormley, *Barriers to Bioweapons*; Jefferson et al. (2014); Vogel, *Phantom Menace or Looming Danger?*
118 Vogel, *Phantom Menace or Looming Danger?*, p. 115.
119 Ibid., p. 57.
120 Kuhn, J. (2008), *Bulletin of the Atomic Scientists* website, 28 February 2008. https://thebulletin.org/roundtable_entry/defining-the-terrorist-risk/
121 Ben Ouagrham-Gormley, *Barriers to Bioweapons*, p. 145.
122 Dance (2021). See also Warmbrod, K., et al. (2021), *EMBO Reports* 22:e53739.
123 Jacobsen, R. (2021), *MIT Technology Review*, 26 July 2021. https://www.technologyreview.com/2021/07/26/1030043/gain-of-function-research-coronavirus-ralph-baric-vaccines/
124 Ball, P. (2004), *Nature* 431:624–6, p. 626.

16 Gods?

1 This idea appeared in the titles of two early books on the topic: Goodfield, J. (1977), *Playing God: Genetic Engineering and the Manipulation of Life* (London, Hutchinson); Howard. T. and Rifkin, J. (1977), *Who Should Play God? The Artificial Creation of Life and What it Means for the Future of the Human Race* (New York, Dell). There have been many more since.
2 It also played a role in the growth of popular interest in computing and the development of various strands of science in the 1970s as the counterculture inspired some researchers, in particular in physics and psychology. Two

snappy academic book titles sum up these links: Fred Turner's *From Counterculture to Cyberculture* and David Kaiser and Patrick McCray's *Groovy Science*. Genetic engineering does not quite fit the bill and is absent from both accounts. For more, see: Turner, F. (2006), *From Counterculture to Cyberculture: Stewart Brand, the Whole Earth Network and the Rise of Digital Utopianism* (Chicago, Chicago University Press); Agar, J. (2008), *British Journal for the History of Science* 41:567–600; Kaiser, D. and McCray, W. (2016), *Groovy Science; Knowledge, Innovation, and American Counterculture* (Chicago, Chicago University Press); Dear, B. (2017), *The Friendly Orange Glow: The Untold Story of the Rise of Cyberculture* (New York, Pantheon); McCray, W. and Kaiser, D. (2019), *Science* 365:550–1.

3 If you want to read some high-octane genetic futurism (it is not to my taste), try Mason, C. (2021), *The Next 500 Years: Engineering Life to Reach New Worlds* (London, MIT Press).

4 One of the first popular expositions of this idea can be found in Church, G. and Regis, E. (2012), *Regenesis: How Synthetic Biology Will Reinvent Nature and Ourselves* (New York, Basic Books).

5 Zimov, S. (2005), *Science* 308:796–8.

6 Lynch, V., et al. (2015), *Cell Reports* 12:217–28; Palkopoulou, E. et al. (2018), *Proceedings of the National Academy of Sciences USA* 115:E2566–74. The exact number of nucleotide differences is not known. I asked on Twitter and got varying replies from people who are working in the area: https://tinyurl.com/mammothdifferences

7 Folcha, J., et al. (2009), *Theriogenology* 71:1026–34.

8 See, for example: *The Guardian*, 13 September 2021; *New York Times*, 13 September 2021.

9 Herridge, V. (2021), *Nature* 598:387.

10 Shapiro, B. (2015), *How to Clone a Mammoth: The Science of De-extinction* (Princeton, Princeton University Press); Shapiro, B. (2017), *Functional Ecology* 31:996–1002.

11 Church and Regis, *Regenesis*, pp. 10–11.

12 Trujillo, C., et al. (2021), *Science* 371:eaax2537. This effect may have been due to the fact that the rest of the genome in the organoids was human; unknown co-adapted Neanderthal alleles required for the normal function of the Neanderthal neuronal genes may have been missing. Although Neanderthals were sufficiently similar to us genetically to produce fertile hybrids (you are probably one of their descendants; I certainly am), they were not identical to modern humans.

13 Farahany, N., et al. (2018), *Nature* 556:429–32.

14 IUCN SSC (2016), *IUCN SSC Guiding Principles on Creating Proxies of Extinct Species for Conservation Benefit. Version 1.0.* (Gland, Switzerland, IUCN Species Survival Commission). For a critique of the initial IUCN position by a member of Restore & Revive, which includes a useful survey of existing projects, see Novak, B. (2018), *Genes* 9:548.

15 Redford, K., et al. (eds) (2019), *Genetic Frontiers for Conservation: An Assessment of Synthetic Biology and Biodiversity Conservation* (Gland, Switzerland, IUCN); Redford, K. and Adams, W. (2021), *Strange Natures: Conservation in the Era of Synthetic Biology* (Yale, Yale University Press), pp. 172–4.

16 McCauley, D., et al. (2017), *Functional Ecology* 31:1003–11; Robert, A., et al. (2017), *Functional Ecology* 31:1021–31.

17 Wray, B. (2017), *Rise of the Necrofauna: The Science, Ethics, and Risks of De-extinction* (Vancouver, David Suzuki Institute/Greystone Books); Seddon, P. (2017), *Functional Ecology* 31:992–5; Lin, J., et al. (2022), *Current Biology* 32:1650–6.
18 Pauling, L., et al. (1949), *Science* 110:543–8.
19 Frangoul, H., et al. (2021), *New England Journal of Medicine* 384:252–260.
20 Urnov, F. (2021), *CRISPR Journal* 4:6–13.
21 Cereseto, A., et al. (2021), *CRISPR Journal* 4:166–8.
22 Komor, A., et al. (2016), *Nature* 533:420–4. The idea was initially developed in 2013 by Komor in an email exchange with Liu when she was a PhD student at Caltech, and then fully developed with Gaudelli in Harvard – Davies, K., et al. (2019), *CRISPR Journal* 2:81–90. This fascinating interview sheds light on both the development of base editing and the process of scientific discovery and publishing.
23 Chu, S., et al. (2021), *CRISPR Journal* 4:169–77.
24 Musunuru, K., et al. (2021), *Nature* 593:429–34.
25 Anzalone, A., et al. (2019), *Nature* 576:149–57.
26 Anzalone, A., et al. (2021), *Nature Biotechnology* https://doi.org/10.1038/s41587-021-01133-w
27 Niu, D., et al. (2017), *Science* 357:1303–7.
28 Yue, Y., et al. (2021), *Nature Biomedical Engineering* 5:134–43. This is not simply a scientific issue, as shown by China's interest in this technology. Faced with a severe shortage of transplant organs, China has allowed people to sell their organs, and until recently has harvested organs from judicial executions, including of prisoners of conscience. Robertson, M., et al. (2019), *BMC Medical Ethics* 20:79.
29 https://www.npr.org/2021/10/20/1047560631/in-a-major-scientific-advance-a-pig-kidney-is-successfully-transplanted-into-a-h
30 *Financial Times*, 2 February 2022; Reardon, S. (2022), *Nature* 601:305–6.
31 Gao, C. (2018), *Nature Reviews Molecular Cell Biology* 19:275–6.
32 Li, S., et al. (2022), *Nature* 602:455–60.
33 Mallaparta, S. (2022), *Nature* 602:559–600.
34 Pixley, K., et al. (2019), *Annual Review of Phytopathology* 57:165–88; *Le Monde*, 30 April 2021.
35 Gao, C. (2021), *Cell* 184:1621–35.
36 Lin, W., et al. (2020), *Nature Biotechnology* 38:582–5.
37 Kolbert, E. (2021), *New Yorker*, 6 December 2021.
38 Shankar, S. and Hoyt, M. (2017), US Patent 2017/034906A1.
39 Voigt, C. (2020), *Nature Communications* 11:6379; Liew, F., et al. (2022), *Nature Biotechnology* https://doi.org/10.1038/s41587-021-01195-w
40 Sadler, J. and Wallace, S. (2021), *Green Chemistry* 23:4665.
41 Galanie, S., et al. (2015), *Science* 349:1095–100.
42 Kromdijk, J., et al. (2016), *Science* 354:857–61; Liu, Y., et al. (2021), *Cell* 184:1636–47.
43 Peplow, M. (2016), *Nature* 530:389–90.
44 Jacob, F. (1998), *Of Flies, Mice, and Men* (Cambridge, MA, Harvard University Press), p. 69. The French edition appeared the year before.
45 Nuffield Council on Bioethics (2015), *Ideas About Naturalness in Public and Political Debates about Science, Technology and Medicine* (London, Nuffield Council on Bioethics); Ducarme, F. and Couvet, D. (2020), *Palgrave*

Communications 6:14; Levinovitz, A. (2021), *Natural: The Seductive Myth of Nature's Goodness* (London, Profile).

46 *Boston Globe*, 17 September 2000; Yetisen, A., et al. (2015), *Trends in Biotechnology* 33:724–34. For Kac's view, see Kac. E. (2016), *Les Actes de colloques du quai Branly Jacques Chirac* 6:1–11.

47 www.glofish.com

48 Magalhães, A., et al. (2022), *Studies on Neotropical Fauna and Environment* doi: 10.1080/01650521.2021.2024054

49 www.multispecies-salon.org/zaretsky/

50 Ledford, H. (2015), *Nature* 524:398–9, p. 398.

17 Perspective

1 Fox Keller, E. (2008), *Historical Studies in the Natural Sciences* 38:45–75.

2 Pauly, P. (1987), *Controlling Life: Jacques Loeb and the Engineering Ideal in Biology* (Oxford, Oxford University Press), p. 51.

3 Watson, J. and Crick, F. (1953), *Nature* 171:964–7.

4 Crick, F. (1958), *Symposia of the Society for Experimental Biology* 12:138–63.

5 Hall, S. (1988), *Invisible Frontiers: The Race to Synthesise a Human Gene* (London, Sidgwick & Jackson), p. 176.

6 Rheinberger, H.-J. (2009), in M. Lock, et al. (eds), *Living and Working with the New Medical Technologies: Intersections of Inquiry* (Cambridge, Cambridge University Press), pp. 19–30, p. 25.

7 Brand, A. and Perrimon, N. (1993), *Development* 118:401–15.

8 Jackson, D. (1995), *Annals of the New York Academy of Sciences* 758:356–65, p. 358.

9 Jamison, A. (2011), in A. Nordman, et al. (eds), *Science Transformed? Debating Claims of an Epochal Break* (Pittsburgh, University of Pittsburgh Press), pp. 93–105, p. 100.

10 Kay, L. (1998), in A. Thackray (ed.) *Private Science: Biotechnology and the Rise of the Molecular Sciences* (Philadelphia, University of Pennsylvania Press), pp. 20–39, p. 32.

11 *Genetic Dreams, Genetic Nightmares* (BBC Radio, 2021), episode 3: www.bbc.co.uk/sounds/play/m000ycvv

12 Annas, G., et al. (2021), *CRISPR Journal* 4:19–24.

13 Jasanoff, S. (2005), *Designs on Nature: Science and Democracy in Europe and the United States* (Oxford, Princeton University Press), pp. 177–9; Drell, D. and Adamson, A. (2001), 'DOE ELSI program emphasises education, privacy: a retrospective (1990–2000)', web.ornl.gov/sci/techresources/Human_Genome/resource/elsiprog.pdf

14 Owen, R., et al. (2021), *Journal of Responsible Innovation* doi: 10.1080/23299460.2021.1948789.

15 McLeod, C. and Nerlich, B. (2017), *Life Sciences, Society and Policy* 13:13; Gold, A., et al. (2021), *Trends in Genetics* 37:685–7.

16 Winner, L. (1980), *Daedalus* 109:121–36.

17 Lentzos, F. (2020), *Nonproliferation Review* 27:517–23.

LIST OF ILLUSTRATIONS

Plates

Text images

INDEX

S

Photo credit: Stuart Phillipson

Matthew Cobb is a professor in the School of Biological Sciences at the University of Manchester. He is the author of numerous books, including *The Idea of the Brain, Life's Greatest Secret, Generation, The Resistance, Eleven Days in August*, and *Smell: A Very Short Introduction*. He lives in England.